ADSORPTION OF INORGANIC IONS ON INORGANIC MATERIALS

ADSORPTION OF INORGANIC IONS ON INORGANIC MATERIALS

S B Kanungo
Former Scientist, Institute of Minerals and Materials Technology
Bhubaneswar, Odisha, India

New Central Book Agency (P) Ltd
LONDON
HYDERABAD ERNAKULAM BHUBANESWAR
DELHI KOLKATA PUNE GUWAHATI

NCBA

REGD OFFICE

8/1 Chintamoni Das Lane, Kolkata 700 009, India
ncbapvtltd@eth.net

OVERSEAS

NCBA (UK) Ltd, 149 Park Avenue North, Northampton, NN3 2HY, UK
ncbauk@yahoo.com

EXPORT

NCBA Exports Pvt. Ltd, 212 Shahpur Jat, New Delhi 110 049
ncbaexports@ncbaindia.com

BRANCHES

208B Peacock Lane, Shahpur Jat
New Delhi 110 049
ncbadel@ncbapvtltd.com

Shop Nos. 3 & 4, Vinayak Towers
681/B Budhwar Peth, Appa Balwant Chowk
Pune 411 002
ncbapune@ncbapvtltd.com

Shop No. 15, Bharati Towers, Ground Floor
Forest Park, **Bhubaneswar** 751 001
ncbabub@ncbapvtltd.com

House No. 3-1-315, 1st Floor
Nimboliadda, Kachiguda, **Hyderabad** 500 027
ncbahydb@ncbapvtltd.com

GSS Shopping Complex, 1st Floor
Opposite College Post Office, Convent Road
Ernakulam 682 035
ncbaernk@ncbapvtltd.com

Radhamayee Complex (Second Floor)
Opp. SBI, Panbazar, Jaswanta Road
Guwahati 781 001
ncbaguwahati@ncbapvtltd.com

ADSORPTION OF INORGANIC IONS ON INORGANIC MATERIALS • S B Kanungo

First Published: 2013

PUBLISHER
New Central Book Agency (P) Ltd
8/1 Chintamoni Das Lane, Kolkata 700 009

PRINTER
New Central Book Agency (P) Ltd
Web-Offset Division, Dhulagarh, Sankrail, Howrah

COVER DESIGNER
Soumen Paul

COVER PRINTER
Techno Process, Kolkata

PROJECT TEAM
Soumi Banerjee, Pradip Biswas, and Pradip Baidya

ISBN 978 81 7381 772 4

Price ₹ 850.00

CONTENTS

FOREWORD

I am extremely happy to write a Foreword for this book authored by Dr Sukriti Bhusan Kanungo who was a most admired and valuable colleague when I spent ten years, as Director, in the Institute of Minerals and Materials Technology, i.e., IMMT (then known as Regional Research Laboratory), Bhubaneswar, Odisha. A first-rate scientist, Dr Kanungo received his early education in what is now Bangladesh and then moved to Kolkata and, finally, Bhubaneswar, where he spent most of his professional life as a scientist in IMMT. He retired as the Head of the Central Physico-chemical and Analytical Facilities Division of the Institute. I am happy that, after retirement, he has found time to delve into his wide experience and knowledge to write this book.

Colloid and surface science finds wide application in many technological fields relating to, for instance, pharmaceuticals and cosmetics, water/waste water treatment, materials development, pollution abatement, oil and natural gas exploration, petrochemicals, biotechnology, agronomy, soil conservation etc. Besides these, to understand certain natural phenomena involving air (e.g., smog), earth (e.g., geochemical processes in aqueous environment), and even products of some biological processes (e.g., milk, latex from rubber plants), it is necessary to invoke the principles of colloid and surface science. An important topic in this field of science is the phenomenon of adsorption, which has now assumed special significance in water and waste water treatment.

The present book deals with the basic aspects of the adsorption behavior of toxic cations and anions on solid particles of inorganic origin in aqueous environment. The author had carried out a good amount of research on this subject during the tenure of his service in the IMMT, Bhubaneswar, Odisha.

The author has aptly and adequately dealt with the subject, starting from the characterization of solid particles to their chemical and electrochemical interactions with ionic species in aqueous medium. To discuss basic aspects of the adsorption phenomenon, it is necessary to use well-characterized materials, and this has been amply emphasized by the author.

I think the book will greatly help the students of physical chemistry, environmental science and technology, soil chemistry, geochemistry and even civil engineering (water treatment).

The author has taken considerable pain in collecting large volume of literature which will be of much help to those who are pursuing research activities on this topic.

I wish the book a wide acceptance among the readers.

Prof. H. S. Ray

Formerly Director, IMMT
Bhubaneswar, Odisha

PREFACE

The phenomenon of adsorption is a widely studied topic in the realms of surface chemistry and related technology. But its ramification has been more widespread than these two fields and now includes environmental chemistry, soil chemistry, geochemistry, material science etc.

The earlier methods of describing adsorption data using the simple partition coefficient and the empirical models like Freundlich and Langmuir equations have given place to rigorous models based on thermodynamic concepts, principles of electrostatics at solid/aqueous solution interface and even quantum chemical principles of bond formation. During this evolutionary process, extensive investigations have been carried out using well-defined/characterised solids of inorganic origin as adsorbents and aqueous solution of inorganic ions as adsorbates. The solids used are either oxides or hydrous oxides or silicates or aluminosilicates which can acquire electrical charge and potential due to acid-base dissociation of their hydroxylated surfaces in aqueous medium. Consequently, the adsorption of inorganic ions is dependent not only upon the concentration of adsorbate, but also upon pH and ionic strength of solution. Besides electrostatic or coulombic interaction, these solids may also interact chemically with the ionic species in solution leading to the formation of certain types of complexes, analogous to the formation of complexes in solution. At this stage, crystalline structure of the exposed surfaces of the solid plays a distinct role in determining the nature of bonding with the adsorbate ions.

Thus, the development of a suitable model—which can adequately describe adsorption data over a wide range of concentration by combining all the relevant factors—is a challenging task. The concept of surface heterogeneity of solid is the direct outcome of the factors mentioned above. This aspect is still being debated. However, the development of different spectroscopic methods, especially X-ray absorption spectroscopy (XAS), has been found to be of great help in understanding the mechanism of adsorption. The author hopes to include a chapter on this topic in the future edition of the book.

With the above background, the objective of writing this book is to present an account of the basic aspects of colloid and surface chemistry that are essential in understanding the phenomenon of adsorption of inorganic ions on well-defined inorganic solids. Several excellently edited books were published by Ann Arbor Science Publishers, Michigan, and American Chemical Society (as symposium series), Washington, USA, in the 1980s.

The topic is of utmost interest to physical and environmental chemists, geochemists and agricultural scientists. However, the scope of the discussion has been consciously confined to the inorganic materials (solids) because of two reasons. First, the objective of the book would have been diluted had the author included heterogeneous, poorly defined, so-called "cheap adsorbents" in the discussion. Such materials belong either to plant products or to industrial and agricultural wastes, many of which are of organic origin. They may have importance from the economic point of view, but not from the point of view of basic research. Secondly, the sheer number of publications that have appeared so far on such adsorbents would entail a separate book.

The chapters of the book have been arranged broadly in line with the evolutionary process of our understanding of the adsorption phenomenon. After a brief introduction to the topic, Chapter 1 provides an outline of the physical characterization of solid adsorbents.

As the hydration of oxide surface is considered the primary step for acquiring charge through subsequent dissociation of the hydroxylated surface, these topics are discussed in Chapters 2 and 3, respectively.

The next chapter deals with the quantitative estimation of surface charge and defines the conditions under which the surface tends to attain zero charge or net zero proton charge. The concept of electrical double layer (EDL) is one of the greatest milestones in the development of colloid and surface chemistry (and also in electrochemistry).

Chapter 5 provides a review of the different models that have been proposed to describe the experimental data on the variation of charge and potential of the oxide-water interface with respect to pH.

The interaction of the electrified surface of solid materials with chemical species in the surrounding medium leads to the process of adsorption. The intrinsic equilibrium constant, which corresponds to the chemical component of the overall free energy of such surface interaction, has to be determined under zero charge and zero potential conditions. Methods which can fulfill these two conditions have been described in Chapter 6. This chapter also describes the principles and methods of the determination of other parameters of the EDL such as surface and diffuse layer potentials, surface charge, etc.

The inorganic constituents of soil are varied. Few literatures provide structural details of these constituents in one place. The oxide constituents of soil can also be prepared in the laboratory and, therefore, their structural and interfacial properties have been extensively studied. Chapter 7 presents structural characteristics of different inorganic constituents of soil.

The interfacial properties of clay minerals are basically different from those of oxide constituents. Consequently, adsorption behavior of clay minerals cannot be modeled in the same manner as those oxides. Such differences have been highlighted in Chapters 8 and 9. However, the influence of organic coatings on the inorganic constituents, though reasonably significant but inadequately defined, has not been included in the discussion.

Chapter 10 provides a detailed description of the adsorption of cations on different inorganic materials. Various factors which influence the adsorption of cations in the absence as well as in the presence of complexing agents have been discussed with extensive literature references.

Similarly, the adsorption characteristics of some selected anionic species including their methods of analysis have been discussed in detail in Chapter 11.

From the contents of the different chapters as briefly presented above, it can be concluded that the book is designed to benefit students (postgraduate), research scholars, scientists, faculty members et al.

Extensive literature references up to 2010 have been made to facilitate further reading. However, the author will greatly appreciate receiving comments and suggestions which will be given due consideration in the future edition of the book.

The author expresses his sincere thanks to all who helped him in his endeavour to write this book. He is particularly thankful to his former students (D. M. Mahapatra, R. Mahapatra, S. K. Mishra, and S. S. Tripathy) who collaborated with him to generate a large volume of experimental results, some of which have been quoted in this monograph. Sincere thanks are also due to Prof. H. S. Ray for writing the Foreword and to the publishers [M/s New Central Book Agency (P) Ltd] for excellent production of the book.

Flat A7/2
Konnagar Abasan
Konnagar 712 235
District Hooghly (West Bengal)
email: sbk_your@yahoo.co.in

Author

CHAPTER 1
INTRODUCTION

1.1 Definition and Origin of Our Present Knowledge on the Adsorption of Inorganics

The phenomenon of adsorption which may be defined as the uptake of dissolved chemical species onto an insoluble substrate by physical and / or chemical forces, has been known to chemists since early nineteenth century. However, its wide and diverse manifestations in chemistry, metallurgy, material science, agriculture, environmental science etc. are being revealed during the last one hundred years. It is now understood that the phenomenon is an inherent part of the natural process that is continuously going on in our surroundings, particularly in natural aquatic system to maintain the balance of inorganic ions between water and sediment. If this balance were not there we would not get fresh water required for our survival. The adsorbents which are ubiquitous in nature are certain hydrous metal oxides of iron, manganese and aluminum etc., siliceous materials of bio-origin, zeolites, clays etc., most of which occur in finely divided state. In aquatic systems these substrates act as scavenger for trace metal ions and also anions. In the underground aquifers sorption process controls the transport of toxic cations and anions by forming secondary repositories in the silt layer containing oxides and clay minerals above water table In soils, some of these hydrous metal oxides occur as coating on other substrates thereby exposing large surface area, a phenomenon though recognized by the soil scientists more than 100 years go, is still being investigated to understand the fertility of different types of soil. Each area of application of the adsorption phenomenon is discussed more elaborately in the next section.

The above adsorbents, especially hydrous oxides of iron, manganese , aluminum , titanium etc. have been widely investigated in connection with their uses in the adsorption of inorganics from aqueous medium . Indeed, most of our present knowledge on the interfacial behavior of solids in aqueous electrolyte solution originates from the extensive research carried out during the last sixty years on the adsorption of inorganics on these substrate. These studies also led to the development of a good number of adsorption models. Some of these models are purely empirical, as they are based on system conditions under study and not on basic principles governing the solid - liquid interface . Others are based on the electrochemical theory of the solid - liquid interface and are applicable even beyond experimental conditions.
The basic fundamental aspects of the development of surface charge and the struc-

ture of electrical double layer at the solid - liquid interface were discovered during the period from 1915 to 1930 by Gouy [1] ,Chapman [2] and Stern [3] and refined later on by Grahame [4] .This concept of electrical double layer subsequently formed the basis of explaining adsorption of inorganics at the solid - liquid interface since nineteen sixties .The investigation that started at that time using pure synthetic hydrous metal oxides of Al , Fe(III) , Mn(IV) ,Ti etc . as adsorbent is still continuing , because to apply the laboratory data to adsorption under natural condition is a challenging job to the surface chemists . However, valuable contributions made since 1960s by well-known research groups led by T.W. Healy (School of Chemistry, Melbourne, Australia), C.P. Huang (Harvard University, USA), J. O. Leckie (Stanford University, USA), F. M. M. Morel (M I T, Cambridge, USA), J. J. Morgan, W. Stumm (C I T, Pasade-na , USA), J W . Murray (School of Oceanography, University of Washington , USA) and many others including those from Dutch school (J Lyklema , van Riemsdijk , T .Hiemstra , L .K . Koopal etc.) to name a few by way of examples from a large community of well-recognized researchers who have advanced our knowledge on the phenomenon of adsorption of inorganics from aqueous medium .

1.2 Scope of the Topic in Different Fields of Science and Technology

1.2.1 Environmental Chemistry

The statement made in the foregoing section also includes environmental aspects of adsorption. Indeed, apart from pure scientific pursuit, environmental consideration is the main driving force behind the growing interest in this field. Our natural water sources, especially fresh water sources, are being polluted by domestic, agricultural and industrial wastes generated by growing human activities. Besides creating awareness among people for protecting our environment by using various corrective measures, it is also necessary to promote self - cleaning of our natural aquatic systems. This is usually carried out through the adsorption process on hydrous oxides or aluminosilicate particles in sediment or in suspended state. If the level of pollution is too high, then the adsorption capacities of these natural adsorbents will be overstressed.Two of the main objectives of the study of environmental chemistry of water are: (i) To describe the nature of occurrence of various chemical species present in the natural water systems and (ii) to characterize both suspended particles and sediments by physico - chemical methods to assess their adsorption capacities. Extensive research is being carried out on these two aspects.

1.2.2 Geochemical Processes in Aqueous Environment

Many geochemical processes in aqueous environment involve adsorption phenomenon. The American Chemical Society organized a symposium in 1985 and its proceedings contain many papers specifically on this topic [5]. Generally, clays and other aluminosilicate minerals like feldspar, zeolites etc. undergo certain structural changes due to adsorption of inorganics from aqueous medium. Even glassy materials of volcanic origins like plagioclase, orthoclase undergo degradation on sea bed due to various ion exchange processes that take place on their surfaces [6,7]. Indeed, the entire system of ionic concentration balance between sediment and water above it in river, lake and sea is controlled by ion exchange through adsorption process [8,9]. Another most relevant example of geochemical process where adsorption plays an important role in aqueous environment is the formation of sea-bed manganese nodules and the occurrence of many trace metals in them [10]. We shall discuss these aspects later on in more details.

1.2.3 Water Purification

The adsorption phenomenon has its most important use in the purification of water including the treatment of sewage. Certain impurities like fluoride [11-15], arsenic (III and IV) [16-26] chromium(VI) [27-29], selenium(IV and VI) [30-32] etc are best removed from water by adsorption technique. During the coagulation of suspended particles and subsequent oxidation step (using chlorine, air etc.) fine particles of hydrous iron and aluminum oxides form flocs which , depending upon the pH of the medium, may adsorb either anions like F^-, AsO_3^{3-} (arsenite)/AsO_4^{5-}(arsenate), $Cr_2O_7^{2-}$ (dichromate)/$Cr_2O_4^{2-}$ (chromate) etc.or toxic cations like Cd(II) [33-38] , Pb(II) [39-45] , Hg(II) [44,46-49] etc. In many cases adsorbents are added from outside to make the process of impurity removal faster and more efficient.

1.2.4 Analytical Chemistry , Process Metallurgy etc

Adsorption is a convenient tool for the separation of cations or oxyanions by suitably adjusting pH , ionic strength and potential of the medium . Some of the most common examples are the separation of ionic (cations & anions) and molecular species using various chromatographic techniques such as ion-exchange, partition (paper & column), thin layer etc wherein, both organic and inorganic ion exchangers or adsorbents have been employed. Another example application of adsorption phenomenon in analytical chemistry is the use of certain ionic dyes (fluorescein, dichlorofluorescein, eosin etc) which undergo sharp change in color when adsorbed as oppositely charged colloidal particles like silver chloride or bromide at the equivalence point of precipitation titration [50] .Certain types of coprecipition that are believed to occur during the gravimetric separation of anions and cations, for example $BaSO_4$, $Mg(HPO_4)$, calcium oxalate etc., are indeed a phenomenon of chemisorption of species of similar structure [50] . This problem is avoided by controlling the pH and sometimes ionic strength of the medium. In mineral processing the phenomenon of adsorption of certain bivalent cations (as activators) or soluble silicates (as depressors) are used to alter the surface charge of a mineral to achieve selective flotation within a particular pH range [51]. In hydrometallurgy, base metals like copper, nickel, cobalt etc. can be separated from ammoniacal solution using sorption on hydrous iron oxide [52], valuable metals like gold, silver, molybdenum etc. are recovered from their aqueous solutions as cyanides using adsorption on activated carbon [53].

1.3 Characteristics of Solid or Adsorbent Phase

1.3.1 Type of Solid Phase

Characterization of solid phase is not only essential but also of foremost importance before studying its adsorption behavior. Solid adsorbent may be either natural or artificial i.e, prepared in the laboratory. In both the cases it is necessary to determine first the chemical composition and identify the crystalline phases present using x - ray diffraction and other appropriate methods. Many naturally occurring adsorbents occur in such a way that the active component or phase forms a coating on another phase of different chemical composition. Crushing and grinding may dislodge coated material or may even form a new mixed phases. Too much grinding leads to extensive alteration in the surface properties. Therefore, both bulk and surface properties have to be studied simultaneously during the preparation of solid adsorbent as elaborated below.

1.3.2 Morphological Studies

Electron microscopy (both scanning and transmission) is the most important tool for morphological study of a solid adsorbent regardless of its origin. The main disadvantage of the method is that during the preparation of the samples for microscopic examination certain changes in size, shape may take place due to which correct picture of the material may not be obtained. Earlier, several researchers have carried out extensive microscopic studies on synthetic oxides, hydroxides, oxy hydroxides etc. of manganese [54,55], zinc [56,57], iron [59], magnesium [58] etc. James and Parks [60] have made a brief review on this aspect which also includes the list of extensive work carried out by Matijevic and his colleagues on the electron microscopic study of the hydrosols of various metal hydroxides during their formation, growth and aging processes. These shall provide ample guidance to those who are interested in this field of particle characterization by microscopic method. However, electron microscopic study coupled with other supplementary methods like coulter counter, light scattering, image shadowing etc. might be useful to get an idea of the extent of morphological alteration / deformation, if any, caused by high vacuum, heat effect due to exposer to electron beam.

1.3.3 Surface area

The interfacial area of a solid when suspended in a solution is the area containing such ionizable sites which can react reversibly with either potential determining ions in solution or adsorb any other ions or molecules at the surface sites. Most of the adsorption models assume planer and non-porous surface. However, an ideal non-porous surface is seldom used in practice as an adsorbent. The interaction of porous surface with ions in aqueous solution leading to development of electrical double layer has been investigated by Lyklema [61], Perram et al [62] and Perram [63]. Though adsorption depends upon pore size distribution , the difference in the results from ideal non-porous surface is not very large . For large pore size (> Debye parameter 1/) the area within the pores acts as good as planer surface. For small pores into which protons and certain metal ions can penetrate , discrepancy /deviation in surface charge versus surface potential or pH-adsorption relationships may appear.

1.3.3.1 Surface Area by Low Temperature Adsorption of gas

Various methods are available to measure the surface areas of finely divided solid materials. Most of these are based on the measurement of sorption under specified conditions , of gas or any other solute molecules, having known dimension or the area occupied by them for flat wise adsorption on a surface . By far the most popular method of measuring surface area for finely divided solid is the low temperature nitrogen adsorption proposed by Brunauer , Emmett and Teller or better known as BET method [64] . The isotherm developed by the authors for multilayer adsorption enables one to estimate the amount of N_2 adsorbed as monolayer using the area occupied by one molecule of N_2 (~ 16.6 A^2) . However, thorough degassing (drying under N_2 flow) of the sample before measurement casts some doubt about the relevancy of the data when the solid material is suspended in water. For example , it has been observed [65] that surface area measured by the BET method shows variation with change in out gassing parameters, such as temperature and period of outgassing, nitrogen flow rate etc . Most importantly, there is no way to verify that entire area is covered with a layer of gaseous nitrogen molecules. Despite this doubt the BET method is still most popular, because of the simplicity of experimental procedure and satisfactory reproducibility under fixed experimental conditions.There are various methods by which the extent of adsorption of gas can be determined. These may be classified into three main groups, namely volumetric, gravimetric and dynamic methods. Each method has its own limits of applicability and precision range [65,66] . In the volumetric method the change in the volume of gas in a calibrated gas burette serves as a measure of the amount of gas adsorbed by the adsorbent. This was the most widely used method in the earlier literature. In the gravimetric method the increase in the weight of the solid adsorbed due to the adsorption of gas is generally measured in a specially designed quartz microbalance. After the development of gas chromatographic technique dynamic method has gained great popularity and most of the modern instruments are based on this technology, as no high vacuum is required in this method. This has greatly reduced the time for measurement of surface area of solid.

Surface area of a solid can also be determined by comparing its isotherm with the isotherm of another solid of known surface area. If the reference solid is non-porous, then we can have additional information on the nature of porosity (micro or meso, see next section for definitions) of the material. Adsorption isotherm of N_2 or Kr on non-porous alumina and silica are available in literature [65]. The quantity adsorbed (in moles/g) on test material is plotted as function of either t or α_t (the ratio of quantity adsorbed at any value of relative pressure i.e. p/p_o to that adsorbed at $p/p_o = 0.4$). The value of t above $p/p_o > 0.4$ is generally independent of the nature of porosity of the adsorbent, but is dependent on the quantity adsorbed. Typical t plots for adsorbents without microporosity (curve I) and with substantial microporosity (curve II) are shown in Fig. 1.1. The multilayer thickness t is calculalated from the following relationship

$$t = (n/n_m)\, t_1 \qquad (1.1)$$

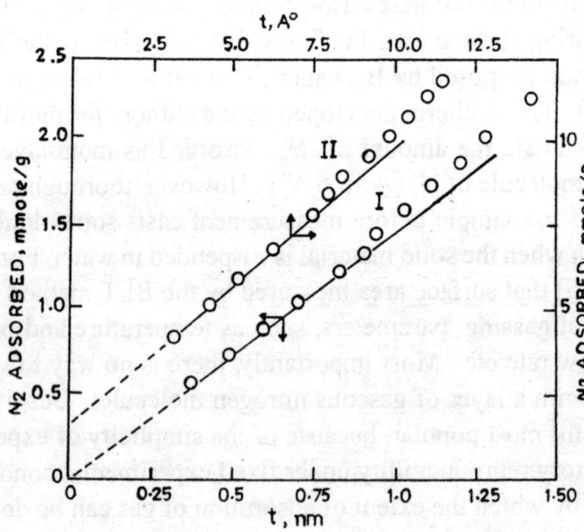

Fig.1.1 Typical nature of the plots of N_2 sorbed vs. thickness of the statistical multilayer (t) for solid adsorbent without micro porosity (curve I) and with micro porosity (curve II). Redrawn using data derived from ref. 249.

where, t_1 is the thickness of monolayer (0.345 nm for N_2), and n and n_m are the quantities adsorbed (in mmole/g) at any value of p/p_o and at monolayer respectively. However, t for any value of n can be obtained directly from the plot of α_t vs. n or reference material at p/p_o >0.35, where the t plot is independent of the nature of adsorbent. Knowing t_1 the value of n_m can be obtained from the slope value of the linear portion of t plot. The surface area is then calculated as follows from the known value of n_m

$$S = n_m \times a_m \times L \tag{1.2}$$

Where a_m is the average area of nitrogen molecule (0.162 nm^2), L is the conversion factor from nm^2/g to m^2/g i.e. 6.02×10^2 . In Fig. 1.1, the deviations of the experimental points at high adsorption from linear behavior represent mesoporosity. For curve II, an estimate of microporosity can be obtained by multiplying the intercept value with molar volume of liquid N_2.

1.3.3.2 Surface area from Liquid Phase Adsorption

An extensive investigation has been carried out on the determination of surface area of solid adsorbent using adsorption of soluble adsorbates like ions , dyes , fatty acids or even surfactant molecules .The advantage of this method over the gas adsorption method is that pre drying of adsorbent is not necessary . The disadvantage of the method is that there is a considerable amount of uncertainty regarding size, configuration and orientation of molecules and the degree to which these molecules might occur in aggregated state during adsorption. Further, these large molecules are not accessible into the pores and therefore actual area is higher than that estimated by adsorption from liquid phase. Orr and Dallavale [67] have discussed various aspects of determining surface area from liquid phase adsorption, especially dyes and fatty acids . Similar discussion is also available in the monograph by Adamson [68]. The use of the adsorption of dyes, especially cationic dyes like methylene blue, malachite green etc .on clays for the determination of surface area , has been in practice for quite a long time . Greenland and Hayes [69] have made a review on the adsorption of organics on clays. Kanungo and De [70] have studied the adsorption of both cationic and anionic dyes on precipitated magnesium hydroxide and magnesium oxide obtained by the calcinations of magnesium carbonate. It has been observed that anionic dyes like orange II and tartrazine exhibit high adsorption while cationic dyes like methylene blue show poor adsorption and consequently low surface area . However, the samples exhibit high adsorption of iodine from aqueous medium and the area obtained by this method is very close to those obtained from low temperature N_2 adsorption .

Another method for the determination of surface area of solid in aqueous suspension is the so called "negative adsorption " [71] . It involves measuring the increase in the concentration of co - ions (same charge as the solid surface) in the bulk solution as a result of their expulsion from the diffuse layer of counter ions immediately adjacent to the solid phase and is therefore proportional to the surface area of the particles .The proportionality constant is derived from diffuse double layer theory which will be discussed in the subsequent section . For the case of bivalent co-ions in aqeous solution at 25°C the authors arrived at the following relationship .

$$S = 1.89 x 10^7 \, V_t \left({\Delta C_i}/{C_i} \right) \sqrt{C_i} \qquad (1.3)$$

where , S= surface area , cm^2

V_t= total volume of liquid , l

C_i = equilibrium concentration of co-ions , moles/l

ΔC_i = increase in concentration of co-ions , moles/l

This technique has the advantage that no pretreatment leading to the alteration of the surface is necessary but for porous solid it often provides much lower values of surface area than other methods. This is attributed to the restricted mobility of the co - ions in porous solid. Because of the uncertainty in the estimated value of surface area, the method has been seldom used in the study of interfacial properties of solid.

1.3.4 Porosity

Porosity represents the fraction of void volume created by pores with respect to the total volume of solid and therefore porosity is a dimensionless quantity. Porosity is a collective concept involving at least three characteristic features of a porous solid. These are: (a) Pore size i.e., diameter of pores, (b) pore volume and (c) area covered by interior of pore surface. Discussion on these three features is beyond the scope of this chapter. Interested readers may consult some excellent monographs [65, 66] on this subject for more information. However, it is worthwhile to distinguish porous solids on the basis of their pore sizes. Thus, pores with diameter less than 5nm, i.e., of molecular dimension are called micropores. Pores with diameter between 5 and 50 nm are called mesopores and those with diameter larger than 50 nm are called macropores. It is micro and meso porosity that contribute mostly to the specific surface area determined by the BET method.

Porosity is directly related to surface area of the porous solid . In gas adsorption method of surface area determination, the nature of the isotherm is affected by the type of pore [65,66] . While planer surface and micropores are completely covered by a layer of adsorbate molecules at pressure much below the saturation pressure of the gas , a capillary condensation in macropores usually occurs at higher pressure where hyterisis during desorption is also observed . Pore size or radius can be determined during the surface area measurement by gas adsorption method under variable adsorbate pressure. A drop in vapour pressure over the cylindrical liquid column (capillary condensation) in a pore of radius r(in cm) is expressed by the Kelvin equation

$$\ln\frac{p}{p_o} = -\frac{2\sigma\cos\theta_w v}{r\,R\,T} \tag{1.4}$$

where,
σ = surface tension ,dynes/cm
v = molar volume of adsorbate, cm^3
R = molar gas constant, atm/mole deg
T = temperature , K
p = equilibrium vapour pressure, torr
p_o = saturation vapour pressure, torr
θ_w = wetting angle between the solid and the liquid

The pressure of equilibrium adsorption in the capillary condensation region is higher than the corresponding pressure of desorption , resulting in hysterisis between adsorption and desorption paths as illustrated in Fig.1.2. In this case desorption occurs from fully filled up capillaries and the wetting angle is equal to zero . It is therefore necessary to affect adsorption up to relative pressure (p/p_o) unity and subsequently the desorption branch of the hysterisis loop of the isotherm is used to calculate the pore radius for every point on the isotherm , since , under this condition , no correc-

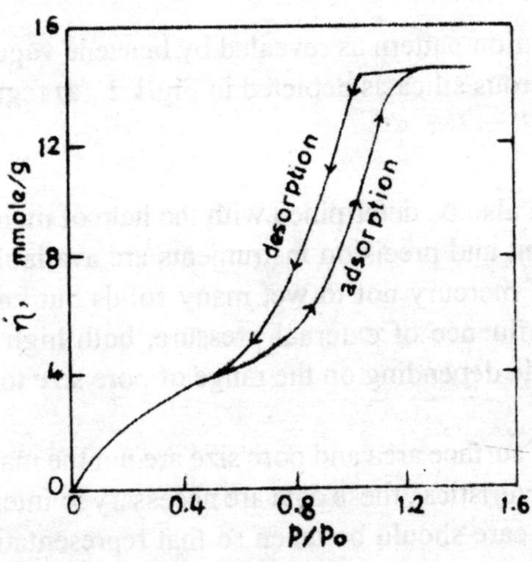

Fig. 1.2 Isotherm for adsorption and desorption of benzene vapor on and from highly porous silica respectively at 20°C. Redrawn on arbitrary scales after I. P. Mukhlyanov et al Catalyst Technology, Mir Publishers,Moscow,1976.

tion for wetting angle is necessary . Each point on the isotherm gives the amount adsorbed which when multiplied by v gives void volume . This is plotted against pore radius calculated at each point to obtain the pore size distribution pattern . Figure 1.3a shows the plot of gradual increment in void volume as the pores are being emptied during desorption as function of pore radius. Figure 1.3b also shows differential increment (dV/dr) which represents the structural characteristics of the porous solid.

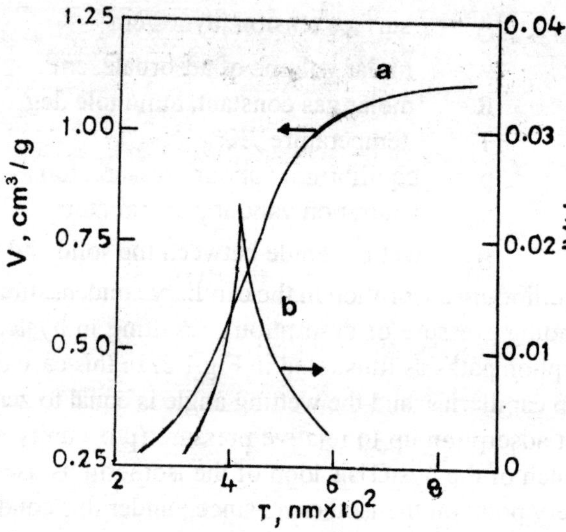

Fig. 1.3 Pore size distribution pattern as revealed by benzene vapour adsorption and desorption on porous silica as depicted in Fig.1.2. (a) regular plot and (b) differential plot.

Pore size distribution can also be determined with the help of mercury porosimeter , for which rapid , efficient and precision instruments are available . The method is based on the property of mercury not to wet many solids but enters even into the finest pores under the influence of external pressure. both high and low pressure porosimeters are available depending on the range of pore size to be studied [65].

Though measurements of surface area and pore size are not the main objective in the study of interfacial characteristics , these data are necessary to interpret the results of such studies .Therefore , care should be taken so that representative values are obtained .

1.3.5 Solubility

Some of the solid adsorbents like oxides or silicates in their finely divided state tend to be dissolved in aqueous solution depending upon the pH of the medium. For example, some dissolution of amorphous hydrous oxide of iron and manganese has been observed at pH below 4.0 [59,72]. Consequently, the initial equilibration of solid and solution may result in over saturation with respect to a more stable particle size or phase , resulting in the precipitation of separate particles of new phase or in the form of coatings different material on the original particles.The process is believed to occur during the diagenesis of minerals or coating of clay particles in soils with hydrous amorphous oxides of iron [73,74].This is a very slow process and therefore requires a very long time to reach equilibrium. Yates [75] argues that dissolution rates in conjunction with other methods of characterization can be used to study penetration of the surface by electrolytes leading to abnormally high double layer capacitance values. Equilibrium solubility data have been compiled exhaustively by several well-known authors [76-79]. Various aspects of the phenomena of dissolution followed by reprecipitation/coprecipitation and their implications on the adsorption behavior have been discussed in chapter ten.

CHAPTER 2

HYDRATION OF OXIDE SURFACE

2.1 Mechanism of the Hydration of Oxide Surface

When a metal oxide is exposed to water or its vapor surface gets hydrated. The degree of hydration depends upon the kinds of bonds broken or exposed during size reduction. If fracture involves breakage of secondary bonds of the van der Waal type, a non- polar or hydrophobic surface results upon immersion in water, e.g. graphite. If the fracture involves rapture of primary ionic or covalent bond, for example, oxides

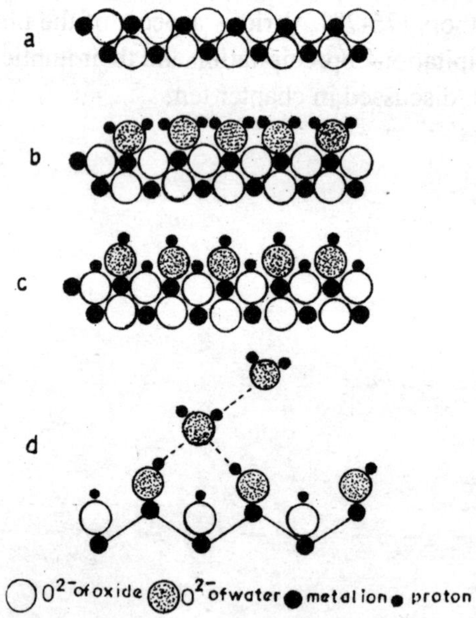

Fig. 2.1 (a) Arrangement of oxygen and metal ions in an idealized oxide structure where metal ions occur as coordinatively unsaturated state. (b) Adsorption of water leads to the saturation of coordination sites. (c) Transfer of proton to the neighboring oxygen of the oxide leads to the formation of hydroxylated surface. (d) Further adsorption of water vapor occurs through hydrogen bonding of singly coordinated O-H sites. Adapted after refs. 27 and 369.

of heavy metals, a polar or hydrophilic surface results. Therefore, the hydration of oxides can be represented by the following general equation:

$$Me_xO_y + yH_2O \leftrightarrow Me_x\,2(OH)_y \tag{2.1}$$

The above equation suggests that every oxygen atom is replaced by two OH groups in the oxide for the sake of charge balance. The widely held mechanism of the dissociative adsorption of H_2O on an idealized oxide surface may be illustrated by the following sequence of figures (Fig .2.1). Figure 2.1a shows the uncoordinated metal ion in an idealized surface layer of metal oxide and water molecules can occupy these vacant coordination sites through chemisorption (Fig. 2.1b) .The surface becomes hydroxylated by proton transfer from the bound water molecules to the neighboring oxide ions (Fig. 2.1c), a process that appears to be energetically favorable because of better charge neutralization or balance in the lattice is achieved [80] . In such a situation two types of O - H groups can be distinguished; those bound to one metal atom and those bound to two metal atoms. However, depending on the crystal morphology of the adsorbent triply coordinated site has also been proposed [81]. Boehm [80] has observed that doubly coordinated OH groups are strongly polarized by the cations, thus loosening the bond to hydrogen and resulting in the acidic character by releasing H^+ and therefore constitute the major cation adsorption sites. For singly coordinated OH groups the polarization is much weaker and these groups exhibit basic character and adsorb anions depending on pH of the solution. Infrared spectroscopy also suggests that H_2O may act as bridge between two metal atoms (M_2OH_2). This site, though contains potentially two ionizable protons, has only one O - H group. Further adsorption of H_2O vapour on hydroxylated surface takes place through hydrogen bonding on singly coordinated O - H sites as illustrated in Fig. 2.1d. All these forms of surface hydration have been verified by infra red spectroscopy, the important findings of which have been summarized by James and Parks [60]. However, some authors [82] view the hydroxylation of oxide surface in a slightly different way. It is believed that when an oxide is immersed in water , instead of adsorption of molecular water, H^+ and OH^- ions are adsorbed on lattice oxygen (Lewis base) and metal ion (Lewis acid) sites respectively through the sharing of electron pairs or simply acid /base bonding . In general, most oxides when exposed to open air will have adsorbed some moisture to form partially hydroxylated surface before being immersed in water. However, prolonged exposure of the surface leads to uniform hydroxylation and such surface when immersed in water the following reactions takes place.

$$H_s^+ + OH^-_{soln} = H_2O \tag{2.2}$$

$$OH_s^- + H^+_{soln} = H_2O \tag{2.3}$$

Where, the subscript s denotes adsorbed surface species. However, reactions (2.2) and (2.3) will tend to produce either net negative or net positive charge depending upon the pH of the solution. Because of the tendency to adsorb more OH^- or H^+ ions the solids sometimes are classified as "solid acid" or "solid base" respectively. The acidic character of doubly coordinated O - H bond and the basic character of singly coordinated O - H bond are sometimes called Lewis acid and Bronsted acids respectively.

The weakly acidic or basic nature of Bronsted sites is also evident from the fact that molecular water is adsorbed by hydrogen bonding on these sites at vapor pressure lower than that required for complete hydroxylation of the surface . For this reason it is difficult or even impossible to prepare a fully hydroxylated surface without adsorbed molecular water .

2.2 Infrared Spectra of Hydroxylated Surface

The infrared spectra of hydrous oxides and silicates have been extensively studied and are well-documented [83,84]. In air, at least three major ranges of absorption bands have been observed in the spectra of the above solids from which various types of OH groups as well as the occurrence of molecular water can be distinguished. Generally, the broad and intense band near 3400 cm-1 represent the O - H stretching of molecular water which tends to obscure the narrow bands occurring in the same region for surface O - H groups . Therefore, to identify the surface O - H groups it is necessary to remove free water as much as possible by prolonged heating at lower temperature (<100°C) with simultaneous evacuation [84]. Using this procedure the broad peak at 3400 cm^{-1} can be eliminated, while the H - O - H bending vibration at 1650 cm^{-1} can be used to detect very trace quantity of molecular water . However, for some solids such as TiO_2, it is extremely difficult to remove molecular water by evacuation at low temperature and requires temperature above 100°C [75] .This water is considered to be coordinately bonded to two Ti atoms . The narrow band near 3700 cm^{-1} observed in some silica or aluminosilicates samples is attributed to OH of the free silanol (SiOH) groups [83]. The band near 3660 cm^{-1} is also attributed to similar group but linked by hydrogen bonding to adjacent oxide or hydroxide or to molecular water. However, more than one band can be observed in many hydrous oxides for free OH groups in the range of 33750 - 3660 cm^{-1} and for the hydrogen bonded OH groups in the range of 3470 - 3660 cm^{-1}. Most of the investigations in this direction have been carried out in connection with the research on the development of oxide and silicate based catalysts [66].

An important objective of the study of infrared spectra of hydrous oxide substrates is the qualitative and quantitative estimation of the acidity (or basicity) of surface OH groups. This is usually carried out by allowing the adsorption of such substances which form complexes by accepting or donating protons to the hydroxylated surface. The shift in O -H stretching vibration frequencies gives the measure of acidity or basicity of the surface . Attempts have also been made to distinguish between surface and bulk OH groups from the shift in absorption band caused by the exchange of surface hydrogen with deuterium or tritium [83]. But it requires considerable experience to extract unambiguous information in this regard. This aspect is briefly discussed in the following section. For more details readers should consult the relevant articles in the monographs by Farmer [83] , Parfitt and Rochester [85], Hair [86] , etc.

2.3 Determination of Density of Surface OH Groups

Many methods have been employed to determine the density of surface OH groups, i.e , the number of OH site per unit area , usually nm^2 . Most of the methods, except IR spectroscopy, involve measurement of maximum sorption of various cations and anions that are capable of interacting chemically with the oxide surface. These are discussed below.

2.3.1 Infrared Spectroscopic Method

The determination of the density of surface OH groups of an oxide adsorbent using infrared spectroscopy requires accurate calibration of the instrument with solids having known amount of adsorbed or desorbed OH groups. This quantity is generally determined by gravimetric or volumetric method and due correction has to made for the presence of molecular water, the quantity of which may be assessed also from the calibrated intensity of H - O - H bending vibration band near 1660 cm^{-1}. Using a pre-calibrated instrument system the integrated intensity of all OH groups (free and combined) is obtained from which the number of OH groups intercepted by the IR beam can be calculated knowing the amount of H_2O vapour adsorbed and Avagadro's number. By following this method Anderson [87] found 4.4OH groups / nm^2 on silica gel surface. Bases like NH_3 and pyridin vapor are used as adsorbents whereby the protons donated by acidic sites combine with these bases. The shift in the characteristic absorption band of NH_3 and pyridin indicate their association with acidic OH groups. Instrument calibrated with known amount of adsorption can provide the measure of surface site density [80,85].

2.3.2 Potentiometric Titration Method

The maximum uptake or release of proton by hydrous oxides at extreme pH values (albeit, without phase change or dissolution of solid) can be obtained from potentiometric or acid - base titration method . The uptake values have been considered to be representative of the total ionizable surface sites. However, certain care has to be taken to extract as much true value as possible. For example, titration carried out within a narrow pH range or at low ionic strength of the solution will provide underestimate for surface site density. At low ionic strength all the surface groups are not fully ionized since high surface potential generated under this condition also inhibit complete ionization. Again, the adsorption of cations and anions (including H^+ and OH^-) is sensitive to pH variation. While for cations the adsorption increases with increase in pH, an opposite effect is observed for anions. The method has been used to determine the site density of the hydrous oxides of iron and manganese [88, 89]. If adsorption is carried out at constant temperature and pH values (adsorption isotherm) instead of acid - base titration, the experiments need to be carried out at pH where the adsorption of ions is maximum [80,90]. Therefore, unless a clear saturation plateau is observed in the potentiometric titration curve, one has to be careful in using the titration data for the estimation of surface site density.

2.3.3 Adsorption of Site Specific Reagents

It has already been stated above that there is variation in the strength of the surface acid (or basic) sites. In other words, the reactivity of the hydroxylated oxide surface towards different reagents which may react with either acidic or basic or amphoteric sites is different. The concentration of OH groups, if measured by the adsorption of suitable reagents capable of reacting with weakest acidic or basic groups, should be equal to the total OH groups present. Based on this approach , Boehm [80] has used several such reagents and measured the total concentration of surface OH groups . Alternatively , these have been characterized and estimated spectrophotometrically by titrating the solid in non-aqueous medium in the presence of a variety of indicator dyes having different pK_a values and noting the change in color of the adsorbed dyes at different pH values. The reagents are capable of reacting with even the weakest acid or basic sites and therefore gives total value of surface OH groups [91] . The method , however, suffers from two major drawbacks: (1) It fails to distinguish OH from adsorbed H_2O. (2) It is difficult to observe the colour change of the dyes with pH in the background of non-white substances like oxides of iron, manganese, chromium, cobalt etc. Evidences have been presented to show that the color change of indicator dyes assumed for acidic OH groups on montmorillonite surface is actually due to ionization of H_2O present in the hydration sheath of exchangeable cations [92]. Despite these, the method has been extensively used for characterizing OH groups in catalysts [91] , clays and clay minerals [93, 94] . All these suggest that the hydroxylated surface contains a variety of OH and H_2O sites which may function as acid or base or both (amphoteric) .

Anions like phosphate, fluoride, selenite and oxalate have also been used as adsorbates for the determination of OH groups [59]. Some authors [80,95] have used Grignard reagent as adsorbent after drying the solid to remove the physically adsorbed or molecular water. Adsorption of methanol has also been used to determine the number of OH sites of a wide range of oxides [96]. Laragne and Brenet [97] developed a method which is based on replacing the H^+ of the OH group with K^+ from KI in the presence of anhydrous dimethyl formamide and small quantity of glacial acetic acid. After 2 - 3 hr of equilibration the suspension is filtered taking enough care not to bring the content any possible contact with water and the liberated iodine is titrated against 0.01M sodium thiosulphate to the disappearance of iodine color. Therefore, in iodine equivalent 1 ml of 0.01M $Na_2SO_3 \equiv 10^{-5}$ mole of OH group . Kanungo and Parida [98] have observed that surface area values calculated from surface OH groups determined by this method for a number of synthetic MnO_2 are in good agreement with those determined by low temperature nitrogen adsorption or BET method, except a few highly porous samples with high surface area values. It has also been shown by the authors that adsorption of Cu^{+2} on these MnO_2 samples show better correlation with areas derived from surface OH groups than those with the BET areas.

However, in practice, wide differences have been observed between the results obtained by different methods due to various factors, such as difference in the sample preparation and subsequent treatment before measurement, adsorbate size and arrangement of functional groups , pH and ionic strength of the medium etc . For example, small ionic size and strong Fe-F bond promote high adsorption of F^- on hydrous iron oxide. Hiemstra and van Riemsdijk [99] did not observe well - defined maxima for F^- ion adsorption and concluded that F^- should not be used as a probe for the density of surface site density. Similarly, phosphate ion adsorption on different iron oxides usually gives lower values owing to the difficulty in reaching site saturation even at very low pH value [100] . Potentiometric titration may underestimate site density because of the accumulation of high surface charge at extreme pH values prevent complete surface ionization due to lateral interaction.

2.3.4 Isotope Exchange Method

Because of several constraints involved in the determination of surface OH groups by chemical reaction , isotope exchange of H with deuterium or tritium has been found to provide better estimate of site density . Using a longer equilibration period it is possible to estimate site density of micro porous solids . For non - porous solids lower equilibration period or rapid isotope exchange may be sufficient . In this method an oxide an oxide sample is labeled with tritium by dispersing it in tritiated water of known activity and agitating for a set period of time. The suspension is then centrifuged, the aqueous phase is removed by decanting and its activity is determined. A known volume of distilled water is added and the new suspension is agitated. At subsequent times the contents are centrifuged to remove supernatant liquid for activity measurement. The number of exchangeable protons is calculated from the total activity removed from the surface [101-104].

2.3.5 Site Density from Crystal Structure

Considerable amount of work has been carried out on the theoretical estimation of surface OH groups based on crystal structure of the oxides [105]. However, to perform such estimation it is necessary that the oxide substrates have well-defined crystal structure. The distribution of surface area among various crystal faces must be known, since in reality, solids generally have a wide range of fractal surfaces. Statistical microscopic observation supported by electron diffraction studies are some of the essential tools required for such studies.

As stated earlier in this section surface OH groups are coordinated to one, two or even three metal atoms in the crystal. There is also another type in which two OH groups are coordinated to one metal atom .The configuration of the various types of OH groups on frequently occurring planes of goethite and hematite surface has been determined by Barron and Torrent [81]. Owing to the differences in the coordination of OH groups to the underlying metal atom, the reactivity of these groups should vary. In contrast to the singly coordinated OH groups which occupy nearly 50% of the total surface sites , the doubly coordinated OH groups on goethite , hematite and lepidocrocite are inert over a wide pH range [106-109] . While in the case of goethite both singly and doubly coordinated OH groups are equally distributed , in case of hematite doubly coordinated OH groups are concentrated on **[001]** plane which remain inactive even at pH 10. Therefore , in case of those two oxides of iron site density has to be determined either at very low or very high pH value. However , the adsorption of impurity ions may alter the adsorption behavior of the inactive crystal planes **[110]**. Jones and Hockey [105] and Yates [75] have shown that for rutile and goethite 60% of the surface sites resides on **[110]** and **[100]** planes respectively (see ref. 59, p223). Table 2 .1 gives surface site density values of a number of solid oxides determined by different methods. However, if the unit cell parameters of a well - crystallized oxide is known, the surface site density may roughly be calculated from 2/3 power of the bulk concentration , i.e. $(n/V)^{2/3}$ where , n is the number of oxygen atoms per unit cell volume V $(A^0)^3$. The values obtained are generally close to the average of the site density values calculated for different crystal planes.

Table 2.1 : Site density (OH groups) of oxides and oxyhydroxides as determined by different methods .

Oxide/Oxyhydroxide	Surface area* (m^2/g)	Site density ($no./nm^2$)	Method of determination	Reference
Goethite	52	16.4	tritium exchange	i , ii
	70	6	acid / base titration	iii

		4	acid / base titration	iv
	51.8	2.6	acid / base titration	v
		1.68	acid / base titration	vi
		7.3	adsorption of F^- at pH 5.5	iv
		5.2	adsorption of F^-	vii
		0.8	phosphate adsorption	viii
		2.9	phosphate adsorption	ix
		1.5	SeO_3^{3-} adsorption	viii
		2.3	oxalate adsorption	x
		2.6	Pb^{2-} adsorption	xi
		7.0	Pb^{2-} adsorption	xi
Amor. iron oxide hydrate		11.4	tritium exchange	i , ii
(ferrihydrite)		10.2	tritium exchange	xii
		1.7 - 2.5 (based on the assumed surface area of 600 m^2/g)	various methods (as compiled by Dzombak and Morel)	xiii
	70	9.3	acid / base titration	xiv
		2.2	F^- adsorption	xiv
Akaganeite	51	15.6	acid / base titration	xiv
Lepidocrocite		8.0	acid / base titration	xv
		9.6	F^- adsorption	xvi
		1.68	acid / base titration	xvii
Hematite		4.4 - 9.1	IR spectra of adsorbed base	xviii
		5.5 - 10	IR spectra of adsorbed H_2O	i , ii
		5.5 - 9	calcd from crystal structure	i , ii
		22.4	tritium exchange	ii
		0.35 - 4.2	phosphate adsorption	xix
Magnetite		5 - 5.2	acid / base titration	xx , xxi
	1.73	14.2	calcd . from crystal structure	xxii
δ - MnO_2	74.1	221	tritium exchange	xxiii
	57.9	30.1	acid / base titration	xxiv
		6.0	acid / base titration	xxv
		8.1	acid / base titration	xxvi
Birnessite	43.6	25.8	acid / base titration	xxiv
α - Al_2O_3	12	30.5	calcd . from crystal structure	xxvii
γ - Al_2O_3	13	8.0	IR study of adsorbed and desorbed CO_2	xxviii
Gibbsite	3.5 -16.8	6.0	fitting of surface titration data	xxix
Quartz	4.2	11.4	tritium exchange	ii
Silica	144 - 390	4.6	tritium exchange	xxx ,xxxi
Anatase	7	15.8	calcd . from crystal structure	xxii

---

* Surface area by low temperature N_2

(i) Yates , D.E. , The structure of oxide/aqueous electrolyte interface . *Ph .D . dissertation Univ. Melbourne, Melbourne, Australia* . (ii) Yates ,D. E.,Greise, F., Cooper, R. and Healy , *T. W. Austr . J . Chem . 30(1977)* , 1656 . (iii) Huang , *C. P., Ph . D. Thesis , Harvard University , 1971* (iv) Sigg , L . and Stumm , *W. Colloids & Surfaces*, 2(1981) , 101 . (v) Balistrieri , L S. and Murray , J. W. , *Am . J . Sci* . 281(1981), 788 .(vi) Lovgren, L ., Sjoberg , S .and Scindler , *P. W . Geochim . Cosmochim . Acta*, 54(1990) , 1301 . (vii) Hingston , F.J. , Atkinson R . J. Posner , A. M . and Quirk , J . P . 9th *Intl . Congr . Soil Sci . Trans* . vol1(1968) , 669 . (viii) Hansmann , D.D. and Anderson , M . A . *Environ . Sci . Technol.* 19(1985), 544 . (ix) Torrent , J., Barron , V. and Schwertmann , U . Soil Sci . Soc . Am . J. 54(1990) ,1007 . (x) Parfitt , R .L ., Farmer, V.C. and Russell , J. D. , *J. Soil Sci.* 28(1977) , 29 . (xi) Hayes , K . F., and Leckie , J. O ., In *Geochemical Processes at Mineral Surfaces* , editors : Davis , J. A. and Hayes K .F. , ACS symposium series 323 , Am Chem . Soc . , Washington D.C. p114,1986, (xii) Hsi , C. D. and Langmuir, D ., *Geochim . Cosmochim . Acta ,* 49(1985), 1931 . (xiii) Dzombak D. A . and Morel , F. M . M . *Surface Complexation Modeling: Hydrous ferric oxide* , John Wiley & Sons , New York , 1990 , 93 .(xiv) Kanungo , S. B. , *J. Colloid Interface Sci* . 162(1994) , 86 . (xv) Waite , T . D . and Morell , F . M. M., *J . Colloid Interface Sci* . 102(1984) , 121 . (xvi) Davis , S . H . R. and Morgan , *J., J .Colloid Interface Sci* . 129 (1989) , 63 . (xvii) Zhang , Y. Charlet, L . and Schindler, P. W. , *Colloids & Surfaces* , 63(1992) , 259 . (xviii) Boehm , P. , *Disc. Farad . Soc.* 52(1971) , 264 . (xix) Barron, V. Herruzo , M . and Torrent , *J., Soil Sc . Soc . Am . J .,* 52(1988) ,647 . (xx) Tamura , H . Majevic , E . and Meites , L ., J . *Colloid Interface Sci* . 92(1983) , 303 . (xxi) Sun , Z . X ., Su , F. W. , Forstling , W. and Samskog , P. O. , *J. Colloid Interface Sci* ., 197(1998) , 151 . (xxii) Sahai , N . and Sverjensky , D. A ., *Geochim . Cosmochim . Acta* , 61(1997) , 2801 , 2827 . (xxiii) Balistrieri , L . S . and Murray , J. W. Geochim . Cosmochim . Acta , 46(1982) , 1041. (xxiv) Tripathy , S . S ., Kanungo, S . B . and Mishra , S . K . , J . *Colloid Interface Sci* . , 241(2001) ,112 .(xxv) Pretorius , P. J. and Linder , P. W . , *Appl . Geochem* ., 16(2001) , 1067 . (xxvi) Fu ,G ., Allen , H . E . and Cowan , C . E ., *Soil Sci* . 152(1991) , 72 . (xxvii) Koretsky , C. M ., Sverjensky , D . A . and Sahai , N., *Am . J. Sci* . , 298(1998) , 349 . (xxviii) Peri , J. B ., *J . Phys . Chem* . , 70(1966) , 3168 . (xxix) MacKinley , J. P. ,Zachara , J.M ., Smith , S . C. and Turner , G . D . , *Clays Clay Minerals* , 43(1995) , 586 . (xxx) Davydov ,V.Y. , Kiserlev , A . V . and Zhuravlev , L. T. , *Trans . Farad . Soc* . 60(1964) , 2254 . (xxxi) Madlev , J. D. and Richmond , R . C. , Z . *Anorg . Allgem . Chem* . , 380(1972) , 92 .

CHAPTER 3
THE IONIZATION OF OXIDE SURFACE

3.1 The Ionization of surface hydroxyl groups

Surface OH groups of hydroxylated oxides undergo amphoteric ionization depending upon the pH of the aqueous medium of constant ionic strength. The situation, in fact, corresponds to adsorption or desorption of H^+ or acid / base equilibria of the oxide surface and can be described by the following reactions:

$$SOH + H^+ \overset{1/K_{a1}^{int}}{\longleftrightarrow} SOH_2^+ \qquad (3.1)$$

and

$$SOH \overset{K_{a2}^{int}}{\longleftrightarrow} SO^- + H^+ \qquad (3.2)$$

where 'S' represents the surface of an oxide. The corresponding equilibrium constants may be written as

$$\frac{1}{K_{a1}} = \frac{\{SOH_2^+\}}{\{SOH\}[H^+]_s} \qquad (3.3)$$

$$K_{a2} = \frac{\{SO^-\}[H^+]_s}{\{SOH\}} \qquad (3.4)$$

where { } denotes the concentration of surface species in moles / kg of solid, [ ] denotes concentration in moles / dm^3 and subscript 's' denotes the term at the surface. The concentration of the surface species may be expressed in the same units as the concentration of the dissolved species by multiplying the former with A / V, e.g.,

$$[SOH] = \frac{A\{SOH\}}{V} \quad \text{moles / dm}^3 \qquad (3.5)$$

where, A = quantity of solid in kg and V = volume of aqueous phase in dm^3. Reaction (3.1) may also be expressed in the reverse order to make the units of both K_{a1} and K_{a2} similar i.e., mole / dm^3. The constants K_{a1} and K_{a2} are known as 'conditional' acidity or stability constants as their values are dependent on the prevailing experimental conditions, such as temperature and ionic strength. In other words , the values depend on the concentration of surface species and consequent generation of surface charge.

The 'conditional' acidity constants can be determined from the potentiometric or acid / base titration of the surface at constant temperature and ionic strength of the medium [111,112]. The concentration of the unreacted OH sites [SOH] is obtained by subtracting the concentration of ionized OH sites at any pH from the total sites i.e., Ns (moles/kg) .The charge on the surface therefore influences the values of Ka_1 and Ka_2. The higher (lower) the positive charge at the surface, the more difficult (easier) is to adsorb (remove) protons. The acidity constants are , therefore , not constant but depend also on the surface charge .

The 'intrinsic' acidity constants, on the other hand, represent the dissociation constants of the OH groups at zero surface charge condition i.e., purely chemical or thermodynamic constants at infinitely low ionic strength . The constants cannot be determined directly. They are, however, can be obtained by plotting the experimentally determined conditional constants against the concentration of the relevant charged species (SOH_2^+ or SO^-) and extrapolating to zero charge condition and at zero ionic strength of the medium. These may be empirically related as follows:

$$K_a(X) = K_a^{int}(X) \exp(-\alpha \{X\}) \tag{3.6}$$

or,
$$\log K_a(X) = \log K_a^{int}(X) - (\alpha / \ln 10)\{X\} \tag{3.7}$$

where , $K_a(X)$ is the conditional acidity or stability constant of any charged surface species ($K_{a_2}(SO^-)$, $K_{a1}(SOH_2^+)$, cation or anion adsorbed species etc.), $K_a^{int}(X)$ is the corresponding intrinsic constant and α is an empirical constant whose value is dependent on the system under consideration. In the subsequent section, method for evaluating α will be discussed. However, experimental results in many instances not only fail to obey the linear behavior of the plot of log $K_a(X)$ against $\{X\}$as envisaged by Eq (3.7), but also different curves are obtained for different ionic strength of the solution . One of the reasons for such deviation is that at very low surface coverage conditional and intrinsic constants are indistinguishable The other factors which are responsible for the observed decrease in stability constants ($K_a(X)$) with decrease in concentration of the charged surface species(X) are: (1) Existence of surface potential developed by adsorbed, charged species. (2) Lateral interactions between adsorbed species, resulting in an increase of their activity coefficients. (3) Surface heterogeneity.

Though all the above factors are somehow interrelated or one may even overlap the other, the influence of surface heterogeneity is probably least understood. This factor varies widely from sample to sample and for the same with the method of preparation. As a result, no comprehensive theory is available for calculating activity coefficients of surface species. Variation in the heat of adsorption of proton gives an idea of the extent of surface heterogeneity. The observable changes in the Gibb's free energy of adsorption (ΔG_{ads}) of proton is the sum of free energies of two components namely, chemical energy and coulombic or electrostatic energy.

$$\Delta G_{ads} = \Delta G_{chem} + \Delta G_{coul} \tag{3.8}$$

where, $\Delta G_{ads} = -RT\ \ln K_a$

$\Delta G_{chem} = -RT\ \ln K_a^{int}$

ΔG_{coul} is the energy required to bring an ion from the bulk of the solution to the solid/liquid interface against the potential ψ_o i.e., $zF\psi_o$. Substituting the free energy values in Eq (3.8) provides a direct derivation relating apparent or conditional constants with surface potential (charge)

$$\log K_a = \log K_a^{int} - (zF\psi_o / 2.303\ RT) \tag{3.9}$$

The above equation may also be derived by replacing the surface concentration of proton $[H]_s$ with its bulk concentration in Eq (3.3) or (3.4) using the following relationship

$$[H^+]_s = [H^+]\exp(-zF\psi_o/RT) \tag{3.10}$$

where, z is the charge of the adsorbed species and F is the Faraday constant in coulomb / mole . Since ψ_o of suspended particles is not easily accessible by experiments, various attempts have been made to determine its value by indirect method. In the following section we shall discuss the relationships between potential, charge and capacitance as envisaged by the simple Helmholtz constant capacitance model. The model proposed by Stumm, Schindler and their co-workers [39,111-113] is based on (i) linear relationship between surface charge (σ_o) and potential (ψ_o) i.e. potential drops linearly with increase in distance from the surface until net charge is zero, or (ii) an empirical correction term to the intrinsic surface acidity constant and is linearly dependent on charge as exemplified by Eq (3.9). According to this model, all ions which are specifically adsorbed contribute to the charge σ_o and the corresponding potential ψ_0. All non- specifically adsorbed counter ions are away from the surface i.e. in the bulk liquid phase (for further discussion , see chapter 5).

3.2 Protonation of Monofunctional Surface Groups :The Concept of One pK_a

Charged latex particles ,e.g . carboxylated polystyrene latexes have one type (usually negative) of surface charge. The adsorption of proton by such solids may be represented by

$$S^- + H^+ \longleftrightarrow SH \tag{3.11}$$

and the concentration of protonated surface is given by

$$[SH] = K_H\,[S^-]\,[H^+]_s \tag{3.12}$$

where K_H is the conditional constant . Rearranging we have

$$\frac{[SH]}{[S^-]} = K_H\left[H^+\right]_s \tag{3.13}$$

Since $N_s = [S^-] + [SH]$ we may replace the concentration terms with the fraction of the relevant species namely $\theta_H = [SH] / N_s$ and $1 - \theta_H = [S^-] / N_s$. Substituting in Eq (3.13)

$$\frac{\theta_H}{1-\theta_H} = K_H\,[H^+]_s \tag{3.14}$$

Assuming the constant capacitance model which is formally consistent with Eq (3.6) to be valid the surface potential ψ_o may be replaced by the ratio σ_0 / C_t where, σ_0 is the surface charge due to S^- $\{-F N_s (1-\theta_H)$, coul /m^2 $\}$ and C_t is the total capacitance (F / m^2) . Therefore ,

$$\psi_0 = \sigma_0 / C_t = \frac{-FN_s(1-\theta_H)}{C_t} \tag{3.15}$$

Replacing the surface concentration of proton with its bulk concentration (see Eq 3.10) and substituting Eq (3.15) for ψ_0 we have ,

$$\frac{\theta_H}{1-\theta_H} = K_H\left[H^+\right]\exp\left\{\frac{zF^2N_s(1-\theta_H)}{C_tRT}\right\} \tag{3.16}$$

The above principle of single acidity constant of latex surfaces has prompted the Dutch school led by van Riemsdijk [114-116] to propose "one pK" model for amphoteric dissociation of the oxide surface. It is assumed that the three positive charges of Al and Fe are equally distributed among six coordinate bonds in oxides so that each M - O bond carries +1/2 charge and each M - OH bond carries -1/2 charge (+1/2 for M-O and -1 for O-H bond). The bond charge has been discussed in more details with illustration later on in this chapter. The protonation reaction for the surface is given by

$$SOH^{½-} + H^{+} \longleftrightarrow SOH_2^{½+} \tag{3.17}$$

Unlike the simple surface protonation reaction (3.11) the reaction (3.17) is able to describe amphoteric character of the oxide surface.

From reaction (3.17) the concentration ratio of the two charged species is given by

$$\frac{\left[SOH_2^{\ 1/2+}\right]}{\left[SOH^{1/2-}\right]} = \frac{\theta_H}{1-\theta_H} = K_H\left[H^+\right]_S \tag{3.18}$$

where θ_H is the fraction of the protonated surface . Though Eq (3.18) is apparently similar to Eq (3.14), the difference between these two is distinct when the charge density is taken into consideration. Thus, from Eq (3.17) total charge for the protonated gibbsite type surface is given by

$$\sigma_0 = zF(-\{SOH^{1/2-}\} + \{SOH_2^{\ 1/2+}\})$$

$$\sigma_0 = 0.5\,F\,(-\{1-\theta_H\} + \theta_H) = FN_s\,(\theta_H - 0.5) \tag{3.19}$$

Consequently, the charge potential relationship is given by

$$\psi_0 = \sigma_0 / C_t = FN_s(\theta_H - 0.5)/C_t \tag{3.20}$$

Replacing the surface concentration of proton in Eq . (3.18) with its bulk concentration (Eq.3.10) and substituting ψ_0 with σ_0 / C_t , we have for gibbsite type surface

$$\frac{\theta_H}{1-\theta_H} = K_H[H]\exp\left(\frac{-0.5F^2N_s(\theta_H-0.5)}{C_tRT}\right) \tag{3.21}$$

From Eqs (3.16) and (3.21), it is clear that the pristine PZC for silica and gibbsite occurs at θ_H value 1.0 and 0.5 respectively. Koopal [117] has shown that by plotting $1-\theta_H$ against $\log K_H[H^+]_s$ the PZC value of silica rises asymptotically and charging becomes relatively difficult with increasing value of $\log K_H[H^+]_s$. On the other hand , for gibbsite PZC occurs at the sharp intersection at the point of zero charge and charging is relatively easy on either side of the PZC.

While deriving Eqs (3.16) and (3.21), if actual concentration terms of the charged surface species are used instead of fractional terms , surface potential ψ_0 is given by

$$\psi_0 = zF\{X^z\}/C_t S \quad \text{(volt)} \tag{3.22}$$

where, S is the area in m² / kg and $\{X^z\}$ is the concentration of the relevant surface species in moles/ kg. Substituting in Eq (3.9) the following linear relationship between apparent acidity or stability constant K_a and concentration of surface species $\{X^z\}$ is obtained as follows

$$\log K_a = \log K_a^{int} - \frac{(zF)^2}{2.303RTC_t S}\{X^z\} \tag{3.23}$$

Comparing Eq (3.7) with Eq (3.23) the slope value α is given by

$$\alpha = \frac{(zF)^2}{RTC_t S} \quad \text{(kg / mole)} \tag{3.24}$$

The above model has been further generalized by Hiemstra et al [108,118,119] who have suggested an integrated approach to surface protonation of oxides based on the following points.

1. The surface chemistry of the oxides is based on the concept of bond valence as enunciated by Pauling from the crystal structures of oxides.
2. Surface oxygens are considered as the sites for proton adsorption and they can be associated with one or two protons.
3. Surface planes may contain more than one type of surface oxygen and therefore different surface groups.
4. As consequence of point 3 crystalline oxides may expose different cyrstal planes and therefore different surface groupes in solutions.
5. A method is provided for *a priori* calculation of intrinsic affinity constants of the surface oxygens for protons on the basis of the behavior of hydrated metal oxide complexes in solution.

Points 1and 2 as stated above, lead to following generalized expressions for surface protonation reactions.

$$Me_n - O^{(n\nu-2)} + H^+ \overset{K_{n.1}}{\longleftrightarrow} Me_n - OH^{(n\nu-1)} \quad (3.24)$$

$$Me_n - OH^{(n\nu-1)} + H^+ \overset{K_{n.2}}{\longleftrightarrow} Me_n - OH_2^{\,n\nu} \quad (3.25)$$

where, ν is the bond valence and n is the number of metal ions coordinated to the oxygen groups; known as singly (n = 1), doubly (n = 2), triply (n = 3) coordinated oxygen etc. The bond valence is the charge of the metal ion divided by its coordination number: an idea based on local charge neutralization [120]. The amphoteric reactions (3.1) and (3.2) representing "two pK" model is the specific case of Eqs (3.24) and (3.25) for n = 1 and ν = 1. On the other hand, the "one pK" model (Eq (3.17)) is consequence of the case for n = 1 and ν = ½.

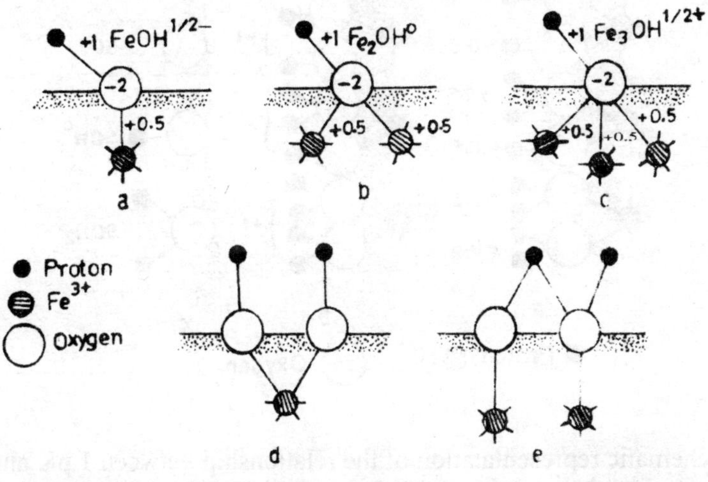

Fig. 3.1 Different types of coordination of O-H groups, namely (a) singly, (b) doubly and (c) triply coordinated O-H groups on a hydrous iron oxide surface. (d) and (e) represent geminal and viscinal OH groups respectively. See the text for explanation of the terms.

Figure 3.1 illustrates the different types of hydration and hydroxylation of metal ion of coordination number six. Thus, a, b and c represent singly, doubly and triply coordinated OH groups respectively depending upon the various crystal faces exposed in the solution phase. As shown in Fig. 3.1 protons attached to doubly coordinated (b - type) OH groups show more acidic character (Lewis acid) than the singly coordinated (a - type) OH groups which show more basic character (Bronsted acid). Therefore , while a - type site can act as either proton acceptor or proton donor (amphoteric) , b - type sites can only release proton . Though the density of total surface OH groups represent all forms of coordination , it is generally observed that site 'a' is more reactive than those of 'b' and 'c' type sites . The other two types of coordination are geminal (Fig. 3.1d) and vicinal (Fig. 3.1e), but they do not contribute significantly to the overall site density.

Figure 3.2 illustrates the relationship between one pK and two pK model as far as surface hydration is concerned. The surface metal ion adsorbs OH or OH_2 groups in different ways to satisfy its coordination. However, OH and OH_2 groups undergo further rearrangement through proton transfer so as to attain net charge which may be either -1 or 0 or +1 as envisaged by two pK model.

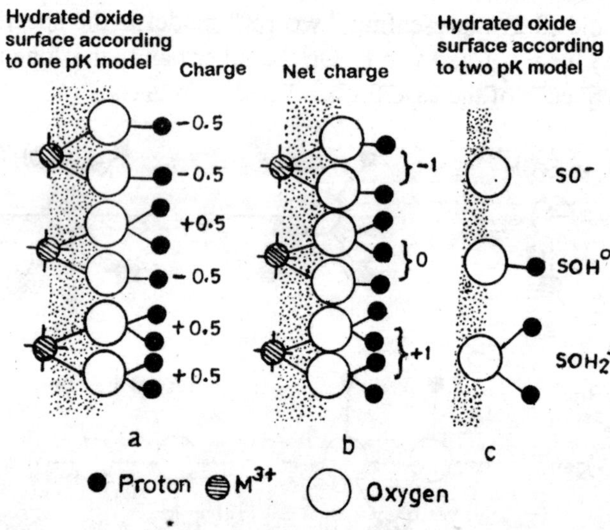

Fig. 3.2 Schematic representatation of the relationship between 1 pK and 2 pK models for a hydroxylated surface.

The equilibrium constants $K_{n.1}$ and $K_{n'2}$ suggest that for only one type of surface oxygen (mono functional surface) , two protonation reactions , in general , are possible. Hence, in principle, each homogeneous mono functional oxide surface can also be a "two pK" surface. In practice, however, it is quite possible that only one of the two reactions is 'active' in the accessible or relevant pH range so that such oxide behaves as a "one pK" surface .The description of the amphoteric oxide by two pK values, according to Koopal [117], is an artifact of "one pK" model since, for a homogeneous surface the difference between $\log K_{n.1}$ and $\log K_{n.2}$ is about 14 $\log K_H$ unit which is not found in actual practice. Therefore, one pK model is more appropriate if the surface is homogeneous and the charges of the surface groups are $1/2^-$ and $1/2^+$. For heterogeneous surface "one pK" model is applicable by considering such a surface as pseudo homogeneous. The 'activity' of the two reactions depends on either of the two $pK_{n.i}$ values. Therefore, *a priori* knowledge of the $pK_{n.i}$ values of different surface oxygen is of great help in describing the charging of oxides in solution. It may be noted that N_s value for "one pK" model is double the quantity assumed for the " two pK " model .

The importance of multisite protonation of oxide surface has been highlighted in points 3 and 4 . This is incorporated in the MUSIC (an acronym for multisite complexation) model proposed by Hiemstra et al [118,119]. This model envisages a method for *a priori* calculation scheme for intrinsic affinity constants. Similarly, the $pK_{n',i}$ values for the protonation of a number of metal oxides serve as reference to provide an alternative method for the determination of intrinsic affinity constants of many oxides , even if MUSIC model is not available.

The method, based on the principle that Gibbs free energy of the protonation of an oxide, is composed of explicit local electrostatic and other unspecified contributions. Hiemstra arrived at the following expression

$$\log K_{n.i} = A - B\,(n\nu / L) \tag{3.26}$$

where, A and B are constants , L is the distance between the metal ion and the adsorbed proton. The significances of n and ν have already been stated earlier. To obtain A and B Hiemstra [118,119] first studied protonation reactions of metal oxides in solution , for which $\log K_H$, n , L and ν values are known. However, a distinction has to be made between oxo and hydroxo complexes, as their respective A values may differ widely. There occurs a definite shift in the value of A from solution complex to surface complex for some oxides / hydroxides. For example, in the case of gibbsite this shift is about 4log K_H unit . This value also holds good for many amphoteric oxides and therefore gibbsite is used as primary reference material. However, in the case of poorly crystalline or near amorphous silica the value of A for both surface and solution complexes are almost similar. Thus knowing the values of A and B, Hiemstra has calculated the K_H values for different surface groups using n , ν and L values from structure and or crystallographic data . A discussion on the method of determining K_H including suggestion for simplification of calculation methods have been presented by Bleam [121]. Although the effect of n is incorporated in the model, it is found that for higher values of n (≥ 3) predicted values of K_H are subjected to approximation of one or two order of magnitude . The difference between upper and lower limits in the values of K_H is particularly higher for oxides with disordered

crystal faces or for oxides deposited on other surfaces .
From the foregoing discussion it may again be inferred that there are broadly two types of oxide surface, namely silica and gibbsite types. For the silica type which follows protonation reaction (3.11), the surface charge σ_0 is implicitly related to pH by substituting θ_H with $[1+(\sigma_0 / FN_s)]$ in Eq (3.14) as shown below

$$\sigma_o = \frac{-F N_s}{1+K_H [H_s]} \tag{3.27}$$

From the Gouy - Chapman - Stern model of the electrical double layer (discussed in detail in chapter 5) at oxide-water interface, it can be shown that

$$1/C_t = 1/C_{st} + 1/C_d \tag{3.28}$$

where, C_t , C_{st} and C_d are the values of total , Stern layer and diffuse layer capacitances respectively . C_d is equal to $\varepsilon_0\varepsilon_1\kappa$, where ε_0 is the permittivity in vacuum (8.854 x 10^{-12} C / V m) , ε_1 is the relative dielectric constant of solution (dimensionless) and κ^{-1} is the Debye length or thickness of the double layer which , in turn , is related to ionic strength . Combining Eqs . (3.10) ,(3.15) and (3.28) we have

$$[H]_s = [H]\exp\left\{\frac{-zF^2N_s}{1+K_H[H]_s}\left(\frac{1}{C_s}+\frac{1}{\varepsilon\varepsilon_o\kappa}\right)\right\} \tag{3.29}$$

From Eqs (3.27) and (3.29) it is therefore evident that the surface charge is related to C_s, N_s and κ (ionic strength of the solution) . The nature of such relationships for silica surface has been discussed by Koopal[117]. The higher the N_s and lower the salt concentration of solution the larger is the deviation from ideal behavior. Again, for low value of C_s the screening of surface charge by electrolyte is very poor and therefore charging of the surface becomes difficult.

Similarly for gibbsite surface the relationship between the surface charge and pH can be obtained by combining Eqs. (3.18) and (3.19)

$$\sigma_o = -0.5FN_s\left(\frac{1-K_H[H]_s}{1+K_H[H]_s}\right) \tag{3.30}$$

The above equation indicates that for gibbsite surface σ_0 is also related to N_s, C_s and κ. It has been observed that for more realistic i.e. , closest to experimental values σ_0 of higher values of N_s (~ 8 - 10 sites/nm^2) and C_s (~ 2 farad / m^2) are necessary. The σ_0 vs pH curves of several hydr(oxide)s such as rutile, hematite and goethite have been described satisfactorily with the simple "one pK " model (108 , 114, 115, 118). This model has the advantage that it is simple and that with a few parameters both surface charge and potential behaviors can be described. The log K_H value represents the overall pH_{PZC}. However, it has to be borne in mind that the model fails to reflect a complex situation like multifunctional surface. Further discussion on surface ionization and complexation has been presented in chapter 6.

CHAPTER 4
THE CONCEPT OF ELECTRIFIED SOLID / SOLUTION INTERFACE

4.1 The Origin and Distribution of Electrical Charge at Solid / Solution Interface

It has already been stated that oxides when immersed in aqueous solution tend to adsorb excess of either H^+ or OH^- resulting in the development of positive or negative charge at the interface depending upon the pH of the solution . At a particular pH, the adsorption of either of these two ions are equal and as a consequence , the solid surface does not acquire any net charge i.e., the net charge and potential also are also zero . This is known as point of zero charge (PZC). However, the most frequently asked question is: "How is the electrochemical equilibrium of a charged chemical species established at the solid / solution interface and how the charge distribution take place between the two phases ? To answer this question, it is believed that the activity and the state of the species in the bulk of the two phases i.e., solid and aqueous solution determine the actual potential difference between the two phases. This potential difference manifests itself in the redistribution of charge at the interface which has led to the development of various models of electrical double layer which will be discussed in the next chapter.

The redistribution of charge may occur in several ways , such as (i) charge transfer across the interface ; (ii) unequal dissolution of oppositely charged ions of which the solid phase is composed ; (iii) unequal adsorption from solution of ions of opposite charges and (iv) adsorption and orientation of dipolar molecules . Surface charge may also originate from lattice imperfection at the solid surface and isomorphpous replacement within the lattice. This is commonly observed in clays and clay minerals. There are basically two types of electrified interface which broadly exemplify the above ways of charge distribution at the interface. These are : (1) completely polarizable mercury - water interface where the potential difference across the interface is established by means of external circuit (e.g. potentiostat and a reference electrode) so that all the charged species have finite concentration by transferring only in one of the two phases i.e., way (i). On the other hand (2), at the completely reversible interface the electrostatic potential is resulted from the chemical interaction of the solutes with the solid surface through the ways (ii) and (iii). An ideal example is silver iodide - water interface, wherein Ag^+ and I^- ions dissolved from the solid phase serve as potential determining ions. Sulphides (e.g., ZnS, PbS etc) - water interface where M^{2+} and S^- act as potential determining ions, also belong to this category. Oxide - water interface though not an ideal example, but belongs to this category, wherein H^+ and OH^- determine the potential across the interface instead of M^{2+} and O^{2-}. The characteristics of the two types of electrified interface, which form the basis of various models of the electrical double layer, are essentially similar. They differ only in the manner in which the potential difference across the interface is established. The reader may consult original references for more details (122-124).

4.2 The Completely Polarized Hg - Water Interface

Different aspects of the Hg-Water interface have been discussed by Grahame [4]. A comparison of the results obtained from the two types of interface have been discussed by Stumm et al [125]. With a polarizable Hg electrode, the change in surface tension of the liquid phase is measured with variation in applied voltage which may be either positive or negative. The interfacial tension can be measured by various means e.g., by measuring the drop weight of the individual drops in the case of dropping Hg electrode or by means of a suitable (Lippman) electrometer. Figure 4.1

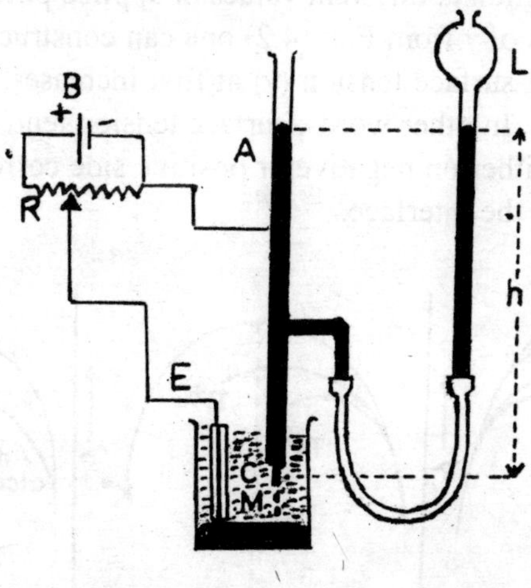

Fig. 4.1 An illustration of a typical dropping mercury electrode system (Lippman electrometer) used for studying the properties of Hg/water interface in preence of soluble ions. A dropping Hg electrode; B battery; C capillary; E reference electrode; L mercury level adjustor; M falling Hg drops; R rheostat.

shows a typical Lippmann electrometer which consists of a long tube A containing mercury and drawn out into a capillary C immersed into an aqueous electrolyte solution. The interface between the solution and the mercury at the capillary tip from where mercury drop emerges serves as a sensor for change in the tension at the interface between the two liquids. The change is manifested in the variation in height h between the upper meniscus and tip of the capillary C of the tube A. By connecting mercury in tube A and mercury in trough M through a reference electrode E to different poles of a battery B a definite potential could be applied and the resulting change in the interfacial tension can be determined from the variation in height h in the level adjustor tube L required to bring the level of mercury in A to its original position i.e., before the addition of potential. The relationship between the two quantities may be written in the following form:

$$2\pi r\gamma = \pi r^2 h \rho g \quad (4.1)$$

or, $$\gamma = (r h \rho g)/2 \quad (4.2)$$

where, r = radius of the capillary, cm

ρ = density of mercury, g / cm^3

g = acceleration due to gravity, 981cm / sec^2

The mercury which has initially a positive potential with respect to solution, is gradually made negative by varying the applied potential (EMF) with the help of rheostat R. By measuring the height h at different values of applied potential and calculating the corresponding values of γ from Eqn (4.2) one can construct the electrocapillary curve. It is observed that surface tension (γ) at first increases, reaches a maximum value and then decreases. In other words, surface tension tends to decrease with increase in applied EMF either on negative or positive side corresponding to adsorption of cation or anion at the interface.

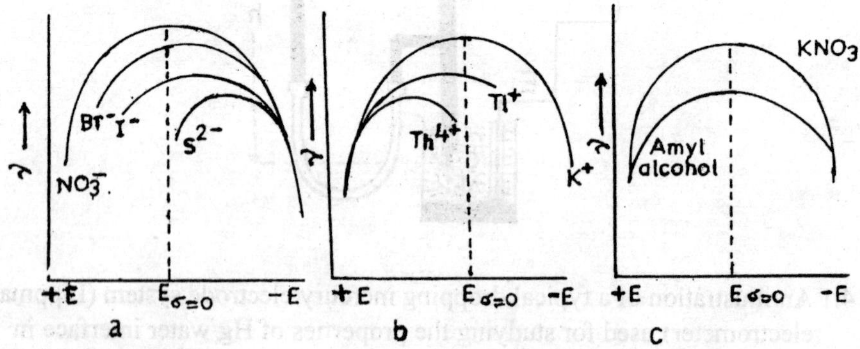

Fig. 4.2 Electrocapillary curves for aqueous solutions containing (a) anions, (b) cations and (c) non-ionic molecular substances like amyl alcohol. Adapted after ref. 127.

The shape of the electrocapillary curves depends on the nature and concentration of electrolyte solution. The presence of NO_3^-, Cl^-, Br^-, I^- and S_2^- ions as their potassium salts shifts electrocapillary maximum (ecm) towards more negative values of potential and lowers surface tension. At sufficiently high negative potential the effects of these anions disappear and electrocapillary curves (ecc) merge together (see Fig. 4.2a). On the other hand, the presence of K^+, Cs^+, Th^{4+} etc. displaces the ecm to more positive values of potential and also reduces the surface tension. The effect of these cations disappears at high positive potential where the curves tend to merge (see Fig. 4.2b). The presence of amyl alcohol, a nonionic molecular substance brings about change in the shape of ecc mainly in the region around ecm (see Fig. 4.2c). At far away from ecm the effect diminishes and the curves coincide with that in pure ionic solution. The substances which exert such an effect are called specifically adsorbed or capillary active substances. Therefore, while the presence of anions affects only the ascending branch (positively charged) the presence of cations affects the descending branch where the mercury surface is negatively charged. The effect of surface active agents on the ecc cannot be attributed to such simple electrostatic principle and chemical forces play distinct role.

The extent of adsorption can be predicted from Gibbs adsorption isotherm which is given by

$$\Gamma_i = -\left(\frac{\partial \gamma}{\partial \mu_i}\right)_{\mu_s, T, p} \tag{4.3}$$

where , Γ_i is the surface concentration of i th component .The right hand side represents the decrease in surface tension γ with variation in the chemical potential of the i th component at constant temperature , pressure and potential of other species . The conditions of applicability of Eq (4.1) to the electrocapillary phenomenon were worked out in detail by Grahame [4] . Thermodynamically , it can be shown [126,127] that the slope value at each point of the electrocapillary curve is equal to the charge density σ_0 on the mercury surface at a given potential Ψ at constant temperature , pressure and chemical potential of other components . Eq (4.3) is therefore modified as follows

$$\sigma_o = -\left(\frac{\partial \gamma}{\partial \psi}\right)_{T,P,\mu_i} \tag{4.4}$$

This is known as Lippmann's first equation. Using this equation one can calculate the charge density at any value of Ψ from the experimental curve and construct another curve of charge vs. potential . From the above equation it follows that on the ascending branch of the electrocapollary curve $\sigma_0 > 0$ i.e., d $\gamma > 0$, whence $d\gamma / d\psi <$

0 or $- d\gamma / d\psi > 0$. On the descending branch , $\sigma_0 < 0$ i.e.,$- d\gamma / d\psi < 0$. By defination , the differential capacity C_d is equal to the derivative of the charge with respect to potential and for electrocapillary curve this is expressed either as the first differential of charge with respect to potential or the second differential of electrocapillary curve . The following equation is also known as the Lippmann's second equation.

$$-\left(\frac{\partial \sigma_o}{\partial \psi}\right)_{T,P,\mu_i} = \left(\frac{\partial^2 \gamma}{\partial \psi^2}\right)_{T,P,\mu_i} = C_d \tag{4.5}$$

Equation (4.5) may also be expressed as follows:

$$d\sigma = C_d \, d\psi \tag{4.6}$$

Since, at the electrocapillary maximum $\sigma = 0$ and $\psi = \psi_{echem}$ integration of Eq. (4.6) under the above limiting conditions yields integral capacity at potential ψ.

$$\int_0^{\sigma} d\sigma = \int_{\psi_{echem}}^{\psi} C_d \, d\psi$$

or,

$$\sigma = C(\psi - \psi_{\sigma=0}) \tag{4.7}$$

Substituting ψ in Eq .(4.6) we have

$$d\{C(\psi - \psi_{\sigma=0})\}/ d\psi = C_d$$

Differentiating the term within the braces, one may arrive at the following relationship between C_d and C

$$C_d = C + (\psi - \psi_0)\, dC_d / d\psi \tag{4.8}$$

Generally, the differential and integral capacities are not identical; except under conditions where $\psi - \psi_{\sigma=0}$ or at $dC_d / d\psi = 0$. The differential capacity can be determined directly, for example by an a.c. (alternate current) method. In order to vary the electrode potential, a d.c.(direct current) voltage is superimposed upon the a.c. component. Figure 4.3 shows typical electrocapillary curves showing charge-potential and differential capacity - potential relationships in the presence of inert electrolyte solution. In the range of potential corresponding to the ascending branch of the curve, the capacity changes comparatively little and varies between 20 and 40 $\mu F / cm^2$. As the electrocapillary maximum approaches, the capacity passes through minimum (which is more pronounced at low concentration of electrolyte) and thereafter rises to a little extent; remains constant at about 20 μF / cm² over a wide range of potential on the descending branch of the curve. The addition of capillary active substances affects not only the electrocapillary curve but also σ - ψ and C_d - ψ curves.

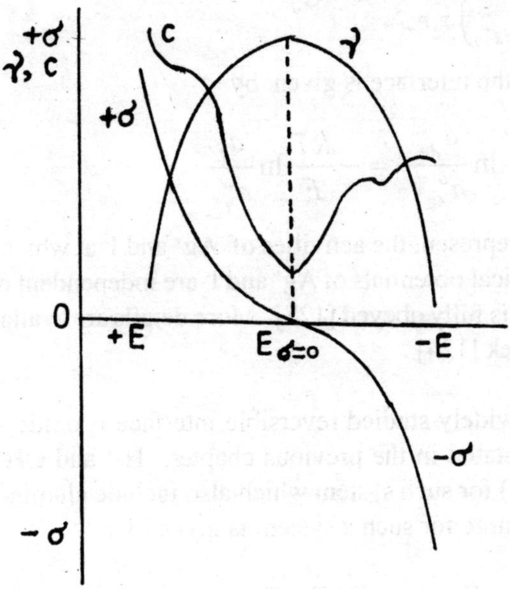

Fig. 4.3 Typical nature of the electrocapillary curves showing the variation of charge(σ), surface tension(γ) and double layer capacity with respect to variation in potential(E) at Hg/aq. KNO_3 solution interface.

4.3 The Reversible Interface

Except AgI - water, many of the solid - solution systems do not approach of ideality to the same extent as the polarized electrode. However, the reversible solid - solution system also provides a convenient tool to explore quantitatively the distribution of charge at the interface, more precisely than the other systems . Thus, in this system the interface is in equilibrium with Ag^+ and I^- depending upon the solution concentration

$$\sigma_o = F(\Gamma_{Ag+} - \Gamma_{I}^{-}) = -F(\Gamma^{+} - \Gamma^{-}) \tag{4.9}$$

where, Γ is the surface excess of either Ag^+ or I^- ion or in this respect any other cation or anion respectively used to vary the electrolyte concentration of the system . The charge - concentration relationship is obtained by determining the extent of adsorption (surface excess) by varying either $[Ag^+]$ or $[I^-]$ concentration in the medium at constant ionic strength. By analogy with the Lippmann's equation for Hg electrode the differential capacity is given by

$$\frac{F}{RT}\left(\frac{\partial \sigma_o}{\partial a_{Ag^+}}\right)_{T,P\,\mu+} = -C_d \qquad (4.10)$$

The potential ψ_0 at the interface is given by

$$\psi_0 = \frac{RT}{F}\ln\frac{a_{Ag^+}}{a^o_{Ag^+}} = -\frac{RT}{F}\ln\frac{a_{I^-}}{a^o_{I^-}} \qquad (4.11)$$

where, a^o_{Ag+} and $a^o_{I^-}$ represent the activities of Ag^+ and I^- at which $\psi_0 = 0$, assuming that the surface chemical potentials of Ag^+ and I^- are independent of their bulk activities in aqueous phase is fully obeyed [128]. More details are available in the paper by Lyklema and Overbeek [124].

However, the most widely studied reversible interface is oxide - aqueous solution system. As already stated in the previous chapter, H^+ and OH^- are the potential determining ions (pdi) for such system which also include aluminosilicate minerals. Therefore, surface charge for such a system is given by

$$\sigma_0 = F(\Gamma_{H^+} - \Gamma_{OH^-}) = -F(\Gamma^+ - \Gamma^-) \qquad (4.12)$$

4.4 Determination of Charge at the Oxide - Water Interface by Potentiometric Titration

The most common and widely used method for measuring proton excess or deficiency is potentiometric or acid - base titration of the oxide surface in the presence of an indifferent or inert electrolyte solution. This means that the ions of electrolyte solution do not contribute to the surface charge of the oxides. Further, titration has to be performed in an atmosphere free from gaseous CO_2 which interfere with the sensitive titration data. Since, during the titration oxide surface either gain or lose protons Eq. (4.12) may also be written as

$$\sigma_0 = F(\Gamma_{H^+} - \Gamma_{OH^-}) = F/A\,S\,\{[SOH_2^+] - [SO^-]\} \qquad (4.13)$$

where A is the surface area in m^2/g and S is the concentration of solid in g / l.

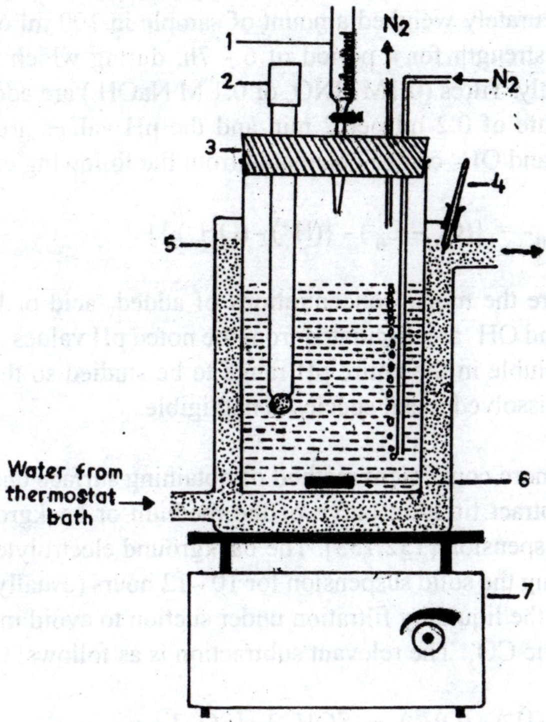

Fig.4.4 Sketch of a jacketed vessel used for rapid potentiometric titration under constant temperature and under inert atmosphere. The significances of the numerals are as follows: 1 burette, 2 combined pH electrode, 3 rubber bunge, 4 thermometer, 5 jacketed vessel, 6 teflon coated magnetic paddle, 7 magnetic stirrer.

4.4.1 Procedure for Rapid Potentiometric Titration

Details of the potentiometric titration are available in the papers by Parks and Debruyn [129], Atkinson et al [130],Yates and Healy [131] etc. These papers provide the design of titration vessel also. However, Fig. 4.4 shows the sketch of the one used by the present author for rapid potentiometric titration. The method is based on the principle of adding small increments of either acid or base of known concentration to a known volume of well stirred suspension of solid and noting the change in the pH after 2 - 3 min. This is known as rapid titration method. The titration cell consists of a wide mouthed jacketed glass vessel of 150 ml capacity . The open end or the mouth of the vessel is closed with a rubber bunge of 10 mm thick, having suitable provisions for inserting a combined glass electrode of the pH meter, nitrogen inlet and outlet tubes, titration burette etc. Water from thermostat bath is allowed to circulate through the jacket and its temperature is noted at the exit point. Titration is carried out after equilibrating an accurately weighed amount of sample in 100 ml of 1:1 electrolyte solution of known strength for a period of 6 - 7h, during which pure nitrogen is bubbled intermittently. Titres (0.1M HNO_3 or 0.1 M NaOH) are added in the stirred suspension at the rate of 0.2 ml per 2 min and the pH values are noted .The net consumption of H^+ and OH^- can be calculated from the following expression .

$$\Gamma_{H+} - \Gamma_{OH^-} = [(C_A - C_B) - \{(H^+) - (OH^-)\}] \tag{4.14}$$

where C_A and C_B are the molar concentrations of added acid or base. The molar concentrations H^+ and OH^- are calculated from the noted pH values . The solid oxide must be highly insoluble in the entire pH range to be studied so that the uptake or release of OH^- by dissolved metal species is negligible.

An alternative and more convenient method of obtaining surface charge due to proton uptake is to subtract titration curve of the medium or background electrolyte solution from the suspension [132,133]. The background electrolyte solution is obtained by equilibrating the solid suspension for 10 - 12 hours (usually overnight) and carefully separating the liquid by filtration under suction to avoid minimum contact time with atmospheric CO_2. The relevant subtraction is as follows:

$$(C_A - C_B)_{susp} = (H^+) - (OH^-) + [SOH_2^+] - [SO^-] + y$$

$$(C_A - C_B)_{blank} = (H^+) - (OH^-) + y$$

$$\Delta(C_A - C_B) = [SOH_2^+] - [SO^-] \tag{4.15}$$

Where, y represents unknown constituents which consume H^+ and OH^- in the system, $[SOH_2^+]$ and $[SO^-]$ are the concentrations of positively and negatively charged surface species in moles / l . Substituting Eq . (4.15) in Eq . (4.13) we have

$$(\Gamma_{H^+} - \Gamma_{OH^-}) = \Delta(C_A - C_B) / A.S \tag{4.16}$$

Instead of calculating $\Delta(C_A - C_B)$ for each pH value , it is more convenient to plot titration data (i.e., volume of either 0.1M strong acid or 0.1M strong base added vs . pH) of suspension as well as blank solution in the same graph . The differences in the volumes of acid or base consumed by the suspension and the blank (Δml) is plotted against the corresponding pH values . The resulting curves generated for different concentrations of electrolyte solution should intersect at a common pH value which is the point of zero charge. The Δml value of acid at the pH of zero charge is noted which is then subtracted from Δml values of acid or added to the Δml value of base consumed at different pH values. These values are then converted to either meq / g or µcoul / cm^2 (since 1 ml of 0.1M acid or base ≡ 0.1 meq ≡ 9.6500 coulomb of charge) if surface area is known, and plotted against pH relative to PZC where all the curves meet. Such type of plots are usually reported in literature [129,130].

4.4.2 Potentiometric Titration by Batch Method

Unlike the rapid method batch titration is essentially an equilibrium method. Accordingly, to a series of well stoppered polyethylene method (125-150 ml capacity) containing appropriate solution of electrolyte, different volumes of 0.1M strong acid (HCl or HNO_3) or 0.1M strong alkali (NaOH or KOH) are precisely transferred in such a way that the total volume of solution in each bottle is exactly same without affecting the ionic strength of the medium. Pure nitrogen is bubbled for a few minutes in each bottle which is then capped immediately. After 2 - 3 hours the pH values of the solutions are noted. Accurately weighed, known quantity of solid is transferred to each bottle. This is followed by bubbling the solutions again with pure nitrgen for 3 - 4 min

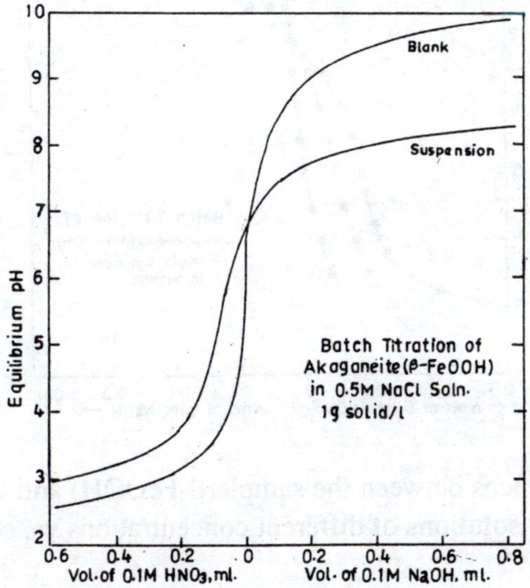

Fig.4.5 Typical batch titration curves (plot of volume of acid or base added vs. pH) of a hydrous iron oxide sample (β-FeOOH) as well as blank solution containing 0.5M NaCl.

after which the bottles are immediately closed. The bottles are then equilibrated for a predetermined period at a constant temperature after which the pH values of the supernatant liquids are noted. Like the rapid titration method volumes of acid or alkali added are plotted against equilibrium pH for both for both suspension and blank on the same graph. This is illustrated in Fig. 4.5 for the titration of akaganeite (β - FeOOH) in 0.5M NaCl solution. The blank solutions in this case are the electrolyte solutions in the bottles whose pH values are noted after adding known volumes of either 0.1M HNO_3 or 0.1M NaOH. Figure 4.6 shows the plot of difference in titre values between suspension and blank (Δml) for three different concentrations of NaCl solution against corresponding pH .It may be noted from the figure that the three curves intersect at pH 7.05 with Δml value of 0.023 ml of 0.1M NaOH. This value is added to every difference (Δml) value in the acid addition region and subtracted from every Δml value in the alkali addition region so as to obtain a revised set of Δml values with respect to pH_{PZC} for each concentration of electrolyte solution. Following the procedure for calculation of surface charge as discussed in the previous section proton adsorption or release values are plotted either as meq/g or as μCoul/cm² against equilibrium pH . This is exemplified in Fig. 4.7. More examples of batch titration are available in references 88 and 89.

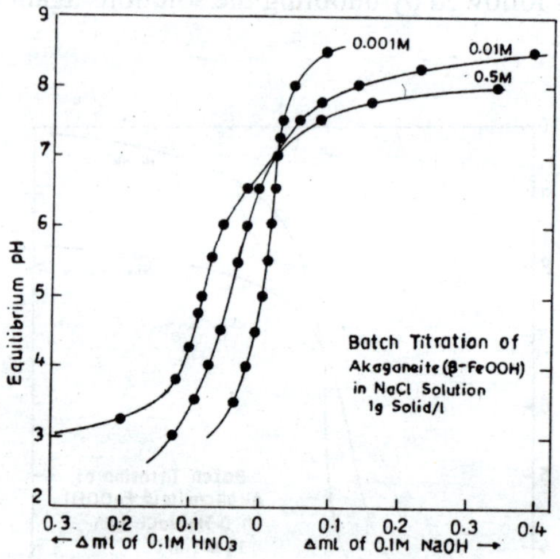

Fig. 4.6 Plots of differences between the sample(β-FeOOH) and the blank titration curves in NaCl solutions of different concentrations vs. corresponding pH values.

4.4.3 Comparison Between Rapid and Batch Method

The starting pH and the rate of titration are the two important factors in the acid - base titration of solid (oxide) surface. The starting pH is the pH at which the solid is in equilibrium with the inert electrolyte solution before the start of titration process. This may lie on the acidic or alkaline side depending upon the nature of the solid

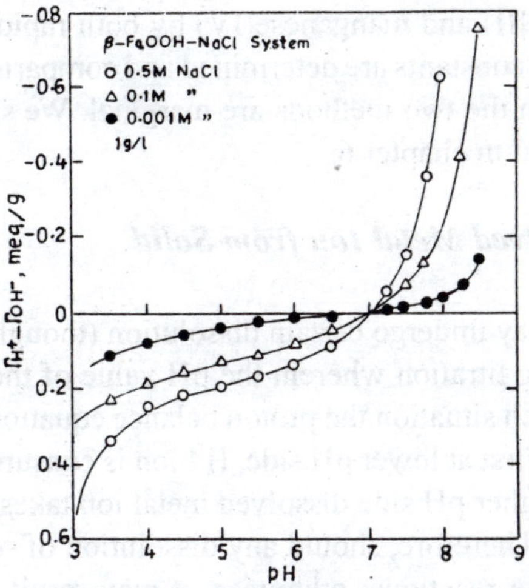

Fig. 4.7 Final representation of the batch titration curves for β-FeOOH in terms of the adsorption of H^+ or OH^- ion with respect to pH_{PZC} vs. pH.

sample and also on the concentration of the solution. Further, kinetics of proton adsorption or desorption is governed by two - step process : a fast initial step followed by a much slower second step. The initial step, wherein the adsorption / desorption process takes place essentially on the surface is complete within few minutes. The second step is generally thought to be the result of exchange within the interior of the solid particles. Rapid titration is therefore employed to suppress this step . If titration of the suspension is not performed rapidly the pH value drifts continuously and the phenomenon of hysterisis between acid and base titration will appear (Ref. 27, p51). Even for fast titration, hysterisis has been observed, more significantly with amorphous than with crystalline oxides [41,75,104].

However, there is no universally agreed view on how fast the titration has to be carried out . The present author has observed that during the titration of hydrous oxides of iron and manganese drift in pH value occurs from the moment acid or base is added to the suspension and continues for about 4 - 5 minutes at the same rate and then at slower rate for more than 15 minutes . Therefore, we followed an arbitrary rate of 0.2 ml per 2 min. An important question that still remains unanswered is how far the acidity constant data obtained from rapid titration is applicable to long term equilibrium state e.g . in natural solid / solution systems . It is also necessary to examine if there is any difference in the surface charge and acidity constant values derived from both rapid and batch titration of the same oxide sample. Kanungo and his coworkers [88,89,90,134] have carried out potentiometric titration of amorphous hydrous oxides of iron(III) and manganese(IV) by both rapid and batch methods. From these data, acidity constants are determined and comparison of the results show that differences between the two methods are marginal. We shall, however, discuss this aspect in more detail in chapter 6.

4.4.4 Effect of Dissolved Metal Ion from Solid

Some hydrous oxides may undergo certain dissolution (though a very small fraction of the total mass) during titration wherein the pH value of the suspension is varied over a wide range. In such situation the proton balance equation may be significantly disturbed in two ways. First at lower pH side, H^+ ion is consumed for the dissolution of solid. Secondly, at higher pH side dissolved metal ion takes up OH^- to form various hydrolytic species. Therefore, should any dissolution of solid occur, it is necessary to account for these reactions; otherwise, it may result in either high or low surface charge estimates. For example, dissolved Al^{3+} from hydrous Al_2O_3 will form several hydrolytic species in solution for which total proton balance equation is modified as shown below [133]:

$$(\equiv AlOH_2^+) - (\equiv AlO^-) = C_A - C_B - \{(H^+) + (OH^-) + AlOH^{2+} + 2Al(OH)_2^+ + 3Al(OH)_3^\circ + 4Al(OH)_4^-\} \quad (4.17)$$

In order to incorporate the required correction for the dissolved metal ion in the titration data, it is necessary to know the concentration of total dissolved metal ion at each point, and the concentration of each species is calculated from literature value of their formation constants. Surface charge is then calculated using equation of the type given above for hydrous Al_2O_3. This is because usual determination of surface charge by subtracting the blank titration curve from suspension titration curve will not be applicable in this case. The suspension blank contains only metal ions dissolved at the initial conditioning pH. It does not account for H^+ produced or released by solid precipitation or dissolution at other pH values.

4.5 Some Other Related Methods for the Determination of Point of Zero Charge of Oxides

A few related methods have been developed for determining pH_{PZC}. Though simple, these methods are not as quantitative as the potentiometric titration. Therefore, they are not used either for the determination of surface charge or for any other interfacial parameters. These are:

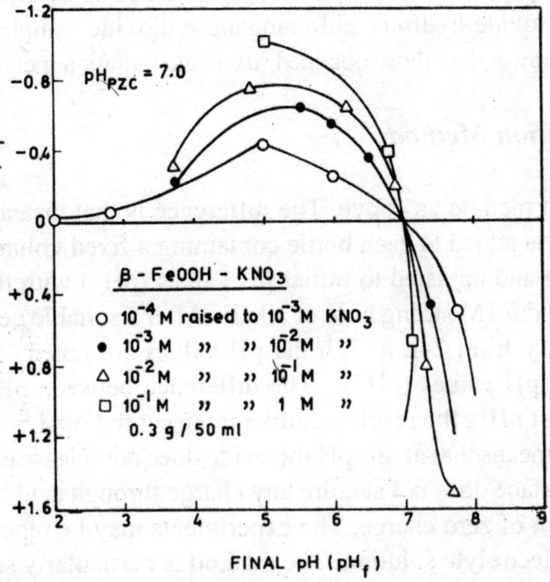

Fig. 4.8 Change in $\Delta pH(pH_i - pH_f)$ of the suspension of a β-FeOOH sample resulting from increase in the concentration of electrolyte solution by adding requisite amount of solid KNO_3 as function of equilibrium pH.

4.5.1 Salt Addition / Titration Method

This is also known as reverse titration method reported first by Muller and Roberts [135]. This is also carried out batch wise in a series of polyethylene bottles in the same manner as described in section 4.4.2 . The initial concentration of inert electrolyte solution is kept as low as feasible, say 10^{-3} - 10^{-4} M. After noting the pH values of the supernatant solutions in each bottle at the end of the requisite period of equilibration, the concentration of the background electrolyte solution is raised by a decade by adding accurately weighed quantity of solid salt to each bottle. After equilibrating (with intermittent shaking) for the same period, pH values are again noted. Similarly the concentration of electrolyte is again increased by a decade and equilibrium pH values are noted. The change (ΔpH) observed for each decade increase in concentration of salt is plotted against equilibrium or final pH . The common point of intersection of these curves at $\Delta pH = 0$ is the point of zero charge of the sample ; since variation in the concentration of indifferent or inert electrolyte does not alter the point of zero charge . A typical example is shown in Fig .4.8 for β-FeOOH - KNO_3 system . It may be noted that variation in the pH value is more sensitive nearer to pH_{PZC} . The occurrence of the distinct point of intersection of all the curves at pH 7.0 encompassing a wide range of concentration change suggests the validity of the method.

Kanungo and Mahapatra [90,133] have used this method to determine pH_{PZC} of several hydrous ferric oxide hydrous and manganese dioxide samples. The values are found to be slightly lower than those obtained from batch titration method (Fig . 4.7) .

4.5.2 Solid Addition Method

This is also a batch method as above. The difference is that instead of adding salt , solid oxide sample is added to each bottle containing a fixed volume of background electrolyte solution and adjusted to initial pH values (pH_i) with the help of either 0.1M strong acid and 0.1M strong base or alkali . After a suitable period of equilibration which may vary from 24h to 72h the pH values are noted . These values are designated as final pH values (pH_f) . The difference between pH_i and pH_f (ΔpH) when plotted against pHi , the resulting curve intersect the $\Delta pH = 0$ line at ascertain value of pHi . This means that at this pH the oxide does not release either H^+ or OH^- in solution i.e., the surface does not acquire any charge through acid base dissociation, a condition for point of zero charge. The experiments may be repeated for different concentrations of electrolyte solution. The method is particularly suitable where the electrolyte solution is complex such as sea water [136,137]. The method is illustrated for an amorphous hydrous iron oxide (HFO) and an amorphous hydrous manganese dioxide (δ - MnO_2) in Figs 4.9a and 4.9b respectively. Figure 4.9a shows that a single curve describes the variation of ΔpH with pHi in the presence of 0.5M and 0.1M NaCl solutions. The point of intersection of the curve with $\Delta pH = 0$ line occurs at pH 6.9 which corresponds with the first charge reversal of the oxide in the relevant electrolyte solutions and therefore represents the point of zero charge. This value is almost same as that determined by potentiometric titration [88] . However , in major ion sea water containing primarily Na^+, Mg^{2+}, Ca^{2+} , Cl^-, and SO_4^{2-} the PZC of the sample shifts to higher pH value (~ 8.2) . this suggests that specific adsorption of Mg^{2+} and Ca^{2+} takes place onto ferrihydrite in sea water medium. It is interesting to note that there is little effect on the variation of ΔpH with pH_i if the concentration of solid in the system is increased from 0.2g /l to 1g /l.

Figure 4.9b shows the plot of ΔpH vs pH_i for δ-MnO_2 sample in three different concentrations of NaCl solution . The nature of the plots for this sample is distinctly different from that for HFO because PZC of δ-MnO_2 lies at pH around 2. Consequently, as the pH_i approaches 3 and below the ΔpH values in the three NaCl solutions tend to follow a single curve and approaches the first charge reversal at pH around 2. Experimentally, however ,it is difficult to achieve the point of intersection of the curve with $\Delta pH = 0$ without appreciable dissolution of the sample .Indeed , δ-MnO_2 begins to dissolve at pH < 3 . Therefore the experimental curve is extrapolated to intersect $\Delta pH = 0$ at $pH_i = 1.8$ which is same as the pH_{PZC} value determined by potentiometric titration [89] . At pH > 3.5 ΔpH vs pH_i curves for the three NaCl solutions gradually separates out and intersect the $\Delta pH = 0$ line at three different pH_i values , clearly indicating that the second charge reversal is not related to pH_{PZC} of the sample .

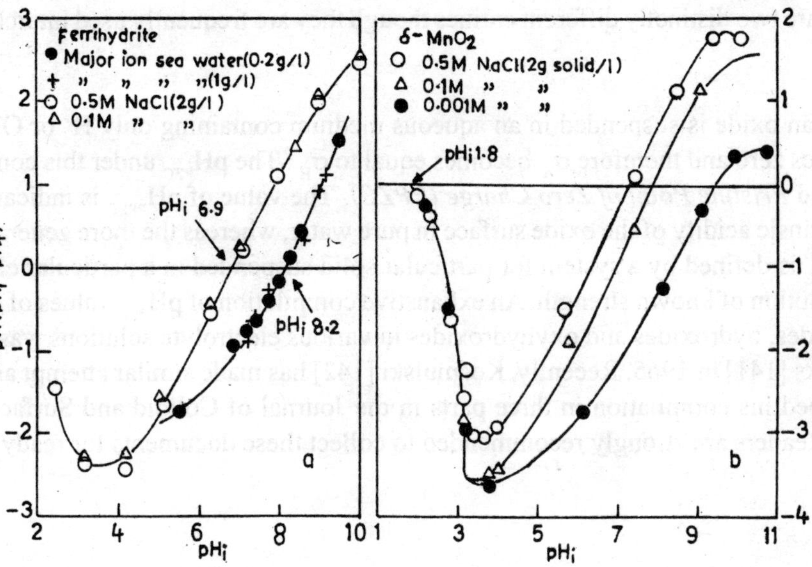

Fig. 4.9 Plots of change in $\Delta pH(pH_i - pH_f)$ values of the suspension resulting from the addition of oxide sample to electrolyte solutions and adjusted to differ ent initial pH values (pH_i) vs. pH_i for (a) ferrihydrite and (b) δ-MnO_2.

4.6 Pristine Point of Zero Charge vs. Point of Zero Charge

Few experimental method other than potentiometric or acid - base titration provide a wealth of information on the interfacial characteristics of oxides in aqueous electrolyte solution . It reveals the manner in which the interfacial charge of an oxide varies with pH, nature and concentration of electrolyte , temperature etc .The unique pH at which a bunch of titration curves for different concentration of electrolyte solutions intersect has been termed by some workers as *point of zero salt effect or* pH_{PZSE} [138-140]. However, any impurity ion or sometimes ions of inert or indifferent electrolyte solution may get adsorbed and shifts the PZC of the sample. In such a situation H^+ or OH^- do not represent the sole potential determiningion, as other ions such as divalent metal ions (M^{2+}) may participate in the coordination reaction with surface OH groups. Thus, the net surface charge so may be written as $\sigma_o = \sigma_H + \sigma_{CC}$, where σ_H is the charge due to proton adsorption or desorption and σ_{CC} is the charge due to formation of coordinative complexes with other ions. From acid base titration of an oxide in true inert electrolyte solution we obtain the pH value at which $\sigma_H = 0$ which may be defined as the *point of zero net proton charge or* pH_{PZNPC}. However, σ_{CC} need not be zero when σ_H is equal to zero and therefore σ_o also need not be zero. Under this condition the definition of pH_{PZC} does not apply. Thus the two terms pHP_{ZNPC} and pH_{PZC} are two distinctly different entities though they are frequently used interchangeably .

When an oxide is suspended in an aqueous medium containing only H^+ or OH^- σ_{CC} becomes zero and therefore σ_o becomes equal to σ_H. The pH_{PZC} under this condition is called *Pristine Point of Zero Charge (PPZC)*. The value of pH_{PPZC} is indicative of the intrinsic acidity of the oxide surface in pure water, whereas the more general term pH_{PZNPC} is defined by a system for particular solid suspended in a particular electro lyte solution of known strength. An exhaustive compilation of pH_{PZC} values of different oxides, hydroxides and oxyhydroxides in various electrolyte solutions was made by Parks [141] in 1965. Recently, Kosmulski [142] has made similar attempt and has published his compilation in three parts in the Journal of Colloid and Surface Science. Readers are strongly recommended to collect these documents for ready reference.

CHAPTER 5
THE ELECTRICAL DOUBLE LAYER

5.1 The Electrokinetic Phenomena and the Concept of Zeta Potential

Before studying the phenomenon of electrical double layer at solid / aqueous solution interface, it is necessary to know the origin of its study. In the early nineteenth century several researchers observed that when water in the two sides of a 'U' tube separated by a porous membrane was connected to an electrical source through two metal electrodes dipped in the water and a suitable EMF was applied, the water began to flow from one side to the other depending upon the direction of the current. The effect, known as 'electro-osmosis ', is one of the four electrokinetic phenomena involving the movement of one phase relative to the other. These are electro-osmosis, electrophoresis, streaming potential, and sedimentation potential. The following table gives the nature and causes of the phenomena.

Table 5.1: Different electrokinetic phenomena , their causes and effects

Phenomenon	Effect	Caused by
Electroosmosis	Movement of liquid, usually aqueous medium along a solid capillary or a bunch of capillaries .	Application of external EMF at both ends of a capillary / capillaries.
Electrophoresis	Movement of fine solid particles dispersed in an aqueous medium .	Application of external EMF at both ends of a capillary .
Streaming potential	Formation of a potential difference (EMF) between up and down stream flow of aqueous solution .	Movement of liquid along capillaries or porous plug generally under pressure.
Sedimentation Potential	Formation of a potential difference (EMF) between top and bottom of the vessel containing solid suspended in water .	Sedimentation of solid particles suspended in an aqueous medium

The occurrence of electrokinetic phenomena is the manifestation of the existence of the electrical double layer at the boundary or interface between a solid and a liquid (generally aqueous solution). This suggests that both solid and liquid phases have to be electrically charged. No movement of either solid or liquid is possible under the influence of electrical field, unless they are charged. Similarly, no EMF is generated at the liquid heads during its flow relative to the solid (streaming potential) unless the liquid is charged or conversely, unless the solid particles are charged when they are allowed to settle in the liquid (sedimentation potential).

The aqueous suspension of solid particles, particularly of oxides and silicates, on the whole is electrically neutral due to the existence of an equivalent charge of sign opposite to that present on the solid surface. The interface comprising of the two electrically charged layers, one on the surface and the other away from the surface i.e., in the bulk of the solution is known as the electrical double layer. Various models of the electrical double layer are discussed in the subsequent sections. The concept of zeta potential arises, if either the solid particles or the liquid in which they are suspended carry some net charge during their relative motion under the influence of applied electrical potential. This means that some extra charges should reside in the solution side beyond the fixed layer of counter ions. Such a proposition suggests the existence of diffuse layer of charge and that the plane of shear lies at a distance beyond the fixed layer of counter ions. The potential at the plane of shear is lower than that of the surface potential (ψ_0). This is known as zeta (ζ) potential which governs the four electrokinetic phenomena as mentioned in Table 5.1. Unlike surface potential, zeta potential is experimentally accessible if the velocity of the relative motion of either solid or liquid is measured in one of the electrokinetic phenomena. We shall discuss these aspects later on in this chapter.

5.2 Parallel Plate Condenser Theory (Model)

The first quantitative theory of the structure of electrical double layer at solid solution interface was put forward by Helmholtz in 1853. According to Helmholtz the double layer as depicted above is closely similar to parallel plate condenser comprising of two oppositely charged layers, one on the solid surface and the other in the liquid phase (see Fig. 5.1). The layers are attracted by the electrostatic forces and are separated by a distance d , which is equal to the radius of the counter ion in the solution phase. Accordingly, the capacity of the condenser is equal to the ratio of the charge(σ_0) to the potential (ψ_0) between the two layers i.e., σ_0 / ψ_0.

This simplest model of the double layer implies the existence of two possible zeta (ζ) potentials or potentials at the slipping planes. It is assumed that all the charges in the solution side are capable of moving together with the liquid, the ζ potential will then

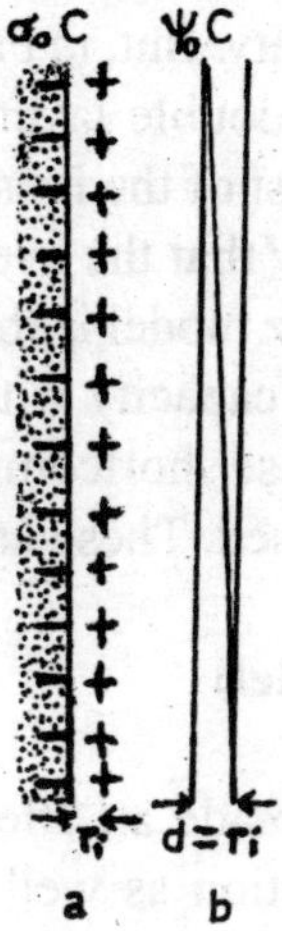

Fig. 5.1 (a) Ionic representation of the Helmholtz model of electrical double layer at the solid/aq. electrolyte solution interface. (b) Sharp drop of potential within the distance between the two oppositely charged layers. The charge on the solid surface is assumed to reside within the solid, so that the distance be tween the two layers is equal to the radius of the ion in solution.

coincide with the surface potential ψ_0 and its variation with electrolyte concentration will follow Nernst equation. If, on the other hand, the charges present in the solution phase are bound along with the solid and move together with it, the ζ potential will be equal to zero. None of these corollaries of the Helmholtz theory is consistent with the experimental observations of the relationship between ψ_0 and ζ potentials unless it is assumed that at very high concentration of electrolyte the value of ζ is zero. At the same time , the double layer capacitance value calculated from the thickness of the parallel plate condenser model is in good agreement with that obtained from experiment and a physically plausible double layer thickness (of the order of 10^{-8} cm) based on the experimental value of capacity . For a parallel plate condenser of thickness d one can write

$$\sigma_0 = (\varepsilon \cdot \varepsilon_0 / 4\pi d)\, \psi_0 = C\, \psi_0 \tag{5,1}$$

where , d is equal to the radius of the ion in solution, $\varepsilon\, \varepsilon_0$ is the dielectric constant (permittivity) of the medium filling the space between the plates of the condenser . In a solution containing no surface active substance or ion the capacity of the Hg electrode determined from the slope of the descending branch of the electrocapillary curve is about 20μF / cm². Since 1μF = $9\text{x}10^5$ esu of capacity, therefore 20μF=18x 10^6 esu . Hence, $d/\varepsilon\, \varepsilon_0 = 1/4\pi C = 1/(4\text{x}3.141\text{x}18\text{x}10^6) = 0.4\text{x}10^{-8}$cm if we assume that the medium between the double layer is in a state close to dielectric saturation, the dielectric constant may be taken as unity. But, in practice the value varies from 2 to 5. In such a case, the thickness of the double layer is of the order of 10^{-8}cm, which corresponds to the radii values of most of the ions. Using the approximate values of d and $\varepsilon\ \varepsilon_0$, back calculation will show that the average C value is about 20μF / cm^2. The main problem with the Helmholtz model is that Eq. (5.1) cannot account for the experimentally observed variation in capacity with ionic (electrolyte) concentration and hence potential . To overcome these shortcomings in the model several improvements have been subsequently proposed. These are discussed below.

5.3 Gouy - Chapman Theory (Model)

Gouy and Chapman independently made an attempt to correlate the double layer properties with electrolyte concentration as well as the effect of temperature. According to them, the fixed array of ions postulated in the Helmholtz model is , in fact, impossible because of the two factors: (a) Electrostatic forces arising from the surface and the ions in solution are not same all over the interface mainly because of the variation in the chemical potential of the surface. (b) The ions in the solution phase are subjected to thermal molecular motion. Because of these two factors, the ions in the solution phase are distributed diffusively relative to the solid surface. This means that the charge density and consequently potential (ψ_0) decreases with increase in the distance from the surface (see Fig .5.2). However, in accordance with the principle

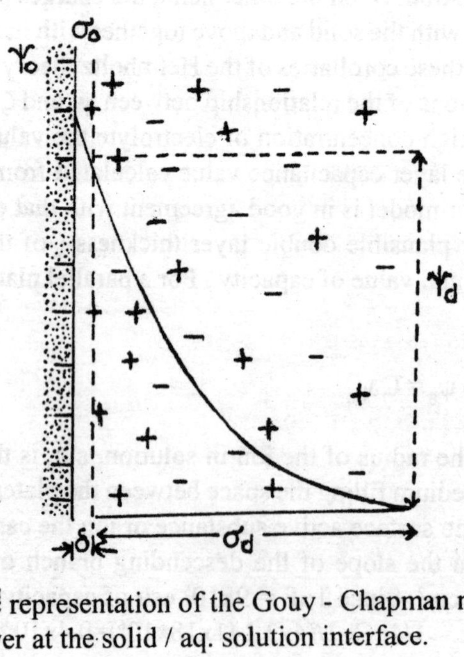

Fig. 5.2 Schematic representation of the Gouy - Chapman model of the electrical double layer at the solid / aq. solution interface.

of electrochemistry the net charge of the system as a whole is zero i.e., charge in the diffuse layer (σ_d) is equal and opposite to the charge on the solid surface (σ_0) or $\sigma_0 + \sigma_d = 0$.

According to the Gouy - Chapman model, ions may be viewed as point charges without any physical dimension or volume and their distribution from the surface to the bulk of the solution phase obeys Boltzmann distribution law. The charge in the diffuse layer is determined by summing all the charges carried by ions present in a liquid column normal to the electrode surface and having a cross section of 1cm^2. This magnitude of charge on the solid surface (σ_0) is also determined in the same way as the charge density of ionic atmosphere calculated from the ionic activity coefficient using the first approximation of Debye - Huckel theory. The starting point in both cases is the Boltzmann - Poisson equation.

$$\nabla^2 \psi = -\frac{e}{\varepsilon \varepsilon_o} \sum z_i n_i \exp\left(\frac{-z_i e \psi}{kT}\right) \tag{5.2}$$

where , ∇^2 = Laplacian operator for rectangular coordinates $\left(\frac{\partial^2}{\partial x^2} + \frac{\partial^2}{\partial y^2} + \frac{\partial^2}{\partial z^2}\right)$

n_i = number of ions of type i per unit volume at a point in the diffuse layer where the mean potential is ψ .
z = valency of the ion
e = electronic charge , 1.602 x 10^{-19} coulomb
k = Boltzmann constant , 1.38 x 10^{-23} J K^{-1} , 10^{-16} erg deg^{-1}
T = absolute temperature (K)
ε = dielectric constant of water (dimensionless)
ε_o = permittivity in free space , 8.854 x 10^{-9} C / V cm

The solution of the above equation has been reviewed by several authors [143-145]. The mathematical details are beyond the scope of the present topic. In determining the charge in the liquid column of unit cross sectional area (1 cm^2) only one coordinate i. e., distance normal to the solid surface is necessary. This means, ∇^2 becomes $\partial^2\psi/\partial x^2$. Using the boundary condition $\psi = \psi_d$ at x = 0 and d ψ / d x = 0 at x = ∞ and for symmetrical electrolyte (z - z) , Eq . (5.2) gives

$$\psi = \frac{2kT}{ze} \ln\left[\frac{1+\gamma \exp(-\kappa x)}{1-\gamma \exp(-\kappa x)}\right] \tag{5.3}$$

where
$$\gamma = \frac{\exp\left(\frac{ze\psi_d}{2kT}\right) - 1}{\exp\left(\frac{ze\psi_d}{2kT}\right) + 1} \tag{5.4}$$

and
$$\kappa^2 = \frac{e^2 \sum n_i z^2}{\varepsilon \varepsilon_o kT} = \frac{2F^2 N_A I x 10^3}{\varepsilon \varepsilon_o RT} \tag{5.5}$$

since, e/kT = F/RT; κ (kappa) is defined as the Debye - Hückel reciprocal length (cm^{-1}) of the ionic atmosphere and in the present case is the reciprocal thickness of the double layer . The expression $\Sigma n_i z_i^2$ represents the ionic concentration of the solution . The number of ions per cm^3 is related to the molar concentration (c_i) by the relation $c_i = (n_i/N_A) x 10^3$ where N_A is the Avagadro's number . Therefore, the quantity $\Sigma n_i z_i^2$ becomes $N_A x 10^3 \Sigma c_i z_i^2$ or $2 N_A x 10^3 I$, where I is the ionic strength of the medium and is equal to ½ $\Sigma c_i z_i^2$. For 1 - 1 electrolyte solution in water I is equivalent to molar concentration c_i. Substituting the numerical values of the constants in Eq (5.5) and assuming 1 - 1 electrolyte in water at 298K when $\varepsilon \varepsilon_0$ is approxmately equal to 78.5.

$$1/\kappa = 3.041 \times 10^{-8}/(c_i)^{1/2} \tag{5.6}$$

The relationship shows that as the electrolyte concentration increases thickness of the double layer decreases .The diffuse layer charge σ_d(coul/m^2), at potential > 0.025V , is given by

$$\sigma_d = -\left\{\left(8\varepsilon\varepsilon_o RTc_i x 10^3\right)^{1/2} \sinh\left(\frac{z\psi_d F}{2RT}\right)\right\} \tag{5.7}$$

At low potential (< 0.025V) Eq.(5.7) may be used in its linearized form which is equivalent to Helmholtz parallel plate condenser or constant capacitance model .

$$\sigma_d = -\varepsilon \varepsilon_o \kappa \psi_d \tag{5.8}$$

Combining with Eq . (5.6) and substituting the values of constants we have

$$\sigma_d = -2.5\,(c_i)^{1/2}\,\psi_d \tag{5.9}$$

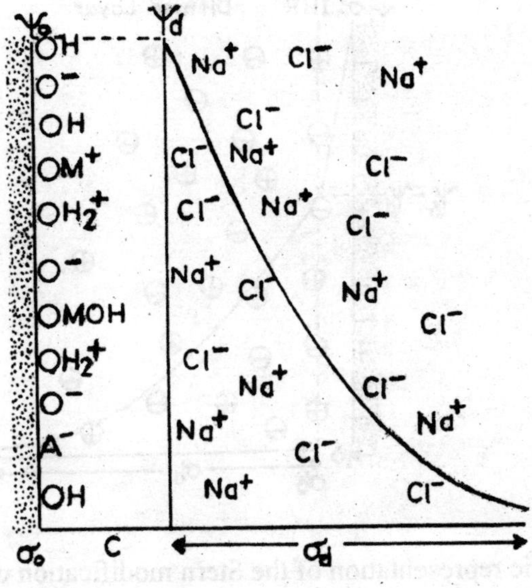

Fig. 5.3 Diffuse double layer model at the oxide / NaCl solution interface.

Similarly, putting the values of the constants and assuming 1-1 electrolyte solution at 298 K Eq .(5.7) may be reduced to

$$\sigma_d = -0.1174\, c^{1/2} \sinh(19.46 \times \psi_d)\ \text{coul./m}^2 \qquad (5.10)$$

The concept of zeta potential does not figure in the diffuse layer theory. Therefore, the potential ψ_d in the above two equations does not differentiate with surface potential ψ_o. However , if one assumes that beyond a certain distance say δ the ions are no longer bound so strongly with the solid surface as to move with the solid phase relative to liquid phase (e.g.,under the influence of electrical field) , the value of the potential corresponding to this distance may be identified with the ζ potential . This value will certainly be less than that of ψ_o (see Fig. 5.2)

Despite certain theoretical limitations / constraints, the Gouy - Chapman diffuse layer model has been successfully used by several authors [27] for the modeling of the adsorption of inorganics at the oxide - water interface. In fact, they prefer to call it two layer model which has been generalized by assigning all adsorbed ions are attached to the surface and counter ions of equivalent charge on the diffuse layer which is considered to be at a certain distance away from the surface layer. Therefore, while considering the surface charge all the ions adsorbed on the surface layer are taken

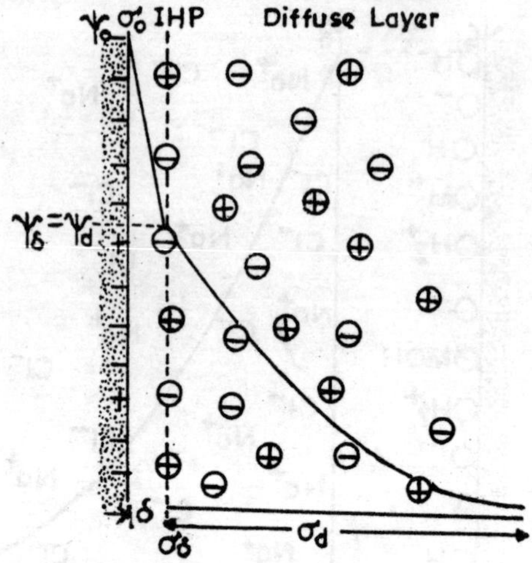

Fig. 5.4 Schematic representation of the Stern modification of the Gouy - Chapman model of the electrical double layer at the oxide / aq. solution interface.

into consideration . In such a situation potential drop is considered to begin from the diffuse layer and not from the surface layer as in this model there is no difference between ψ_0 and ψ_d. Fig . 5.3 illustrates the diffuse layer model for an oxide / aq. NaCl solution interface.

5.4 The Stern Model

The neglect of ionic dimension in the Gouy - Chapman diffuse layer model leads to unsatisfactory agreement between experimental and theoretical values of surface charge and potential . In 1924 Stern [3] suggested that the ions adsorbed closed to the surface are of definite size and his work formed the basis of modern theory of electrical double layer. According to Stern, double layer consists of (i) those ions which constitute the surface charge (σ_0) and potential (ψ_0) and (ii) those ions in solution which are adsorbed by specific chemical interaction energy (ϕ) or partly by electrostatic force at a distance δ away from the surface corresponding to the average radii of the ions. The plane at which such adsorption takes place is called 'Stern plane' or inner Helmholtz plane (IHP) and corresponds to potential ψ_δ and charge σ_d . Between the immobile IHP and the bulk of the solution, ions are distributed diffusively and the charge (σ_d) and potential (ψ_δ) relationship follows Gouy - Chapman theory . Such a distribution is proportional to the "sites" still free in the IHP and the bulk solution multiplied by the Boltzmann factor for the particular chemical and electrostatic energy ($zF\psi_\delta$) [146] . Therefore, Stern model is essentially a combination or synthesis between Helmholtz parallel plate condenser and Gouy - Chapman diffuse layer models. The original or basic Stern model is illustrated in Fig. 5.4. Stern derived the following equation for the charge on inner Helmholtz plane

$$\sigma_\delta = F\Gamma_{max}\left[\frac{1}{1+\frac{1}{c}\exp\frac{\phi_+ + F\Psi_\delta}{RT}} + \frac{1}{1+\frac{1}{c}\exp\frac{\phi_- - F\Psi_\delta}{RT}}\right] \tag{5.11}$$

where , Γ_{max} = maximum number of ions capable of being adsorbed on an unit area

ϕ_+ , ϕ_- = specific adsorption potential of ions . Negative value of potential suggests condition for favourable adsorption .

c = concentration g. ion / l

Generally , the adsorption of one ionic species predominates so that Eq .(5.11) becomes

$$\sigma_\delta = \frac{FT_{max}}{1+\frac{1}{c}\exp\frac{\phi + F\Psi_\delta}{RT}} \tag{5.12}$$

The diffuse layer charge (coul / m^2) is given by the Gouy - Chapman model

$$\sigma_d = -(8\,\varepsilon\,\varepsilon_o\,RT\,c_i \times 10^3)^{1/2}\sinh(zF\psi_d/2RT) \tag{5.13}$$

The condition for electro-neutrality is given by:

$$\sigma_o + \sigma_\delta + \sigma_d = 0 \tag{5.14}$$

The integral capacity of the Stern plane is given by

$$C_{St} = \frac{\sigma_o}{\Psi_o - \Psi_\delta} = \frac{\varepsilon\varepsilon_o}{\delta} \tag{5.15}$$

Rearranging and substituting for σ_o

$$\psi_o - \psi_\delta = (-\delta/\varepsilon\varepsilon_o) \times (\sigma_\delta + \sigma_d) \tag{5.16}$$

Combining Eqs . (5.12), (5.13) and (5.16)

$$\psi_o - \psi_\delta = \delta / \varepsilon \varepsilon_o \{(8 \varepsilon \varepsilon_o RT c \times 10^3)^{1/2} \sinh (zF\psi_d / 2RT) - \frac{FT_{max}}{1 + \frac{1}{c}\exp\frac{\phi + \Psi_\delta}{RT}}\} \quad (5.17)$$

It follows from Eq . (5.17) that as c → 0 the right hand side also tends to zero. Since, according to Stern assumption $\psi_\delta = \psi_d$, then under the above condition $\psi_o = \psi_\delta = \psi_d$. Therefore, in an infinitely dilute solution all the charges in the solution side tend to be randomly distributed in the double layer, according to Gouy - Chapman model. Conversely, as the concentration increases the degree of randomness will diminish since σ_δ is directly related to c (first power) .On the other hand, σ_d is related to $c^{1/2}$, suggesting that increase in σ_d is much slower than that of σ_δ with increase in c. Consequently, for concentrated solution σ_d and ψ_d may be disregarded compared to high charge (σ_δ) and potential (ψ_δ) in Stern layer . This picture is closely related to the Helmholtz parallel plate or constant capacitance model. Similar conclusion may be reached from the consideration of total integral capacitance (C_t) of the double layer which may be expressed as follows:

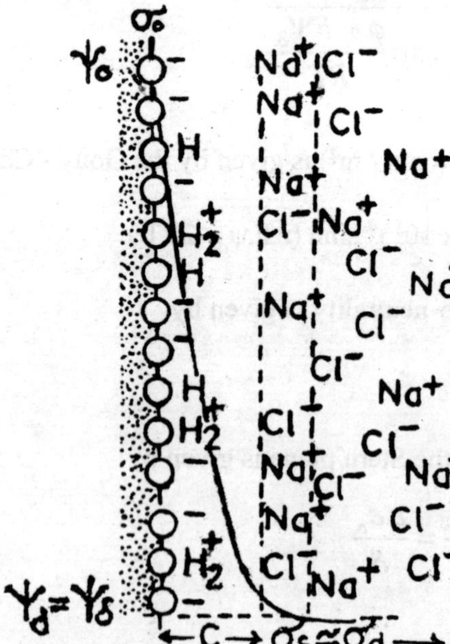

Fig. 5.5 The constant capacitance model of the electrical double layer at the oxide / water interface in the presence of high electrolyte concentration.

$$1 / C_t = 1 / C_{St} + 1 / C_d \tag{5.18}$$

At low ionic strength and relatively low potential C_d is much less than C_{St}, the total capacitance is approximately equal to C_d, a situation akin to the diffuse layer model. At high ionic strength C_d is much greater than C_{St}, the total capacitance is equal to C_{St}. This is analogous to the Helmholtz model of double layer with constant capacitance value as illustrated in Fig. 5.5 for an oxide suspended in aqueous NaCl solution.

One of the important outcome or corollaries of the Stern model is its ability to explain the phenomenon of charge reversal. In solution containing strong surface active ions, adsorption may not be equivalent to the surface charge, but may exceed it by a certain amount depending upon the properties of both ions and the surface. The reversal of charge occurs when counter ions are adsorbed so strongly in the inner Helmholtz plane (specific adsorption) that sign of the surface charge get reversed. Such specific adsorption involves some form of chemical interaction with the surface and therefore differs from the usual electrostatic force by which oppositely charged ions are adsorbed close to the surface. Consequently, specific adsorption needs suitable free energy change (ϕ) so as to favor the formation of chemical bond between the surface and the adsorbing ions and the more negative the value is, the stronger is the bond. The phenomenon is schematically illustrated in Fig. 5.6.

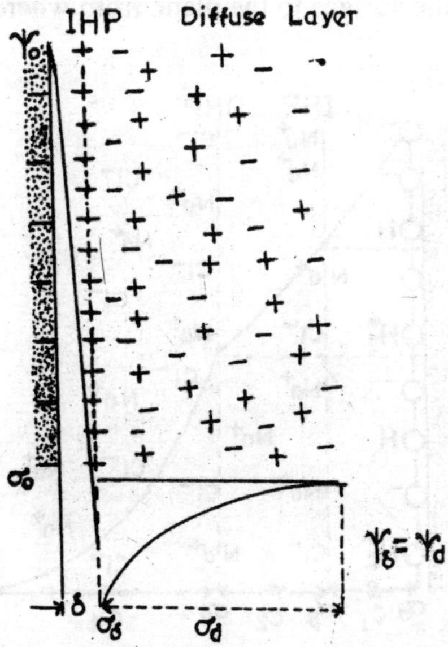

Fig. 5.6 Schematic representation of the phenomenon of charge reversal that occurs due to high concentration of counterions close to the solid surface.

5.5 The Extended Stern Model (Grahame Modification)

Grahame [4] introduced an important refinement in the basic Stern model by recognizing the significance of the hydration state of ions and their plane of adsorption, which, according to him, lies at closest distance from IHP. This he designated as outer Helmholtz plane (OHP) which forms the innermost plane of the diffuse layer. It is assumed that no specific or chemical adsorption takes place at OHP. The region between the surface and the IHP and the region between the IHP and the OHP are considered to behave as two parallel plate condensers having two capacitance values, say C_1 and C_2. The second Helmholtz plane was proposed because the double layer capacity at positive potential (negative counter ion) was found to be grater than at negative potential (positive counter ion) indicating that the distance of closest approach for anions was less than that of cations, since capacitance is inversely proportional to the distance between the two charged layers. If it is considered that ions of the same size or even same ions adsorbed in the OHP may present in the IHP, they are probably dehydrated, at least in the direction of the surface, thus allowing closer approach to the surface than in the OHP [147]. Figure 5.7 illustrates the schematic representation of the extended Stern model for an oxide - electrolyte interface, where β and γ denote the distances between the surface and the IHP and between the IHP and the OHP respectively. The sum of these two distances, as denoted by d, represents the distance from the surface to the plane from where the diffuse layer begins.

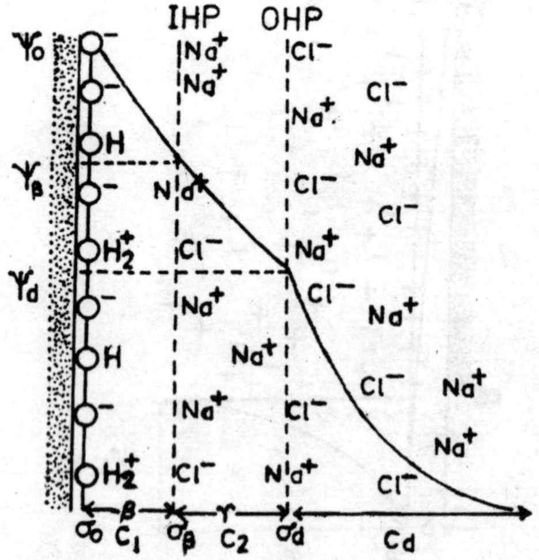

Fig. 5.7 Schematic representation of the extended Stern model of the electrical double layer with two different capacitance values (also called triple layer model).

Accordingly, σ_o and ψ_o ; σ_β and ψ_β ; σ_d and ψ_d are the charge densities and potentials at the surface, compact Stern (IHP), and diffuse (OHP) layers respectively. For overall neutrality of the electrical double layer

$$\sigma_o + \sigma_\beta + \sigma_d = 0 \tag{5.14}$$

The capacitance - charge relationships are given by

$$\psi_o - \psi_\beta = \sigma_o / C_1 \tag{5.19}$$

$$\psi_\beta - \psi_d = (\sigma_0 + \sigma_\beta) / C_2 = -\sigma_d / C_2 \tag{5.20}$$

The charge- potential relationship is, however, same as that for the basic Stern model

$$\sigma_d = -0.1174\, c^{1/2} \sinh(19.46 \times \psi_d) \quad \text{coul. /m}^2 \tag{5.10}$$

While the consideration of an additional Helmholtz plane is important for understanding the mechanism of adsorption, it does not alter the Stern's fundamental picture of the interface. Stern also observed that the error introduced would be small if the second capacitance C_2 was neglected and the potential at OHP was set equal to IHP i.e., $\psi_\beta = \psi_d$. He used this simplification to calculate the total capacitance and potential values in the OHP . This simplified model with only one Helmholtz plane may also be regarded as " Basic Stern model " (Fig. 5.8) and the model with two distinct Helmholtz planes , as discussed above , as the " Extended Stern model (Fig . 5.7).

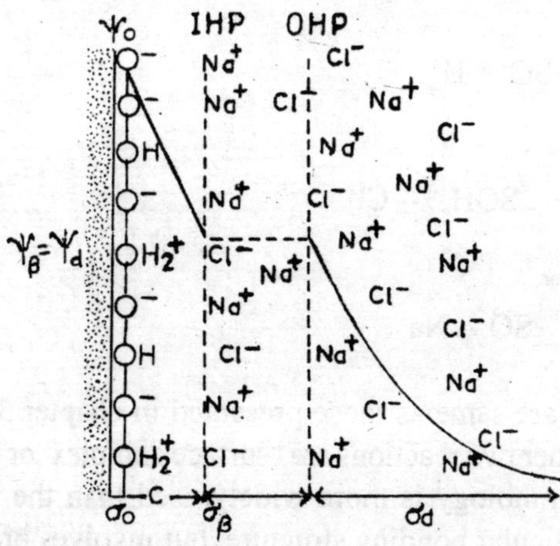

Fig. 5.8 Basic Stern model of the electrical double layer with single capacitance value.

5.6 Site - binding model

In principle , the site binding model is closely similar to the "extended Stern model." or the Grahame's modification of the basic Stern model as discussed in the preceding paragraphs and illustrated in Fig .5.7. It has been stated that all the charged species which are adsorbed by electrostatic or coulombic forces contribute to the surface charge (σ_0). Consequently, the adsorption of supporting electrolyte ions on the surface, if any, will contribute to the surface charge. It has been observed [123,125,148] that surface charge density and differential capacity for many oxide / solution interfaces are much larger than the completely polarized mercury and reversible AgI surfaces. Despite the high surface charge, electrokinetic and other colloid stability measurements show that diffuse layer charge is relatively low , seldom exceeding 20% of the surface charge [131,149] . The difference in charge must be balanced by specific adsorption of electrolyte ions within the region inside the OHP or the shear plane of the diffuse layer. Yates et al [148] proposed that ions other than those of potential determining ions (pdi) i.e., H^+ and OH^-, are adsorbed specifically as counter ions to charged surface sites on a plane between the surface and the OHP and called it " site binding model ". Thus, if three types of sites , namely SOH, SO^- and SOH_2^+ are considered to be present on the oxide surface suspended in a supporting electrolyte say NaCl solution, there are three possible solute species present in the system. These are H^+, Na^+ and Cl^- and their interactions with the three surface sites are as follows:

$$SOH_2^+ \overset{Ka_1^{int}}{\longleftrightarrow} SOH + H_s^+ \tag{3.1}$$

$$SOH \overset{K_{a2}^{int}}{\longleftrightarrow} SO^- + H_s^+ \tag{3.2}$$

$$SOH_2^+ + Cl^- \overset{K_{Cl^-}^{int}}{\longleftrightarrow} SOH_2^+ \text{--} Cl^- \tag{5.21}$$

$$SO^- + Na^+ \overset{K_{Na+}^{int}}{\longleftrightarrow} SO^- \text{--} Na^+ \tag{5.22}$$

The first two reactions are same as those presented in chapter 3 and therefore only reproduced here. The other two reactions are "surface complex"or "ion pair"formation. Though the former terminology is more widely used than the latter, this does not necessarily imply a particular bonding structure, but involves broad physico-chemical interaction of non - coulombic nature. Therefore, this model is also known as triple layer model to emphasize the presence of a separate layer of specifically adsorbed ions which may be identified with IHP in the Stern model .Thus, the innermost layer consists of only H^+ and OH^- ions which contribute to the surface charge σ_0 and experience the potential ψ_0. Separated from the surface lies the plane of specifically adsorbed counter ions contributing to charge σ_β and experiencing potential ψ_β. The charge and potential at which diffuse part of the dissociated counter ions occur (OHP) are σ_d and ψ_d respectively . The capacitance values in the two intervening regions are as usual C_1 and C_2 respectively. The concentrations of ions in the regions of electrical double layer is related to their bulk concentrations by the Boltzmann law and is represented by the general form as follows :

$$[i^z]_s = [i^z] \exp(-z e\psi_\beta / kT) \tag{5.23}$$

where, the superscript z and subscript s denote ionic charge and surface respectively. Accordingly, the concentrations of Na^+ and Cl^- in the β plane may be expressed as follows

$$[Na^+]_s = [Na+] \exp(-e\psi_\beta / kT) \tag{5.24}$$

$$[Cl^-]_s = [Cl^-] \exp(e\psi_\beta / kT) \tag{5.25}$$

One of the important features of triple layer model which distinguishes it from the Stern - Grahame model is that the concentration of surface species can be obtained from the equilibrium constants of the reactions (3.1) , (3.2) , (5.21) and (5.22) which are applicable over a wide range of ionic strength . Thus from Eqs . (5.21) and (5.22) and replacing surface concentration of $[Na^+]$ and $[Cl^-]$ with their bulk concentration in solution

$$K_{Cl^-}^{int} = \frac{\{SOH_2^+ - Cl^-\}}{\{SOH_2^+\}[Cl^-]_S} = \frac{\{SOH_2^+ - Cl^-\}}{\{SOH_2^+\}[Cl^-]} \exp\left(\frac{-e\Psi_\beta}{kT}\right) \tag{5.26}$$

$$K_{Na^+}^{int} = \frac{\{SO^- - Na^+\}}{\{SO^-\}[Na^+]_S} = \frac{\{SO^- - Na^+\}}{\{SO^-\}[Na^+]} \exp\left(\frac{e\Psi_\beta}{kT}\right) \tag{5.27}$$

It is however more appropriate to write Eqs. (5.21) and (5.22) as direct complexation reaction with neutral surface species , SOH .

$$SOH + H^+_s + Cl^-_s \overset{1/{}^*K_{Cl^-}^{int}}{\Longleftrightarrow} SOH_2^+ \text{--} Cl^- \tag{5.28}$$

$$SOH + Na^+_s \overset{*K^{int}_{Na+}}{\longleftrightarrow} SO^- \text{--} Na^+ + H^+_s \tag{5.29}$$

The equilibrium constants of these two reactions are, in fact, combination of Eqs.(3.3) and (3.4) with Eqs . (5.26) and (5.27) respectively .

$$^*K^{int}_{Cl^-} = \frac{\{SOH\}[H^+]_s[Cl^-]_s}{\{SOH_2^+ - Cl^-\}} = \frac{Ka_1^{int}}{K^{int}_{Cl^-}} \tag{5.30}$$

and

$$^*K^{int}_{Na^+} = \frac{\{SO^- - Na^+\}[H^+]_s}{\{SOH\}[Na^+]_s} = Ka_2^{int} \; x \; K^{int}_{Na^+} \tag{5.31}$$

Converting the surface concentration of the soluble species in the above two equations into their bulk concentrations by invoking Eqs . (3.10) , (5.24) and (5.25) we have

$$^*K^{int}_{Cl^-} = \frac{\{SOH\}\exp\left(-e(\Psi_o - \Psi_\beta)/kT\right)[H^+][Cl^-]}{\{SOH_2^+ - Cl^-\}} \tag{5.32}$$

and

$$^*K^{int}_{Na^+} = \frac{\{SO^- - Na^+\}[H^+]\exp\left(-e(\psi_o - \psi_\beta)/kT\right)}{\{SOH\}[Na^+]} \tag{5.33}$$

All the mass law quotients are thermodynamic or intrinsic equilibrium constants if all the concentration terms including those of surface species are expressed in terms of their activities. As stated in chapter 3, if the constants are evaluated using only the concentration terms by neglecting activity coefficients of the surface species as well as the electrostatic terms associated with the conversion from surface to bulk activities in case of dissolved species, they will not be intrinsic constants, but will vary with pH and electrolyte concentrations i.e., surface charge and potential. These constants are therefore "apparent constants", whereas 'intrinsic constants" are those which are evaluated under the condition of zero charge and potential. We shall, however, discuss these aspects in more detail in the next section.

The formation of surface complexes readjusts the acid - base equilibrium and therefore affects the surface charge. The surface charge of amphoteric oxides represents the net number of protons released or consumed by all surface reactions and not just the formation of ionized species like SOH_2^+ and SO^-. Increase in electrolyte concentration results in additional binding of counter ions until the equilibrium is reestablished, depending upon the effects of electrostatic field. In this manner, surface complexation provides a mechanism for the development of surface charge in addition to the role played by proton uptake and release during acid - base dissociation of surface sites [150,151] . With the surface species defined, it is possible to write charge balance equations on the surface (σ_0), on the plane of specific adsorption of counter ions (σ_β) and on diffuse layer (σ_d) in an electrolyte, say NaCl solution.

$$\sigma_0 = B[\{SOH_2^+\} + \{SOH_2^+ - Cl^-\} - \{SO^-\} - \{SO^- - Na^+\}] \quad (5.34)$$

$$\sigma_\beta = B[\{SO^- - Na^+\} - \{SOH_2^+ - Cl^-\}] \quad (5.35)$$

$$\sigma_d = B[\{SO^-\} - \{SOH_2^+\}] \quad (5.36)$$

The surface species distributed among the total number of sites available i.e., Ns is given by

$$Ns = B[\{SOH\} + \{SOH_2^+\} + \{SOH_2^+ - Cl^-\} + \{SO^-\} + \{SO^- - Na^+\}] \quad (5.37)$$

where , B is the conversion factor from moles/kg to $\mu C/cm^2$ and is expressed as follows

$$B = (F/S) \times 10^2 \quad (5.38)$$

S = surface area of solid (m^2 / kg)

F = Faraday constant (coulomb / mole)

{ } = concentration of surface species in mole / kg

The condition of electroneutrality and the charge - potential relationship in the diffuse layer are similar to Eqs . (5.14) and (5.10) respectively .

During the last thirty years since the publication [150,151] of the above reformulated model of the triple layer, a large number of authors [88,89,152 -155] have made attempts to describe various oxide / solution interfaces. However, one of the main problems of the site binding or triple layer model is that it involves multiple parameters, the determination of which by different methods gives different results [146,156]. With many adjustable parameters it is , indeed , possible to fit any experimental data

The problem has been studied by Westall and Hohl [146] who have observed that a wide range of parameters values yield optimal fit, so that it is very difficult to separate adsorption energy unambiguously into electrostatic and chemical components. Indeed, all models can represent the data equally well, but the values of the corresponding parameters in different models are different. According to them, the model may be viewed as being the mathematical form to present or describe the data, but not necessarily be an accurate description of the physical picture of the interface.

5.7 Thermodynamic Basis of Chemical and Electrostatic Interactions at the Solid/Aqueous Solution Interface

The foregoing discussion on the surface ionization and complexation reaction of solid hydr(oxide) in aqueous electrolyte solution is based primarily on the principles of coordination chemistry and the theory of electrical double layer, blended suitably to explain experimental observation on the variation of surface charge and potential with pH. However, the equilibrium reactions are based strictly on the principles of chemical thermodynamics. The thermodynamic basis of surface ionization and complexation reactions has been discussed by many authors [128,140,157-160]. Hayes and Leckie [160] have dealt on this aspect and the present discussion is based on their treatment.

Let us start with the example of the well - known surface protonation reaction

$$SOH_2^+ \longleftrightarrow SOH + H^+ \tag{3.1}$$

The following basic relationship of chemical thermodynamics may be applied to every species of the above reaction

$$\mu_i = \mu_i^o + RT \ln [i] + RT \ln \gamma_i \tag{5.39}$$

where, μ_i and μ_i^o are the chemical and standard chemical potential respectively of the ith species ; [i] the molar concentration of species i ; γ_i is the activity coefficient of species i ; and is defined in such a way that at reference state condition (normally, at 1M concentration for both surface and soluble species) $\gamma_i = 1$. Substituting for all the species in Eq . (3.1) and using the subscript o, + and H^+ to represent SOH, SOH_2^+ and H^+ respectively, we have

$$\mu_o = \mu_o^o + RT \ln [SOH] + RT \ln \gamma_o \tag{5.40}$$

$$\mu_{H^+} = \mu_{H^+}^o + RT \ln [H^+] + RT \ln \gamma_{H^+} \tag{5.41}$$

$$\mu_+ = \mu_+^o + RT \ln [SOH_2^+] + RT \ln \gamma_+ \tag{5.42}$$

Adding Eqs (5.40) and (5.41) and subtracting Eq (5.42) from the total

$$\Delta\mu_i = \Delta\mu_i^o + RT\ln\frac{[SOH][H^+]}{[SOH_2^+]} + RT\ln\frac{\gamma_o\gamma_{H^+}}{\gamma^+} \tag{5.43}$$

where , $\Delta\mu_i = (\mu_o + \mu_{H+} - \mu_+)$ and $\Delta\mu_i^o = (\mu_i^o + \mu^o_{H+} - \mu_+^o)$. At equilibrium $\Delta\mu_i = 0$, and since - $\Delta\mu_i^o$ /RT may be substituted by ln K_{al}^o we get from Eq (5.43)

$$K^o{}_{al} = \frac{[SOH][H^+]}{[SOH_2^+]}\frac{\gamma_o\gamma_{H^+}}{\gamma_+} \tag{5.44}$$

where K_{al}^o is the surface protonation constant at the standard state; γs are the activity coefficients of the relevant species . Equation (5.44) is the thermodynamic macroscopic description of Eq (3.1) .

The unit for surface species is immaterial in this case, as the concentration term of the surface species occurs in both numerator and denominator. However, for a surface reaction it is difficult to specify any meaningful reference states which may allow γ_i to be measured unambiguously. An alternative approach to Eq (5.39) is to define the chemical potential of the surface species differently from that of dissolved species such as H^+. This is because charged species on the surface can exert more significant influence on the electrical properties of the solution over a greater distance than the ions present in the solution itself. Therefore, as stated briefly in chapter 3 the chemical and coulombic contributions to the overall chemical potential of the surface species need to be separated as follows:

$$\hat{\mu}_i = \hat{\mu}_i^o + RT\ln[i] + RT\ln\gamma^s_i + zF\phi \tag{5.45}$$

Proceeding in the same way as above for reaction (3.1) we get

$$K_{al}^O = \frac{[SOH][H^+]}{[SOH_2^+]}\frac{\gamma_o^s\ \gamma_{H^+}}{\gamma_+^s}\exp\left(\frac{-zF\psi_o}{RT}\right) \tag{5.46}$$

Where, $\hat{\mu}_i$ and $\hat{\mu}_i^o$ are electrochemical and standard electrochemical potentials respectively ; γ^s_i is the surface activity coefficients due to chemical factor ; the exponential term represents activity correction due to coulombic factor i.e. the difference in potential between solution and the surface and z is the charge of the solution species . This separation or differentiation of activity coefficients between the chemical and coulombic factors of the surface species form the basis of triple layer model (TLM) of Davis et al [150] . In true sense, such a separation is nothing but using extra - thermodynamic approach to address the electrostatic aspect of the problem. Consequentially, it makes use of the electrical double layer theory to estimate the activity coefficient of the two factors in the form of the following expression.

$$K_{a1}^{int} = \frac{[SOH][H^+]}{[SOH_2^+]} \frac{\gamma_o^s \ \gamma_{H^+}}{\gamma_+^s} \exp\left(\frac{-zF\psi_o}{RT}\right) \tag{5.47}$$

The significances of the symbols are same as discussed above or elsewhere in this chapter.

In a subsequent development of the TLM, the separation of chemical and coulombic factors has been done away with and is replaced by a new definition of chemical potential where the activity the activity coefficient due to chemical factor is replaced by the coulombic or electrostatic term for both surface and dissolved species. This has been made possible by redefining the reference state of the chemical potential term of every species. This is expressed as follows:

$$\mu_i = \mu_i^o + RT \ln [i] + zF\phi \tag{5.48}$$

The last term of the right hand side of the above equation represents the free energy required to bring a species with charge z from the reference state potential to the given potential ϕ. The reference state is chosen for all the species as infinite dilution relative to the aqueous phase and also zero charge condition of the surface. The latter condition ensures that there are no ionic interactions affecting either solution or surface species in the standard and reference states and allows thermodynamic constants to be obtained by extrapolation to zero surface charge and solution concentration. This is distinctly the case where extra thermodynamic assumptions have been invoked. However, it will be shown below that these are coherent with thermodynamic basis.

Applying Eq (5.48) to surface protonation reaction (3.1) for all the species and noting that z is equal to zero for the neutral SOH and proceeding as in the case of Eqs (5.40) - (5.44):

$$K_{a1}^{O} = \frac{[SOH][H^+]}{[SOH_2^{\ +}]}\exp\left(\frac{\Phi_{H^+} - \Phi_{SOH_2^+}}{RT}\right) \tag{5.49}$$

Since the potential difference between the bulk of the solution phase (H^+) and the surface species SOH_2^+ is equivalent to the TLM potential ψ_o , Eq (5.49) may be written as follows:

$$K_{a1}^{O} = \frac{[SOH][H^+]}{[SOH_2^{\ +}]}\exp\left(\frac{-F\Psi_o}{RT}\right) \tag{5.50}$$

Comparing Eq (5.44) with Eq (5.50) it is evident that

$$\frac{\gamma_o \gamma_{H+}}{\gamma_+} = \exp\left(\frac{-F\Psi_o}{RT}\right) \tag{5.51}$$

This shows that thermodynamic parameters are equivalent to coulombic or extra thermodynamic parameters. This is what is envisaged in the TLM or for that matter in any other surface complexation model where the following relationship holds good

$$\gamma_i^{\ s} = \gamma_i^{\ s}(o)\exp\left(\frac{zF\Psi}{RT}\right) \tag{5.52}$$

Typically, $\gamma_i^s(o)$ represents the activity coefficient in the standard state, which is assumed to be equal for all the surface species, thereby rendering Eq (5.47) valid for all interfacial reactions. Since the product $\gamma_{H+}[H^+]$ is equal to the activity (a_{H+}) which may directly be converted to the pH value for experimental purpose , we are left with the ratio $\gamma_o^{\ s} / \gamma_+^{\ s}$. Chan et al [161] have shown from insightful physical reasoning that the ratio of the surface activity coefficients is equal to unity. Therefore, $K_{al}^{\ o}$ in Eq (5.50) may be considered as thermodynamic or intrinsic constant. Some additional arguments to demonstrate that the experimentally (extrapolated) determined constants are equal to thermodynamic constants are as follows:

In the TLM the surface charge σ_o is related to surface potential ψ_o and the plane of adsorption of ions is given by

$$\sigma_0 = C_1 (\psi_o - \psi_\beta) \tag{5.53}$$

where, C_1 is the capacitance between the surface plane (o) and the plane of electrolyte ion adsorption (β) . Under the condition of extrapolation to zero ionic strength $\psi_\beta = 0$ and hence $\psi_o = 0$ since σ_0 is also zero. Since the activity coefficient term or the exponential term also becomes unity under this condition (since $e^0=1$) the extrapolated intrinsic constants are equal to thermodynamic ones.

Following the above arguments, it is also possible to derive similar relationships between intrinsic constants for the formation of surface complexes and the activity coefficients of the complexated species as envisaged in Eqs. (5.28) and (5.29). Thus

$$^*K_{Cl^-}^{\text{int}} = \frac{\gamma_o^s\,[SOH]a_{H^+}^s\,a_{Cl^-}^s}{\gamma_{+-}^s\,[SOH_2^+ - Cl^-]} \tag{5.54}$$

$$^*K_{Na^+}^{\text{int}} = \frac{\gamma_{-+}^s\,[SO^-Na^+]a_{H^+}^s}{\gamma_o^s\,[SOH]a_{Na^+}^s} \tag{5.55}$$

By substituting the activities of soluble species on the surface with their corresponding bulk activities by applying Eqs (5.24) and (5.25) we have

$$^*K_{Cl^-}^{\text{int}} = \frac{\gamma_o^s\,[SOH]a_{H^+}\,a_{Cl^-}}{\gamma_{+-}^s\,[SOH_2^+ - Cl^-]}\exp\left[\frac{-e(\Psi_o - \Psi_\beta)}{kT}\right] \tag{5.56}$$

$$^*K_{Na^+}^{\text{int}} = \frac{\gamma_{-+}^s\,[SO^- - Na^+]a_{H^+}}{\gamma_o^s\,[SOH]a_{Na^+}}\exp\left[\frac{-e(\Psi_o - \Psi_\beta)}{kT}\right] \tag{5.57}$$

where the symbols have their usual significances ; the superscript 's' signifies surface species. For bulk concentration no subscript is used . The above equations may therefore be written as

$$^*K_{Cl^-}^{\text{int}} = {}^*Q_{Cl^-}\frac{\gamma_o^s}{\gamma_{+-}^s}\exp\left[\frac{-e(\Psi_o - \Psi_\beta)}{kT}\right] \tag{5.58}$$

$$ {}^{*}K^{\,\text{int}}_{Na^+}{}' = {}^{*}Q_{Na^+}\,\frac{\gamma^{s}_{-+}}{\gamma^{s}_{o}}\exp\left[\frac{-e(\Psi_o-\Psi_\beta)}{kT}\right] \tag{5.59} $$

where ${}^{*}Q_{Cl^-}$ and ${}^{*}Q_{Na+}$ are the complexation quotients and are given by

$$ {}^{*}Q_{Cl^-} = \frac{[SOH]\,a_{H^+}\,a_{Cl^-}}{[SOH_2{}^{+}-Cl^{-}]} \tag{5.60} $$

and

$$ {}^{*}Q_{Na+} = \frac{[SO^{-}-Na^{+}]\,a_{H^+}}{[SOH]\,a_{Na^+}} \tag{5.61} $$

Therefore , important differences arising out of the introduction Eq (5.48) in place of Eq (5.33) are worth noting here : (i) The reference and the standard states for both surface and dissolved species are same, keeping in view that surface complexation in TLM or the site binding model treat surface functional groups as constrained ligands in solution . (ii) The activity coefficients of surface species are not separated into chemical and coulombic terms. (iii) The effect of charge and potential of surface species on the activity of solution species is implicitly incorporated in the thermodynamic treatment.

5.8 Parallel capacitor model

This model of the electrical double layer was proposed by Smit [162] to explain the very low experimental value of outer layer (between IHP and OHP) capacitance observed for the single crystal oxide semiconductor in aqueous electrolyte solution. Subsequently, Smit [163] incorporated some modifications in the model to address certain discrepancies like true thermodynamic character of the equilibrium constants, high inner layer and low outer layer capacitance values etc. The abnormal capacitance values led the author to suggest parallel capacitor model of the double layer. Although the physical picture of the model has not received much acceptance among the colloid and surface chemists, it is discussed below as an interesting attempt to address certain discrepancies in the experimental results. For an interface of 1 cm^2 area the author made two common assumptions; (i) only two types of charged species namely, SOH_2^+ and SO^- can influence the charge in the diffuse layer and (ii) the charges of the species SOH_2^+- A^- and SO^--C^+ are shielded by the adsorbed ions. It is

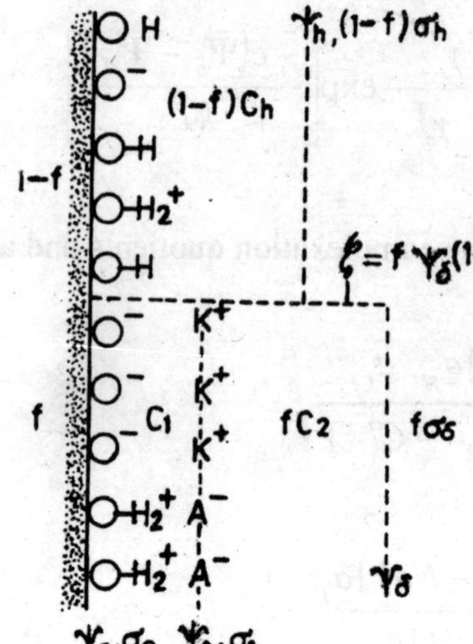

Fig. 5.9. A schematic representation of the parallel capacitor model of the electrical double layer.

further assumed that the above two complex species or ion pairs move to one part of the surface and the other sites, mainly SOH_2^+, SO^- and SOH to the remaining part. This results in two parallel capacitor branches of the charged interface. The fraction 'f' occupied by the complexated sites is given by

$$f = B[\{SOH_2^+ - A^-\} + \{SO^- - C^+\}] / N_s = \alpha_+ + \alpha \qquad (5.62)$$

where N_s has the same unit as the surface species in coulomb per unit area. The fraction f depends upon pH and electrolyte concentration. When f = 1 the outer capacitance will be C_2. The planes ζ and ψ_δ are considered different, the former corresponds to the charges SOH_2^+ and SO^- and the latter to the complexated sites. Similarly, the outer Helmholtz plane (OHP) is considered to lie nearer to to the surface and its capacitance is not differentiated from IHP in the uncomplexated side. Thus when f = 1 the charge on the diffuse layer corresponds to its potential ψ_δ. When f = 0 the diffuse layer coincide with the OHP and the charge (σ_h) corresponds to the potential ψ_h. Therefore, when $f \neq 0$ the Helmholtz layer capacitance in the uncomplexated side is (1-f) C_h and the charge is $(1\text{-}f)\sigma_h$. Under this condition the inner and outer layer capacitances C_1 and C_2 become fC_1^* and fC_2^* respectively. Figure 5.9 illustrates the schematic diagram of the parallel capacitor model and its components as envisaged by Smit. The charge balance equation is given by

$$\sigma_o + \sigma_\beta + f\sigma_\delta + (1\text{-}f)\,\sigma_h = 0 \qquad (5.63)$$

According to this model the potential at different planes are assigned as follows;

$$\psi_o - \psi_\beta = -\sigma_\beta / C_1 = B[\{SOH_2^+ - A^-\} - \{SO^- - C^+\}] / C_1 \qquad (5.64)$$

Combining with Eq (5.62)

$$\psi_o - \psi_\beta = N_s(\alpha_+ - \alpha_-) / fC_1^* = [N_s(\alpha_+ + \alpha_-)] / (\alpha_+ - \alpha_-) C_1^* \qquad (5.65)$$

$$\psi_\beta - \psi_\delta = -\sigma_d / C_2 = -\sigma_d / fC_2^* \qquad (5.66)$$

where σ_d is defined below

$$\psi_o - \psi_h = B[\{SOH_2^+\} - \{SO\text{-}\}] / C_h = B[\{SOH_2^+\} - \{SO^-\}] / (1\text{-}f)C_h^* \qquad (5.67)$$

and $$\zeta = f\,\psi_\delta + (1\text{-}f)\,\psi_h \qquad (5.68)$$

In the above equations capacitors marked with asterisk represent values for modified or parallel capacitor model and these when multiplied by f give values for unmodified or series capacitor model . When f = 0 the charge in the diffuse layer is given by

$$\sigma_h = -0.1174c^{1/2} \sinh(19.47z\,e\,\psi_h / 2.3kT) \quad \text{coul}/m^2 \qquad (5.69)$$

when f = 1 the charge in the diffuse layer is given by

$$\sigma_d = -0.1174c^{1/2} \sinh(19.47z\,e\,\psi_\delta / 2.3kT) \quad \text{coul}/m^2 \qquad (5.70)$$

The concentration of all the surface species including the complexated ones are calculated in the same way as in the series capacitor model. Therefore, there are as many unknown / adjustable variables as the number of equations which are then solved by using Newton - Raphson iteration method. However, all the intrinsic constants need to be corrected for activity coefficients which mean that they need to be true thermodynamic constants. Smit [163] proposed a new method to evaluate true constants as discussed below.

Taking logarithm on both sides of equations (5.60) and (5.61) and rearranging we have

$$p^*Q_{A^-} = pH - \log a_{A^-} + \log\,[SOH_2^+ - A^-] / [SOH] \qquad (5.71)$$

$$p^*Q_{C+} = pH + \log a_{C+} - \log [SO^- - C^+] / [SOH] \quad (5.72\}$$

Assuming that both SOH_2^+- A^- and SO^- - C^+ are present on either side of PZC, Eqs (5.71) and (5.72) may be expressed as follows:

$$p^*Q_{A-} = pH - \log a_{A-} + \log [\alpha_+ / 1 - (\alpha_+ + \alpha-)] \quad (5.73)$$

$$p^*Q_{C+} = pH + \log a_{C+} - \log [\alpha- / 1 - (\alpha_+ + \alpha)] \quad (5.74)$$

The slope values of the plots of $p^*Q_{A^-}$ and p^*Q_{C+} against α_β or $(\alpha_+ - \alpha_-)$ are not equal to zero. Thus, when $C_1 = fC_1^*$ the ratio of the activity coefficients of the two surface species is not independent of either α_+ or α_-. Indeed, $\log(\gamma_o^s / \gamma_{+-}^s)$ or $\log(\gamma_o^s / \gamma_{-+}^s)$ are linearly related to α_β, i.e.,

$$\log(\gamma_o^s / \gamma_{+-}^s) = b(\alpha_+ - \alpha_-) / 2.3 \quad (5.75)$$

and

$$\log(\gamma_o^s / \gamma_{-+}^s) = -a(\alpha_+ - \alpha_-) / 2.3 \quad (5.76)$$

Combining Eqs (5.75) and (5.76) with (5.58), (5.59) and (5.65) we have

$$p^*Q_{A-} = p^*K_{A-}(tr) + b(\alpha_+ - \alpha_-) / 2.3 - \frac{eN_s(\alpha_+ - \alpha_-)}{2.3kT(\alpha_+ + \alpha_-)C_1^*} \quad (5.77)$$

$$p^*Q_{C+} = p^*K_{C+}(tr) - a(\alpha_+ - \alpha_-) / 2.3 - \frac{eN_s(\alpha_+ - \alpha_-)}{2.3kT(\alpha_+ + \alpha_-)C_1^*} \quad (5.78)$$

where (tr) represents true or thermodynamic constants . Since at pH away from PZC only one complexated species (either α_+ or α_-) predominates the last term of the equations (5.77) and (5.78) is essentially constant. Thus, when the linear portion of the plots of either p^*Q_{A-} or p^*Q_{C+} vs $(\alpha_+ - \alpha_-)$ are extrapolated to $(\alpha_+ - \alpha_-) = 0$ the intercept values will yield true thermodynamic constants from the following relationships.

$$p^*K_{A^-}(\text{extr}) = p^*K_{A-}^{\text{int}}(tr) - \frac{eN_s}{2.3kT\,C_1^*} \quad (5.79)$$

$$p^*K_{C+}^{int}(\text{extr}) = p^*K_{C+}^{int}(\text{tr}) + \frac{eN_s}{2.3kTC_1*} \qquad (5.80)$$

The above relationships show that true or thermodynamic constants are different from those of the extrapolated values as the second term on the RHS of the equations (5.79) and (5.80) is almost equal to one. Therefore, intrinsic constants as envisaged in Eqs (5.32) and (5.33) are related to true constants as follows

$$p^*K_{A^-}^{int} = p^*K_{A^-}^{int}(\text{tr}) + \log(\gamma_o^{s}/\gamma_{+-}^{s}) \qquad (5.81)$$

and

$$p^*K_{C^+}^{int} = p^*K_{C^+}^{int}(\text{tr}) - \log(\gamma_o^{s}/\gamma_{-+}^{s}) \qquad (5.82)$$

The above relationships show that intrinsic constants become true constants only when the ratio of the activity coefficients of the relevant surface species is unity. Therefore, combining all the relevant equations namely, (5.75), (5.76), (5.79), (5,80), (5.81) and (5.82)we get complete relationships between intrinsic constant and extrapolated constant

$$p^*K_{A^-}^{int} = p^*K_{A^-}^{int}(\text{extr}) + [b(\alpha_+ - \alpha_-)/2.3] - \frac{eN_s}{2.3kTC_1*} \qquad (5.83)$$

$$p^*K_{C^+}^{int} = p^*K_{C^+}^{int}(\text{extr}) + [a(\alpha_+ - \alpha_-)/2.3] - \frac{eN_s}{2.3kTC_1*} \qquad (5.84)$$

Smit determined the parameters a and b from the slope values of the plots of p^*Q_{C+} vs α_- and p^*Q_{A-} vs α_+ at pH above and below PZC respectively . The adjustable parameters of the model are C_I^* and C_2^* and C_h^* . Smit [163] obtained excellent fitting of the experimental data for the variation of σ_o and ψ_d as function of pH and electrolyte concentration with those of calculated for TiO_2 / NaCl solution using parallel capacitor model . Kanungo [164] also observed better fitting of the data for hydrous MnO_2 / NaCl system using this model.

5. 9 Discreteness of the Charge Effect

In the above discussion it is assumed charges on the surface as well as on the IHP are uniformly distributed or "smeared out". But in actual or physical picture the charges carried by the adsorbed ions on these planes are not distributed uniformly, but in "discrete patches ". This arrangement will impose a self - atmosphere potential on the adsorbed ions which is the two dimensional analogue of the three dimensional self-atmosphere potential occurring in the Debye - Huckel theory of electrolyte solution. This discreteness of charge effect or discrete ion effect was introduced by Esin and Shikov [165] on the suggestion originally made by Esin and Markov [166] to explain the shift in PZC of Hg / solution interface in the presence of surface active agents.

The importance of this effect in the stability of colloid has been stressed by Levine et al [128,167,168] and has also been reviewed by several authors [143,147]. Due to physical discreteness of the charge, the surface layer experience a micro potential which is not equal to the mean "smeared out" potential ψ_o resulting from charge σ_0. Similarly, the smeared out potential ψ_β will be different at the β plane. However, in the diffuse layer the discreteness of charge effect is assumed to be absent. Thus, for ions adsorbed on the surface plane the coulombic attraction force becomes [156]:

$$\gamma_c(o) = \exp\{-ze(\psi_o + \varphi_o)/kT\} \tag{5.85}$$

where, φ_o is the self atmosphere potential due to discreteness of charge effect [167]. Similarly, for ions at the β plane the coulombic attraction is given by

$$\gamma_c(\beta) = \exp\{-ze(\psi_\beta + \varphi_\beta)/kT\} \tag{5.86}$$

where , φ_β is the self atmosphere potential due to discreteness of the charge effect in that plane .

The values of φ_o and φ_β are found to be dependent on the surface charge and electrolyte concentration [167,168]. However, the effect is often assumed to be small or insensitive compared to variation in surface charge and electrolyte concentration so that the effect is included in the intrinsic equilibrium constant. According to Koopal et al [156] the terms $\gamma_c(o)$ and $\gamma_c(\beta)$ are considered " as weighting factors , weighting the probability of finding an ion at the given plane in the absence of specific interaction relative to the probability of finding it in the bulk solution ". It will be shown in the subsequent chapter that $\gamma_c(o)$ and $\gamma_c(\beta)$ are directly related to the activity coefficients of the soluble charged species incorporated in the intrinsic equilibrium constants of various surface reactions .

Another consequence of the discreteness of the charge effect in colloid chemistry is that it predicts , under certain conditions , ψ_d can reach maximum as ψ_o is increased , so that further increase in ψ_0 leads to decrease in ψ_d and therefore to the decrease in the inter particle attraction.

5.10 Porous Double Layer Model

From the study of variation of σ_0 and ψ_o values with respect to pH for a number of oxide / electrolyte suspensions, Lyklema [61] has observed that (i) there is a characteristic property of steep rise in σ_0 with increase in pH i.e., slope values much higher than that predicted by the Nernst equation ; (ii) the greater the porosity the greater is the surface charge, the absolute value of which sometimes exceeds the total number of OH groups per unit area and (iii) notwithstanding the high value of σ_0 the experimental values of zeta potential are not particularly high. These observations have lad the author to postulate porous double layer model. The essential feature of the model is that the surface charge (σ_0) is not confined to the surface property but is distributed within some depth behind the surface layer. Since counter ions are also envisaged to pass through the surface, the σ_0 and ζ (zeta) potential values tend to be low on the solution side.

A major problem in the use of this model is the lack of knowledge on the depth to which the counter ions can penetrate. Breeuwsma and Lyklema [149] termed this as "chemical porosity" to differentiate it from physical porosity which can be quantified by gaseous adsorption studies. High values of surface charge reflect extensive counter ions penetration within the pores which is determined more from geometric consideration than from its affinity towards solid. Thus, Breeuwsma and Lyklema [149] have explained the high sorption of Li^+ on a α - Fe_2O_3 due to similar ionic radii of Li^+ (0.68 A°) and Fe^{3+} (0.65 A°). But, there is a considerable controversy as to how Li^+ gets dehydrated before entering the porous layer.

Ideas similar to the conceptual model of Lyklema were also developed by Perram et al [62]. Improving upon the concept of porous oxide surface, the authors postulated the occurrence of boundary gel layer at the interface. A gel layer of 20 - 40 A° has been determined quantitatively for oxides with higher values of surface potential, This surface layer is believed to be distorted and deformed due to the penetration of various ions and is a characteristic feature of oxide / solution interface [62] . This gel layer effectively shields to a certain extent the effect of surface charge on the diffuse layer in the bulk of the solution. On the other hand, several authors [169-171] have suggested that the occurrence of structured water close to the interface. The order decreases exponentially with the distance from the surface due to lesser influence of charged field and greater influence of the thermal motion. However, neither the porous double layer nor the gel layer theory has been able to predict adsorption behavior of inorganics at the oxide water interface and therefore has found little practical use.

CHAPTER 6

EVALUATION OF THE EQUILIBRIUM CONSTANTS OF REACTIONS OCCURRING AT SOLID / SOLUTION INTERFACE AND RELATED DOUBLE LAYER PARAMETERS

6.1 Surface Charge as a Function of pH and Electrolyte Concentration

It has been stated in the preceding chapter that irrespective of the double layer model the surface charge is determined from the net number (or quantity) of protons released or consumed during the formation of ionized surface species i.e., acid - base dissociation of oxide surface . The site - binding or triple layer model further suggests that the formation of surface complex readjusts the acid - base equilibrium and consequently affects the surface charge. Therefore , the total charge on the oxide surface corresponds not only to the ionized species like SOH_2^+ and SO^- , but also their complexated species with anion (SOH_2^+ - A^-) and cation (SO^- - C^+) respectively present in the swamping electrolyte solution . In the earlier literature, not only this aspect but also the electrostatic factor (between the surface and the solution phase) was not taken into consideration while estimating the equilibrium constants. Therefore, the total ionizable surface sites was represented as follows:

$$N_s = \{SOH\} + \{SO^-\} + \{SOH_2^+\} \qquad (6.1)$$

It was assumed that at pH << pH_{PZC} the surface charge (σ) derived from potentiometric titration was equal to $\{SOH_2^+\}$. Therefore,

$$\{SOH\} = N_s - \{SOH_2^+\} = N_s - \sigma \qquad (6.2)$$

where, σ_0 is the titratable surface charge in moles / kg . Similarly, when pH >> pH_{PZC}, the titratable charge was assumed due only to be $\{SO^-\}$. Consequently,

$$\{SOH\} = N_s - \{-SO^-\} = N_s + \sigma \qquad (6.3)$$

From Eqs. (3.3) , (3.4) and (3.10) it can be easily shown that

$$pK_{a1}^{int} = pH + \log\frac{\{SOH_2^+\}}{\{SOH\}} + \frac{e\Psi_o}{2.3kT} = pK_{a1} + \frac{e\Psi_o}{2.3kT} \quad (6.4)$$

$$pK_{a2}^{int} = pH - \log\frac{\{SO^-\}}{\{SOH\}} + \frac{e\Psi_o}{2.3kT} = pK_{a2} + \frac{e\Psi_o}{2.3kT} \quad (6.5)$$

However, combining Eqs.(6.2) and (6.3) with Eqs. (6.4) and (6.5) respectively and expressing concentration in terms of fractional ionization of the total surface sites i.e., $\alpha_{\pm} = \pm\sigma_0 / N_s$, the above two equations may be written as

$$pK_{a1}^{int} = pH + \log\frac{\alpha_+}{1-\alpha_+} + \frac{e\Psi_o}{2.3kT} = pK_{a1} + \frac{e\Psi_o}{2.3kT} \quad (6.6)$$

$$pK_{a2}^{int} = pH - \log\frac{\alpha_-}{1-\alpha_-} + \frac{e\Psi_o}{2.3kT} = pK_{a2} + \frac{e\Psi_o}{2.3kT} \quad (6.7)$$

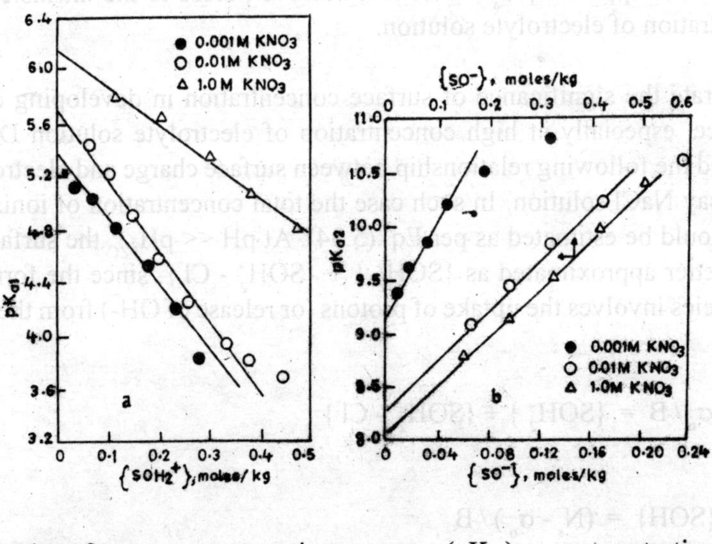

Fig. 6.1 **(a) Plot of apparent protonation constant (pK_{a1}) vs. concentration of protonated hydrous ferric oxide(HFO) surface species. (b) Plot of apparent deprotonation constant(pK_{a2}) vs.concentration of deprotonated species. Both the figures indicate that the intrinsic constants depend not only upon zero charge condition of the surface, but also upon the zero concentration of elec trolyte solution i.e., the condition of zero surface potential.**

where, pK_{a1} and pK_{a2} represent apparent ionization quotients of the reactions (3.1) and (3.2) respectively . Attempts were made by the earlier authors [112,125,172] to obtain the values of intrinsic ionization constants by plotting pK_{a1} and pK_{a2} against the corresponding concentration of either $\{SOH_2^+\}$ or $\{SO^-\}$ and extrapolating the linear region to zero concentration values . This is also equivalent to plotting pK_{a1} and pK_{a2} against α_+ or α_- and extrapolating to $\alpha_\pm = 0$ to obtain the so called intrinsic constants under zero charge condition [150]. However, it was observed that the values were dependent upon the concentration of electrolyte solution. In other words, pK_a values, under this condition, were not extrapolated to zero potential condition . Figure 6.1a and 6.1b illustrate such plots for amorphous hydrous oxide of iron in KNO_3 solution. The figures clearly suggest that intrinsic constants are dependent on electrolyte concentration even though they are obtained under zero charge condition. Similar results were also obtained by Davis et. al.[150] for the experimental data of Yates [75] for the acid - base titration of TiO_2 in $LiNO_3$ solution and Hingston et al [173] for the data on titration of goethite in NaCl solution. Blesa et al [174] also observed similar effect of the electrolyte concentration in their plots of pK_a vs.α_+ or α_- for the suspension of magnetite in KNO_3 solution . The differences are sometimes too large to be explained by the activity correction. This implies that the mechanism of charging of oxide / solution interface involves more than the simple amphoteric dissociation. The apparent pK_a values are, however, close to the intrinsic values at low concentration of electrolyte solution.

To demonstrate the significance of surface concentration in developing charge on oxide surface, especially at high concentration of electrolyte solution Davis et al [150] derived the following relationship between surface charge and electrolyte concentration, say NaCl solution. In such case the total concentration of ionizable surface sites should be estimated as per Eq .(5.34). At pH << pH_{PZC} the surface charge should be better approximated as $\{SOH_2^+\} + \{SOH_2^+ - Cl^-\}$, since the formation of both the species involves the uptake of protons (or release of OH-) from the solution. Thus,

$$\sigma_o / B = \{SOH_2^+\} + \{SOH_2^+ - Cl^-\} \tag{6.8}$$

Therefore,

$$\{SOH\} = (N_s - \sigma_o) / B \tag{6.9}$$

Combining Eq (5.26) and Eq (6.8)

$$\sigma_o / B = \{SOH_2^+\} + \{SOH_2^+\} K_{Cl^-}^{int} [Cl^-]_s \tag{6.10}$$

Rearranging

$$\{SOH_2^+\} = \frac{\sigma_o}{B\left(1 + K_{Cl^-}^{\text{int}}\,[Cl^-]_s\right)} \quad (6.11)$$

Combining Eqs . (3.3), (6.9) and (6.11) we get

$$K_{a1}^{\text{int}} = \frac{\frac{N_s}{B}\left(1 - \frac{\sigma_o}{N_s}\right)\left[H^+\right]_s\left(1 - K_{Cl^-}^{\text{int}}\left[Cl^-\right]_s\right)}{\sigma_o} \quad (6.12)$$

Dividing both numerator and denominator by N_s / B and converting the surface concentration of all soluble species into their bulk concentration

$$K_{a1}^{\text{int}} = \frac{\left(1 - \alpha_+\right)\left[H^+\right]\exp\left(\frac{-e\Psi_o}{kT}\right)\left[1 + K_{Cl^-}^{\text{int}}\,[Cl^-]\exp\left(\frac{e\Psi_\beta}{kT}\right)\right]}{\alpha_+} \quad (6.13)$$

Taking negative logarithm on both sides and rearranging we have

$$pK_{a1}^{\text{int}} = pH + \log\frac{\alpha_+}{1 - \alpha_+} + \frac{e\Psi_o}{2.3kT} - \log\left[1 + K_{Cl^-}^{\text{int}}\,[Cl^-]\exp\left(\frac{e\Psi_o}{kT}\right)\right] \quad (6.14)$$

Similarly , at pH >> pH_{PZC} we may assume that the surface charge is the sum of the following species

$$-\sigma_0 / B = \{SO^-\} + \{SO^- - Na^+\} \quad (6.15)$$

and $$\{SOH\} = N_s / B - [\{SO^-\} + \{SO^- - Na^+\}] \quad (6.16)$$

Proceeding as above by invoking Eqs (3.4) and (5.27)

$$pK_{a2}^{\text{int}} = pH - \log\frac{\alpha_-}{1 - \alpha_-} + \frac{e\Psi_o}{2.3kT} + \log\left[1 + \left[Na^+\right]K_{Na^+}^{\text{int}}\exp\left(\frac{-e\Psi_\beta}{kT}\right)\right] \quad (6.17)$$

Equations (6.14) and (6.17) clearly indicate that surface ionization constants are dependent on electrolyte concentration. In other words, these two equations also suggest that the plots of pH + log (α_+ /1- α_+) vs α_+ or pH - log (α_- /1- α_-) vs α_- should not only be extrapolated to zero charge condition ($\alpha\pm = 0$) but also the ionization quotients should be estimated at infinitely low concentration so that the last term in Eqs (6.14) and (6.17) becomes zero . Under these conditions both σ_0 and ψ_β are zero and therefore ψ_o also becomes zero.

Let us now examine the surface charge condition at high salt concentration i.e., the condition at which surface complexation is primarily responsible for experimentally observed surface charge. Therefore, for a positively charged surface

$$\sigma_0 / B = \{SOH_2^+ - Cl^-\} \tag{6.18}$$

and for a negatively charged surface

$$-\sigma_0 / B = \{SO^- - Na^+\} \tag{6.19}$$

Consequently, $\{SOH\} = [N_s - |\sigma_0|] / B$ (6.20)

Substituting in Eq (5.32)

$$*K^{int}_{Cl^-} = \frac{(N_s - |\sigma_o|)\,[H^+][Cl^-]\exp\left[\dfrac{-e(\Psi_o - \Psi_\beta)}{kT}\right]}{\sigma_o} \tag{6.21}$$

Dividing both numerator and denominator by N_s and taking logarithm

$$\log(*K^{int}_{Cl^-}) = \log\frac{1-\alpha_+}{\alpha_+} + \log[H^+] + \log[Cl^-] - \left[\frac{e(\Psi_o - \Psi_\beta)}{kT}\right]$$

Reversing the sign on both sides

$$p^*K^{int}_{Cl^-} = pH + \log\frac{\alpha_+}{1-\alpha_+} - \log[Cl^-] + \left[\frac{e(\Psi_o - \Psi_\beta)}{2.3kT}\right] \tag{6.22}$$

$$\text{or,}\quad p^*K_{Cl^-}^{int} = p^*Q_{Cl^-} + \left[\frac{e(\Psi_o - \Psi_\beta)}{2.3kT}\right] \tag{6.23}$$

Similarly, from Eq. (5.33) one can derive the following equation

$$p^*K_{Na+}^{int} = pH - \log\frac{\alpha_-}{1-\alpha_-} + \log[Na^+] + \frac{e(\Psi_o - \Psi_\beta)}{2.3kT} \tag{6.24}$$

$$\text{or,}\quad p^*K_{Na+}^{int} = p^*Q_{Na+} + \left[\frac{e(\Psi_o - \Psi_\beta)}{2.3kT}\right] \tag{6.25}$$

where, $p^*Q_{Cl^-}$ and p^*Q_{Na+} are the complexation quotients for anion and cation respectively . Davis et al [150] have shown from earlier literature that when $p^*Q_{Cl^-}$ and p^*Q_{Na+} values are plotted against α_+ or α- respectively for different concentrations of electrolyte solution, a single straight line can describe most of the data. Some deviations, however, occur at low charge and at low ionic strength where the approximations for Eqs. (6.18) and (6.19) lead to error because ionized surface sites SO^- and SOH_2^+ also contribute significantly to the surface charge. However, extrapolation of the plots at higher concentration such as 1M solution to $\alpha_\pm = 0$ gives values close to intrinsic complexation constants.

Using the potentiometric titration data of different oxide- solution systems obtained by different workers, Davis et al [150] calculated intrinsic ionization and complexation constants for these systems. From these constants the authors calculated surface charge and zeta potential (assuming $\zeta = \psi_d$) and satisfactory agreement between the experimental and calculated values was observed. The only adjustable parameters that were not accessible by experiment are inner and outer layer capacitance values. The inner layer capacitance (C_1) i.e., capacitance between surface and IHP was assumed to be 140 μF / cm² for $LiNO_3$ and 100 μF / cm² for KNO_3 solution . These values are higher than those measured from classical colloid systems indicating small distance of about 0.1 nm between the planes. On the other hand , the capacitance between IHP(β - plane) and the OHP(diffuse layer) was assumed to be 20 μF / cm² and was in accordance with the experimental values observed for either Hg / water or AgI / water systems .

6.2 Determination of Intrinsic Ionization Constants by Double Extrapolation Method

The method developed by James et al [175] and discussed further by James and Parks [60] is based on simultaneous extrapolation of charge ($\alpha\pm$) and electrolyte concentration to zero condition . In this method the ionization quotients (pK_{a1} and pK_{a2}) is plotted against $\alpha_{\pm}$ plus k times the function of electrolyte concentration . The value of k is chosen in such a way that data of one concentration are sufficiently separated from the data of other concentration and at the same time concentration function becomes zero at zero electrolyte concentration. This was found to be achieved best by using $kc^{1}/_{2}$. The value of k may be anything except zero and has to be determined by trial and error and lies usually around 0.1 - 0.3. There are two routes that can be used for double extrapolation.

1. For each ionic strength a smooth curve through the experimental points is extrapolated to intersect the line vertical to the x - axis at $\alpha_{\pm}= 0$ i.e., $\alpha_{\pm}+ kc^{1/2} = kc^{1/2}$, where c is the electrolyte concentration. Now, a smooth curve can be drawn through the $\alpha_{\pm} = 0$ points for each ionic strength and extrapolated to $\alpha_{\pm}+kc^{1/2}= 0$ where c must also be zero. The intercept value on the pK_a axis corresponds to intrinsic constant.

2. Using the same plot there is another way of estimating the intrinsic constants. For each ionic strength a set of vertical lines corresponding to the arbitrary values of $\alpha_{\pm}$ is drawn along the curve. The intersections of these lines with the smoothened curve for the appropriate concentration identify the points of $\alpha_{\pm}$ along the curve. A smooth curve is now drawn connecting the points of common a± and extrapolated to intersect the curve at c = 0 which usually lies very close to the curve for low ionic strength (say, 10^{-3} M). This curve is again extrapolated to $\alpha_{\pm}+ kc^{1/2} = 0$ at which α , σ_0 and c = 0 . The intercept value corresponds to intrinsic constants. It may be mentioned here that intercept values obtained by both the methods should be same. Figures 6.2 through 6.4 illustrate some typical examples of the determination of intrinsic constants (pK^{int}_{a2}) for hydrous iron oxide (HFO), hydrous manganese dioxide (birnessite) and a clay mineral (illite) respectively, using the rapid potentiometric titration data generated in the author's laboratory.

6.3 Determination of Intrinsic Complexation Constants by Double Extrapolation Method

Equations (6.23) and (6.25) suggest that $p^{*}Q_{Cl^-}$ and $p^{*}Q_{Na^+}$ values for an oxide suspended in strong NaCl solution when plotted against $\alpha_{\pm}$ - k log c and the resulting curves when extrapolated to $\alpha_{\pm} = 0$ and c = 1M , the intercepts on the axis of $p^{*}Q_{Cl^-}$ and $p^{*}Q_{Na^+}$ give the intrinsic constants . The value of k is so chosen by trial and error that the curve for one ionic strength is separated from the other at a reasonable distance. Like intrinsic ionization constants, there are also two methods by which intrinsic complexation constants can be determined. These are:

1. For each ionic strength, a smooth curve drawn through the experimental points is extrapolated to intersect the line at $\alpha_{\pm}$ - k log c = k log c . At this point both $\alpha_{\pm}$ and c must be zero. Now, a smooth curve can be drawn through the $\alpha_{\pm} = 0$ and this line, in turn, when extrapolated to $\alpha_{\pm}$ - k log c = 0 i.e.,$\alpha_{\pm} = 0$, and c = 1M the intercept value on the p*Q axis gives the intrinsic constant, since at this point the electrostatic term in Eqs. (6.23) and (6.25) are also zero.

2. For each ionic strength, a set of vertical lines corresponding to suitably chosen arbitrary values of a± is drawn. The intersection of these lines with the experimental curve for each ionic strength are joined by smooth line corresponding to c = 1M for that value of $\alpha_{\pm}$. A smooth curve joining the points of intersection with c = 1M when extrapolated to $\alpha_{\pm}$- k log c = 0 gives the intrinsic constant values. Figures 6.5

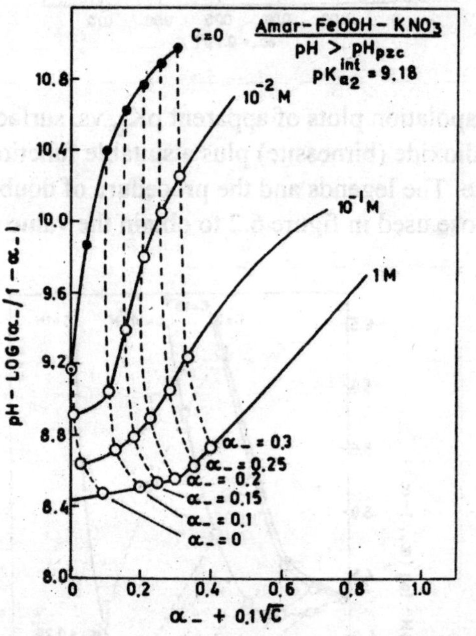

Fig. 6.2 Double extrapolation plots of apparent pK_{a2} vs. surface charge of HFO plus a suitable function of concentration of electrolyte (KNO_3). Open circles represent the values of pK_{a2} calculated at constant fractional charges (α_-) which are then extrapolated by smooth curves to intersect the theoretical curve rep resenting the pK_{a2} values for C = 0. These points of intersection (filled circles) on further extrapolation to zero charge condition (α_-=0) gives the value of intrinsic constant ($p^*K^{int}_{a2}$).

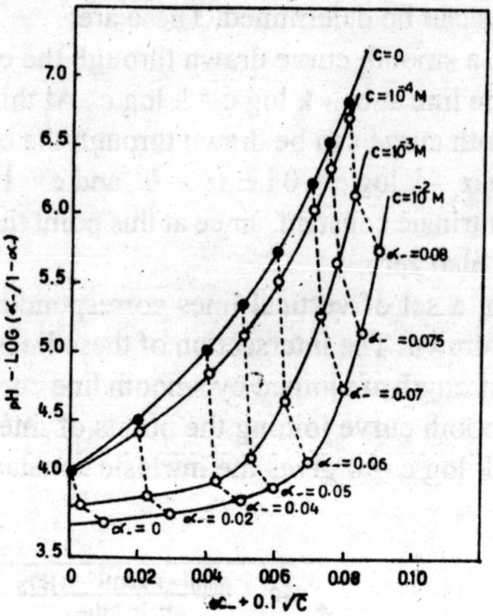

Fig. 6.3 Double extrapolation plots of apparent pK_{a2} vs. surface charge of a hydrous manganese dioxide (birnessite) plus a suitable function of the concentration of electrolyte. The legends and the procedure of double extrapolation are similar to those used in figure 6.2 to obtain the value of intrinsic constant.

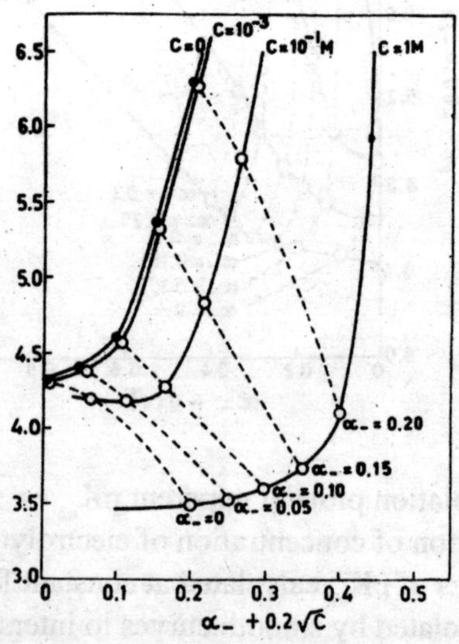

Fig. 6.4 Double extrapolation plots for the determination of pK_{a2}^{int} value of an illite sample in NaCl solution.

and 6.6 illustrate some typical examples of such plots for the determination of $p^*K_{NO3-}^{int}$ (in KNO_3 solution) and $p^*K_{Na+}^{int}$ (in NaCl solution) for HFO and β-FeOOH samples respectively. The success of the double extrapolation method, however, depends upon the precision of the titration data.

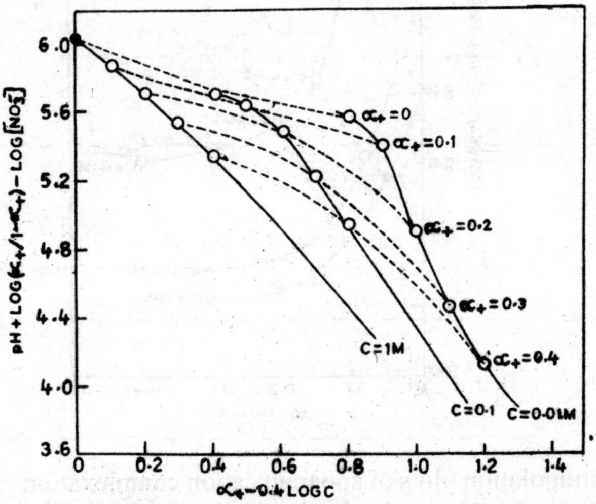

Fig. 6.5 Double extrapolation plots of apparent anion complexation constant (pQ_{NO3-}) of a synthetic akaganeite (β-FeOOH) sample vs. fractional surface charge (α_+) and a suitable logarithmic function of KNO_3 concentration (α_+ - 0.4logC). Open circles represent pQ_{NO3-} values for different fractional charges at 0.1M and 0.01M concentrations which are connected by smooth curve to the plot representing 1M KNO_3. The points of intersection on 1M curve when extrapolated to zero charge condition give the intrin sic complex ation constant $p^*K^{int}_{NO3-}$.

6.3.1 Double Extrapolation to pH_{PZC}

Sprycha and Szczypa [176,177] suggested a double extrapolation method wherein they plotted k(pH) + $c^{1/2}$ against the corresponding pK_a values . As followed in the previous methods each vertical line for different ionic strength is extrapolated to kpH_{PZC} + $c^{1/2}$ and a vertical line joining these extrapolated points (i.e., $k(pH_{PZC}) + c^{1/2}$ values) is again extrapolated to zero concentration i.e., only $k(pH_{PZC})$ values where the surface charge is assumed to be zero. The value of k is so chosen that the vertical line for one ionic strength is separated from the other by a reasonable distance. The value of ionization quotient corresponding to the doubly extrapolated point represents intrinsic constant . The intrinsic constants derived by Sprycha [177] for γ - Al_2O_3 in NaCl solution are also similar to those obtained by James and Parks [60] using the double extrapolation method of James et al [175], described in section 6.2.

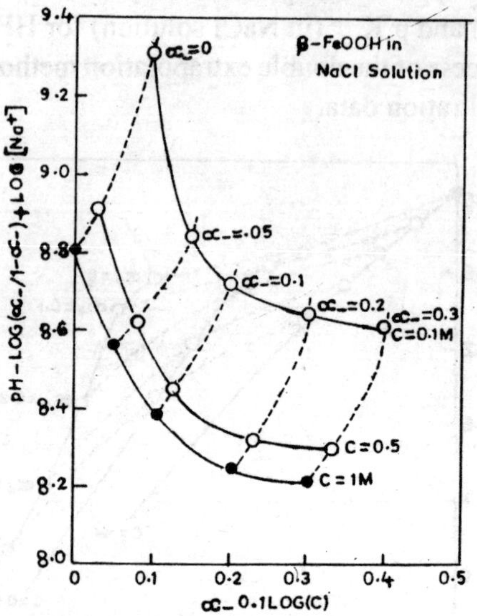

Fig. 6.6 Double extrapolation plots of apparent cation complexation constant (pQ_{Na^+}) of a synthetic akaganeite(β-FeOOH) sample vs. fractional surface charge (α_-) and a suitable logarithmic function of KNO_3 concentration (α_- - 0.1logC). The remaining procedure to obtain the value of intrinsic constant is similar to that given in figure 6.5.

But, in case of anatase in KNO_3 solution considerable differences were observed [175,178]. Indeed, these authors observed that the accuracy of James and Parks method is rather poor because the extrapolated smooth curves lie asymptotically to the vertical lines, in contrast to the straight line in the present method. Thus, any inexactitude in tracing the curves may show large differences in the values of constants.

6. 4 Numerical Method of Optimization

Numerical or computer aided nonlinear method of optimization has always been preferred by the surface chemists over the graphical or analytical method for evaluating intrinsic constants of surface reactions because of the following reasons. Graphical methods are time consuming or tedious, less precise , difficult to apply to complex systems and above all, often applicable to limited range of experimental data . In many instances extrapolated values are greatly influenced by individual biasness of the lines drawn. Though the results appear to be accurate, conflicting data are, sometimes, generated among different workers. In contrast, computer programs are fast, bias free, applicable to complex systems and cover a wide range of experimental data. Another advantage of the statistical optimization is that the error range for every estimated parameter can be obtained. However, computer programs cannot distinguish between right and wrong input data. Input of any erroneous data will result in wrong and sometimes fictitious output. On the other hand, such wrong and undesirable data can be screened out during the operation of graphical method. Secondly, solution of multi parameter (or multi adjustable parameters) requires complicated algorithm which sometimes is either difficult to handle or is not accessible to many researchers.

A number of computer programs have been developed during the last thirty years by those researchers who have been working in the solution of problems related to chemical equilibrium in aqueous system. However, many of the programs have been developed are, in fact, designed to solve a particular type or class of problems. But, there are several programs which are designed to solve general problems of chemical equilibrium, though they are less rigorous as far as the thermodynamic database is concerned. The REDEQL by Morel and Morgan [179] and refined by McDuff and Morel [180] and WATEQL by Truesdel and Jones [181] are some of the earlier general programs based on which most of the subsequent popular programs such as MINEQL and MICROEQL by Westall et. al. [182,183] and GEOCHEM by Sposito and Mattigold [184] have been developed. While both WATEQL and GEOCHEM emphasize thermodynamic database for equilibrium calculations in natural water and other geochemical systems, MINEQL is found to be more applicable to equilibrium calculation for laboratory experiments on adsorption. Later on, U.S. Environmental protection agency has developed a new program called MINTEQ [185] by adopting essentially important features from both MINEQL and WATEQL. Excepting MICROEQL which is written in BASIC, the other programs are written in FORTRAN using Newton-Raphson method of iteration for minimizing square root error.

Simultaneously with the development of MINEQL and MICROEQL , Westall developed another program named FITEQL [186] which is an iterative , nonlinear least square optimization program based on Gauss method for unconstrained problem , in the sense that unknown parameters are free to assume any non - zero values . To solve a general equilibrium problem in aqueous medium let us take , for example , an acid base titration data set of an oxide . The first step in the FITEQL optimization procedure is to set up a multi component chemical model for the system . Then we solve the equilibrium model at each point of titration. For this purpose it is useful to arrange the stoichiometric relationship between the components necessary to form the relevant species in a matrix or tabular form (Tableau). Before we make a Tableau it is worthwhile to have a clear distinction between components and species as shown below.

Species is defined as every chemical entity which is considered in the solution of equilibrium problem. Components are defined as the entities which when involved in a chemical reaction yield a species as product and no component can be written as the product of a reaction involving the other components. Every component is associated with one material balance equation with two variables : free concentration (X) and the total concentration (T) . Depending upon the available experimental data a component in an equation can be classified into three groups as follows :

Group I. Components for which T is known . For example,

T I : Experimental value
X I : Guess for XI to be calculated

Group II. Components for which both X and T are known. For example,

T II : Experimental value
X II : Experimental value

Group III. Components for which only X is known . For example ,

X III : Experimental value
T III : Zero is the convenient value , although not obligatory. If TIII = 0 then the calculated value
YIII : is the total amount of component III dissolved in the liquid phase .
YIII : (residual concentration of component III) will be defined later on. If X and T for a given component are adjustable parameters, the component is assigned to as if X and T were known . In this regard the optimization procedure should be considered as ad justment of known value rather than determination of unknown values.

6.4.1 Tableau for Acid / Base Titration: An Example of Problem Formulation

Having defined species and components we are now in a position to formulate equilibrium problem in the form of tableau. A chemical tableau is a convenient and systematic means of properly organizing an equilibrium problem. A basis set of species is selected from all chemical species in the system . For the set of chemical species in a particular aqueous solution, a proper basis or component set is such that is such that every species can be represented as the product of reaction involving only the components . The example discussed below is the hydrous iron oxide suspended in NaCl solution and titrated by either strong base (NaOH) or strong acid (HCl) . The species H^+ , $FeOH^O$, Na^+, Cl^- and H_2O provide a convenient and appropriate set of components from which all the species in the system (H_2O, H^+, OH^-, $FeOH^o$, $FeOH_2^+$, FeO,

Na^+, Cl^-) can be expressed in unique way . This is shown in Table 6.1 where the stoichiometric coefficient of each species corresponding to each component is given on the horizontal row. Except for species that are also component a mass law equation can be written for each line of tableau using the stoichiometric coefficients as exponents for free concentration of components. For example,

$$[\,OH\,] = [\,H+\,]^{-1}\,k_w \tag{6.26}$$

$$[\,FeOH_2^{\,+}\,] = [\,H^+\,]\,[FeOH\,]\,/\,K_{a1} \tag{6.27}$$

In the above mass law equations activity coefficients are taken as unity and [ ] represents molar concentration. For each column of the tableau a mole balance equation can be written using the stoichiometric coefficient of each species. For example,

Table 6.1 Tableau for Acid - Base Titration of Hydrous Ferric Oxide (HFO) in NaCl Solution

Species / Component	H^+	$>FeOH^o$	Na^+	Cl^-	$P = \exp(-F\psi/RT)$	K
H^+	1					
OH^-	-1					K_w
Na^+			1			
Cl^-				1		
$>FeOH_2^{\,+}$	1	1			1	$1/K_{a1}^{int}$
$>FeOH$		1				
$>FeO^-$	-1	1			-1	K_{a2}^{int}
NaCl			1	1		$(NaCl)_T$
NaOH	-1			1		$(NaOH)_T$
HCl	1			1		$(HCl)_T$
>FeOH		1				$(>FeOH)_T$
					1	$(AS/F)_\delta$

$$[FeOH]_T = [FeOH_2^{\,+}] + [FeOH^o] + [FeO^-] \tag{6.28}$$

$$[H^+]_T \;=\; [H^+] - [OH^-] + [FeOH_2^{\,+}] - [FeO^-] = [HCl]_T - [NaOH]_T \tag{6.29}$$

where $[HCl]_T$ and $[NaOH]_T$ are the total concentration of strong acid and base respectively added to the system . The mass law equations corresponding to the rows in

Table 6.1 and the mole balances corresponding to the columns constitute a complex set of equations for the chemical equilibrium problem under consideration, assuming that concentrations of all the added "ingredients " in the system such as salt, acid, base, and solid adsorbent are known . It is also necessary that equilibrium constants of the reactions involving components are known or their initial values guessed . However, the values of apparent equilibrium constants of the surface reactions are usually not known *a priori*, because of the dependence of coulombic term $\{\exp(-e\psi_o / kT)\}$ on the equilibrium sorption of surface charge. This problem has been solved by including explicitly the coulombic term as a dummy component (P) in the system and assigning nominal charge of the surface species by expressing the mass law equilibrium in the appropriate form by incorporating P term in the RHS of Eq. (6.27). The mole balance corresponding to the component P is simply the total surface charge.

$$(P)_T = [FeOH_2^+] - [FeO^-] = (S_v / F)\,\sigma_0 \quad \text{moles / l} \tag{6.30}$$

where, S_v = area per unit volume of suspension , m^2 / l
F = Faraday constant , C / mole
σ_0 = surface charge , C / m^2

For diffuse layer model Eq. (5.10) holds good and substituting it in Eq (6.30) we have for 1:1 electrolyte solution at 25°C

$$(P)_T = S_v / F \{0.1147 c^{1/2} \sinh(19.46\psi_d)\} \tag{6.31}$$

The above calculation is, however, valid only for diffuse layer model. The inclusion of water as a component, though implicitly necessary for completeness of the tableau, does not yield any useful information. This is because, activity of water in dilute solution is unity and therefore does not appear in the mass law equation. Therefore, in practice, H_2O as a component is omitted.

6 .4.2 Tableau Preparation for the Adsorption of M^{2+} on HFO as Function of pH

The formulation of equilibrium problem becomes little more complex if we consider the adsorption of a divalent metal ion on hydrous iron oxide in a medium of constant ionic strength, but pH value is varied by adding strong acid and base. Let us assume that the metal ion is added to the system as chloride (MCl_2) so that M^{2+} and Cl^- are also included as components. The systematic tableau format (see chapter 8, Table 8.1) allows us to write both mole balance and mass law equations as follows for diffuse layer model.

Mole balance equations

$$[H^+]_T = [H^+] - [OH^-] + [FeOH_2^+] - [FeO^-] - [FeOM^+] - [MOH^+] - 2[M(OH)_2] = [HCl]_T - [NaOH]_T \quad (6.32)$$

$$[FeOH]_T = [FeOH_2^+] + [FeOH^o] + [FeO^-] + [FeOM^+] \quad (6.33)$$

$$[Cl^-]_T = [Cl^-] + [MCl^+] + [MCl_2^o] \quad (6.34)$$

$$[M^{2+}]_T = [M^{2+}] + [MOH^+] + [M(OH)_2] + [FeOM^+] + [MCl^+] + [MCl_2] \quad (6.35)$$

$$(P)_T = [FeOH_2^+] - [FeO^-] + [FeOM^+] = (S_v / F)\,\sigma_0 \quad (6.36)$$

Mass law equations

$$[OH^-] = [H^+]^{-1} \quad 10^{-14} \quad (6.37)$$

$$[MOH^+] = [H^+]^{-1}\,[M^{2+}] \quad K_1 \quad (6.38)$$

$$[M(OH)_2] = [H^+]^{-2}\,[M^{2+}] \quad K_2 \quad (6.39)$$

$$[FeOH_2^+] = [H^+] \quad [FeOH^o] \quad \exp(-e\psi_o / kT) \quad (K_{a1}^{int})^{-1} \quad (6.40)$$

$$[FeO^-] = [H^+]^{-1} \quad [FeOH^o] \quad \exp(e\psi_o / kT) \quad K_{a2}^{int} \quad (6.41)$$

$$[FeOM^+] = [H^+]^{-1}\,[M^{2+}]\,[FeOH^o] \quad \exp(-e\psi_o / kT) \quad K_{1M}^{int} \quad (6.42)$$

Total surface charge

$$\sigma_0 = (F / S_v)(P)_T = (F / S_v)\{[FeOH_2^+] - [FeO^-] + [FeOM^+]\} \quad (6.43)$$

No differentiation has however been made between strong and week sites in this tableau

6.4.3 Mathematical Representation of the Equilibrium Problem in FITEQL

In the above section, the procedure for making tableau containing species, components and constants for presentation of stoichiometry of various equilibrium reactions has been described . The mathematical representation of the equilibrium problem presented below (Table 6.2) forms the background of numerical solution using

FITEQL or any other similar computer program. In this table the terms within parentheses after description represent indices or scalar quantities which, in turn, are again represented by the corresponding matrix or vector terms. Further, i denotes species and j denotes any component .

Table 6.2: Mathematical description of chemical equilibrium problem**

Description	Matrix or Vector
Stoichiometric coefficient of component j in species i (a_{ij})	A (matrix)
Free concentration of species i (C_i)	**C** (vector)
------ do------- in logarithm ($\log C_i$)	**C*** (vector)
Stability constant of species i (K_i)	**K** (vector)
-------do-------in logarithm ($\log K_i$)	**K*** (vector)
Total analytical concn . of component j (T_j)	**T** (vector)
Free concentration of component j (X_j)	**X** (vector)
-------do-------- in logarithm ($\log X_j$)	**X*** (vector)
Residual in material balance equation for component j (Y_j)	**Y** (vector)
Partial derivative of Y_j with respect to X_j ($\partial Y_j / \partial X_j$)	**Z** (Jacobian)

Mass law equation

$\log C_i = \log K_i + \sum_i (a_{ij} \log X_j)$ $\qquad$ $\mathbf{C^* = K^* + A\,X^*}$

Mass balance equation

$Y_j = \sum_i (a_{ij} C_i) - T_j$ $\qquad$ $\mathbf{Y = \tau_{AC} - T}$

Iteration Procedure

$Z_{jk} = (\partial Y_j / \partial X_k) - \sum_i (a_{ij}\, a_{ik}\, C_i / X_k)$ $\qquad$ $\mathbf{Z\,\Delta X = Y}$ where $\mathbf{\Delta X = X_{original} - X_{improved}}$

** Adapted from ref . 187

6 .4 .4 Criteria for goodness of Fit of FITEQL Output

The criterion for the goodness of FITEQL output is given by the overall variance Vy which is the weighted sum of squares of residuals (SOS) divided by the degrees of freedom (DF)

$$SOS / DF = Vy = \frac{\sum\left[Y_{R(m)}/S_{R(m)}\right]^2}{n_p n_R - n_U} \tag{6.44}$$

where, $Y_{R(m)}$ is the residual of the mass balance equation weighted for each data point 'm' to reflect the uncertainties of the data. If this is done, residuals for more precise data have a greater influence on the objective function than do residuals for less precise data. In the above equation n_p is the number of data points, n_R is the number of R components or the number of Gr. II components, n_U is the no. adjustable parameters. In FITEQL relative and absolute errors estimates entered for the experimental data are used to calculate corresponding error in $Y_{R(m)}$ via propagation of error analysis, and these errors are used to weigh the residuals [27]. Hence the objective function is actually the sum of the squares of weighted residuals over all R components and all data points. Therefore,

$$SOS = \sum\left[Y_{R(m)}/S_{R(m)}\right]^2 \tag{6.45}$$

where, $S_{R(m)}$ is the error calculated for $Y_{R(m)}$ from the experimental error estimates. The goodness of fit of the proposed model to the data is indicated by the proximity of Vy to1. The value of Vy depends on the experimental error estimates and according to Westall [187] the value of Vy between 0.1 and 2.0 indicates a reasonably good fit.

The details of the theoretical basis for error estimation and optimization procedure for unknown parameters can be obtained from either the original FITEQL manual (document) version 3.2 [187] or from ref 27. The manual along with the CD may be available from the Chemistry Dept., Oregon State University, Corvallis, OR, US. Both MINEQL and MICROEQL versions may be available from Prof. David. Dzombak, Dept of Civil Engineering, Carnegie Mellon University, Pittsburg, PA 15213 - 3890, USA.

6.5 Significance of Intrinsic Surface Ionization Constants

The intrinsic surface ionization constants, namely K_{a1}^{int} and K_{a2}^{int} are the two important interfacial parameters of oxides and silicates in aqueous medium. They indicate the degree with which the surface can undergo ionization (diprotonation and deprotonation) under the influence of electrostatic field of the surrounding ions, especially H^+ or OH^-. The degree of ionization depends upon the nature of solid oxides and silicates, because the essential precondition for ionization is that surface must get hydrated and hydroxylated which are the potential sites for acid / base dissociation depending upon the pH of the solution. The difference between the negative logarithm of the two ionization constants (ΔpK_a) or, in other words the ratio of K_{a1}^{int} / K_{a2}^{int} provide many information on the oxide solution interface. Some of these are as follows:

(1) It can be shown from Eqs (3.3) ,(3.4) and (3.10) that at pH_{PZC} where $\{SOH_2^+\} = \{SO^-\}$ the following relationship holds good .

$$[H^+]_{PZC} = (K_{a1}^{int} \times K_{a2}^{int})^{1/2} \tag{6.46}$$

$$pH_{PZC} = \tfrac{1}{2}(pK_{a1}^{int} + pK_{a2}^{int}) \tag{6.47}$$

From the above equation it can also be deduced that

$$pK_{a1}^{int} = pH_{PZC} - (\Delta pK_a / 2) \tag{6.48}$$

and

$$pK_{a2}^{int} = pH_{PZC} + (\Delta pK_a / 2) \tag{6.49}$$

Therefore, if ΔpKa and pH_{PZC} are independently accessible by experiments, one can estimate both the ionization constants . Again, from Eqs . (3.3) and (3.4)

$$K_{a1}^{int} / K_{a2}^{int} = \frac{\{SOH\}^2_{PZC}}{\{SOH_2^+\}^2_{PZC}} = \frac{\{SOH\}^2_{PZC}}{\{SO^-\}^2_{PZC}} \tag{6.50}$$

since, at PZC $\{SOH_2^+\} = \{SO^-\}$

$$\text{Or, } \{SO^-\}_{PZC} = \{SOH_2^+\}_{PZC} = (K_{a2}^{int} / K_{a1}^{int})^{1/2}\{SOH\}_{PZC} \tag{6.51}$$

For $\{SO^-\}_{PZC}$ and $\{SOH_2^+\}_{PZC}$ to be very small , say smaller than 1% of $\{SOH\}_{PZC}$ i.e., $10^{-2} \times \{SOH\}_{PZC}$ the ratio $K_{a2}^{int} / K_{a1}^{int}$ should be smaller than 10^{-4}. This means that the equilibrium constant for acid dissociation (K_{a1}^{int}) should be at least ten thousand times stronger than the base dissociation constant of the surface (K_{a2}^{int}). In other words, the value of ΔpKa should be equal or larger than 4. Once the concentration of the charged species is very small its corresponding ion pair formation is also small at pH around PZC . Thus, at pH<< pH_{PZC} we may neglect the concentration of $\{SO^-\}$ and $\{SO^- - Na^+\}$. Similarly, at pH >> pH_{PZC} one may neglect $\{SOH_2^+\}$ and $\{SOH_2^+$-

$Cl^-\}$ provided ΔpK_a is very large. This has been demonstrated by Koopal et al [159] by calculating the values of pK_{a1}^{int} and pK_{a2}^{int} from the experimental data of Yates [75] for TiO_2 - KNO_3 solution system using the James and Parks [60] method of double extrapolation (large ΔpK_a) and the Westall and Hohl [146] method of numerical optimization (small ΔpK_a). Therefore, for small ΔpK_a the approximation of neglecting $\{SO^-\}$ and $\{SOH_2^+\}$ at pH << pH_{PZC} and at pH >> pH_{PZC} respectively is invalid and introduce significant error while estimating ΔpK_a value by graphical extrapolation method. However, at low electrolyte concentration one may neglect $\{SO^- - C^+\}$ at pH << pH_{PZC} and $\{SOH_2^+ - A^-\}$ at pH >> pH_{PZC} because such assumption leads to an error of less than 2 - 3 %. By introducing an additional parameter δ, Eq. (6.51) may be written as

$$\{SO^-\}_{PZC} = \{SOH_2^+\}_{PZC} = (\delta / 2)\{SOH\}_{PZC} \tag{6.52}$$

or,
$$\frac{\{SOH\}_{PZC}}{\{SO^-\}_{PZC}} = \frac{\{SOH\}_{PZC}}{\{SOH_2^+\}_{PZC}} = \frac{2}{\delta} \tag{6.53}$$

where,
$$\delta = 2\,(K_{a2}^{int} / K_{a1}^{int})^{1/2} = 2\times10^{-\Delta pK_a/2} \tag{6.54}$$

(2) Another manifestation of ΔpK_a value of an oxide / solution interface is its influence on the ψ_o vis-à-vis concentration of electrolyte solution. Using the triple layer model several authors [60,151,152,156,188] have shown that ψ_o should depend little on the concentration of electrolyte solution. Such a phenomenon is distinctive of system or interface with low ΔpK_a. This is because, for low value of ΔpK_a, any change in the concentration of charged species (e.g., variation in pH) on either side of pH_{PZC} does not result in any significant change in the ratio of SOH_2^+ / SO^- at pH_{PZC}. Consequently, the surface around the PZC behaves like a pseudo Nernstian with slope value around 59 mV/pH. For large ΔpK_a value the concentration of charged species around PZC is very small. As a result, any change in the concentration of either of the charged species brings about significant change in the ratio of SOH_2^+/ SO^- at pH_{PZC}, thereby causing large deviation from the Nernst relationship. Any change in the ionic strength of the medium will result in the deviation of slope values, especially near PZC. For small ΔpK_a system, ψ_o - pH relationship at different ionic strength is not only linear, but also very close to each other.

6.6 Estimation of ψ_o and ψ_β and Their Relationships with Other Interfacial Parameters

As discussed in chapter 5, ψ_o and ψ_β are the two important parameters of the electrical double layer at oxide solution interface. Several indirect methods have been suggested [177,178,188] to estimate these two parameters. None of these two potentials can be determined experimentally by any independent method for an oxide / water colloid system. Since, both the potentials are interlinked, determination of ψ_o helps in the determination of ψ_β. Since, unlike AgI / water system oxide / solution system does not behave reversibly, Nernst equation is not applicable directly to the system. However, the problem has been tackled by invoking hydrolysis of the constituent ions of an oxide i.e., O^{2-} and M^{n+} to form OH^- and H^+ respectively in aqueous medium [147]. The following "Nernstian" relationship is then justified by attributing these two ions as potential determining ions at oxide / water or even for silicate / water interface.

$$\psi_o = \frac{RT}{F}\ln\frac{a_{H^+}}{a_{H^+\,\Psi=0}} = \frac{2.3RT}{F}\left(pH_{\Psi=0} - pH\right) \quad (6.55)$$

where, $pH_{\psi=0}$ = pH at which $\psi = 0$ or pH_{PZC}.

The concentration of potential determining ions in oxide/water system are controlled by adjusting the pH value of the system through external addition of acid or base . In this process the solid surface adsorbs excess H^+ or OH^- to form charged species. The above relationship is then modified as follows .

Multiplying Eqs (3.3) and (3.4) and substituting $[H^+]_s$ by its bulk concentration $[H^+]$ we have

$$K_{a1}^{int} \text{ x } K_{a2}^{int} = \frac{[H^+]^2\{SO^-\}}{\{SOH_2^+\}}\exp\left(\frac{-2e\Psi_o}{kT}\right) \quad (6.56)$$

Combining Eq .(6.56) with (6.46) and rearranging

$$\frac{[H^+]^2_{PZC}}{[H^+]^2} x \frac{\{SOH_2^+\}}{\{SO^-\}} = \exp\left(\frac{-2e\Psi_o}{kT}\right)$$

Taking logarithm and rearranging we have the general equation correlating ψ_o with pH of the aqueous medium.

$$\frac{e\Psi_o}{kT} = 2.303(pH_{PZC} - pH) - \frac{1}{2}\ln\frac{\{SOH_2^+\}}{\{SO^-\}} \tag{6.57}$$

The above equation suggests the importance of the role played by the ratio of the two charged species on the ψ_0 - pH relationship. The dependence of ψ_0 on the last term of the equation (6.57) becomes less significant and the effect of ionic strength becomes more significant in the vicinity of PZC . If the assumption that $\sigma_0 \approx B\{SO^-\}$ for pH >> pH_{PZC} and $\sigma_0 \approx - B\{SOH_2^+\}$ for pH << pH_{PZC} are valid, the variation of surface charge with ΔpH i.e. (pH_{PZC} - pH) is large . Such a situation arises when ΔpK_a is large (> 4.0) and therefore the concentrations of the charged species are low around PZC. Therefore, even a small change in their concentrations brought about by variation of either pH or ionic strength of the medium may lead to significant variation in the ratio of $\{SOH_2^+\}/\{SO^-\}$. This results in the deviation of the slope value of the plots of ψ_0 vs pH as predicted by Eq . (6.55) . With decrease in the value of ΔpK_a, the concentration of the charged species in the vicinity of PZC also increases. Consequently, the deviation from the Nernstian slope also decreases as the influence of the last term of the Eq. (6.57) diminishes, because $\{SOH_2^+\}/\{SO^-\}$ tends to be 1. At the same time, the variation of ψ_0 value with ionic strength of the medium also diminishes so that ψ_0 - pH plots at different concentrations of electrolyte solutions tend to merge.

However, the estimation of the absolute concentrations of the charged species, especially near the vicinity of PZC is rather difficult without making certain assumptions. Therefore, attempts have been made to relate the concentration ratio with more precise experimentally accessible terms, which, in turn, are related to ψ_0 through electrostatics of the double layer. The procedure followed by Bousse and Meindle [188] in this regard is worth considering and is therefore discussed below.

At the outset it is convenient to convert the concentration terms of all the surface species into fractional or dimensionless terms by dividing them with the total surface sites i.e. N_s.

$$\alpha_+ = \{SOH_2^+\}/N_s, \quad \alpha_- = \{SO^-\}/N_s, \quad \alpha_n = \{SOH\}/N_s \tag{6.58}$$

Consequently, $\alpha_+ + \alpha_- + \alpha_n = 1$ (6.59)

By expressing the net surface charge $B[\{SOH_2^+\} - \{SO^-\}]$ in the fractional form , we may introduce a new function χ (chi)

$$\chi = \alpha_+ - \alpha_- \tag{6.60}$$

The last term of the RHS of Eq. (6.57) is expressed in terms of the another function F

$$F = (\alpha_+ / \alpha_-)^{1/2} \tag{6.61}$$

From Eqs . (6.60) and (6.61) we may express

$$\alpha_+ = \frac{F^2 \chi}{F^2 - 1}, \quad \alpha_- = \frac{\chi}{F^2 - 1} \tag{6.62}$$

A similar relationship for α_n can be derived as follows

$$K_{a1}^{int} / K_{a2}^{int} = \frac{\{SOH\}^2}{\{SOH_2{}^+\}\{SO^-\}}$$

$$\text{or,} \quad \{SOH\} = [(K_{a1}^{int} / K_{a2}^{int})\{SOH_2^+\}\{SO^-\}]^{1/2} \tag{6.63}$$

Substituting the term δ as proposed in Eq.(6.54) and α_+ and α_- terms as given in Eq. (6.62) in the above equation one gets the corresponding relationship for α_n .

$$\alpha_n = \frac{2F\chi}{\delta(F^2 - 1)} \tag{6.64}$$

The term δ introduced first by Healy et al [189], gives an indication of the strength of the tendency to ionize the surface sites. The higher the value of δ, the higher is the tendency of the surface to become charged through reactions (3.1) and (3.2). Substituting Eqs. (6.62) and (6.64) in Eq. (6.59).

$$\frac{F^2 \chi}{F^2 - 1} + \frac{\chi}{F^2 - 1} + \frac{2F\chi}{\delta(F^2 - 1)} = 1$$

simplifying

$$F^2(\chi - 1) + (2/\delta)\chi F + (\chi + 1) = 0 \tag{6.65}$$

Taking the negative root of the above quadratic equation for the sake of convenience in simplification

$$F = \frac{-2\chi/\delta - \left[(4\chi^2/\delta^2) - 4(\chi+1)(\chi-1)\right]^{1/2}}{2(\chi-1)}$$

Changing the sign and dividing both numerator and denominator by 2 the value of F is reduced to

$$F = \frac{\chi/\delta + \left[1 + (\chi/\delta)^2(1-\delta^2)\right]^{1/2}}{(1-\chi)} \tag{6.66}$$

Since , F is equal to $[\{SOH_2^+\}/\{SO^-\}]^{1/2}$ we may substitute Eq . (6.6) in Eq. (6.57)

$$e\psi_o / kT = 2.303\,(pH_{PZC} - pH) - \ln F \tag{6.67}$$

This is the desired equation relating ψ_o and pH provided that the discreteness of the charge effect is neglected i.e., charge is uniformly smeared out on the surface. The above relationship is better applicable at lower ionic strength. At higher ionic strength (> 10^{-2} M) surface complexation takes place and F value is modified as follows.

In this modification the nature and significance of χ and F functions in Eq. (6.60) and (6.61) remain unaltered, only the number of fractional surface species is increased by two complexated species in the mass balance equation, for example, when NaCl solution is used as electrolyte

$$\alpha_+ + \alpha_- + \alpha_{Cl^-} + \alpha_{Na^+} + \alpha_n = 1 \tag{6.68}$$

where , $\alpha_{Cl^-} = \{SOH_2^+ - Cl^-\}/N_s$ and $\alpha_{Na^+} = \{SO^- - Na^+\}/N_s$

The net surface charge is therefore given by

$$\sigma_0 = B[\{SOH_2^+\} + \{SOH_2^+ - Cl^-\} - \{SO^-\} - \{SO^- - Na^+\}]$$

The concentration ratio of the surface species in Eqs (5.26) and (5.27) may be expressed in fractional terms.

$$n = \frac{\alpha_{Cl^-}}{\alpha_+} = [Cl^-]\,{}^*K_{Cl^-}^{int}\exp\left(\frac{e\Psi_\beta}{kT}\right) \tag{6.69}$$

$$p = \frac{\alpha_{Na^+}}{\alpha_-} = [Na^+]^* K^{int}_{Na^+} \exp\left(\frac{-e\Psi_\beta}{kT}\right) \quad (6.70)$$

where n and p are the ratios of the fractional concentrations of anion complexated and cation complexated species which along with their respective uncomplexated species constitute the mass balance equation. Therefore, Eq. (6.68) can now be written as

$$(1+n)\,\alpha_+ + (1+p)\,\alpha + \alpha_n = 1 \quad (6.71)$$

Substituting Eqs . (6.62) and (6.64) in Eq (6.71)

$$(1+n)\frac{F^2\chi}{F^2-1} + (1+p)\frac{\chi}{F^2-1} + \frac{2F\chi}{\delta(F^2-1)} = 1 \quad (6.72)$$

On rearranging the quadratic equation for function F takes the following form

$$F^2\{\chi(1+n)-1\} + (2/\delta)F\chi + \{(1+p)\chi + 1\} = 0 \quad (6.73)$$

Proceeding as before the value of F obtained from the negative root is as follows

$$F = \frac{\chi/\delta + \left[(\chi/\delta)^2 + \{1-\chi(1+n)\}\{1+(1+p)\chi\}\right]^{1/2}}{1-\chi(1+n)} \quad (6.74)$$

The value of F can be substituted in Eq.(6.57) in the case where complexated species contribute significantly to the surface charge. However, one may neglect $\{SOH_2^+\text{-}Cl\}$ i.e., the term n at pH >> pH_{PZC} and $\{SO^- - Na^+\}$ i.e., the term p at pH << pH_{PZC}. Again, if δ is small $1 - \delta^2 \approx 1$. Tripathy et al [89] have used Eq.(6.57) to calculate ψ_o values as function of pH for two types of hydrous MnO_2 / solution system by neglecting surface complexation since for a low ΔpK_a system, as in this case, variation in the concentration of either of the charged species does not effect the slope value of ψ_o- pH plot .

6.6.1 *Estimation of ψ_β*

The potential generated at the plane of either coulombic or specific adsorption of counter ions (β - plane) can be estimated using Eqs (6.23) and (6.25). Indeed, the difference between intrisic complexation constants ($p^*K_{Cl^-}^{int}$ or $p^*K_{Na+}^{int}$) and their corresponding complexation quotients (p^*Q_{Cl-} or p^*Q_{Na+}) can be used as a measure of (ψ_o - ψ_β). This is obtained by differentiating the two equations with respect to pH [178] as follows

$$\frac{-p^*Q_{Cl^-}}{pH} = \frac{e}{2.3kT}\left(\frac{\Psi_o - \Psi_\beta}{pH}\right) \tag{6.75}$$

and

$$\frac{-p^*Q_{Na^+}}{pH} = \frac{e}{2.3kT}\left(\frac{\Psi_o - \Psi_\beta}{pH}\right) \tag{6.76}$$

For both the equations (ψ_o - ψ_β) = slope x (2.3 kT /e). The other method is to plot p^*Q_{Cl-} or p^*Q_{Na+} vs. fractional concentration of either SOH_2^+ - Cl^- ($\alpha_{\beta+}$) or SO^- - Na^+ ($\alpha_{\beta-}$) at higher ionic strength. Partial differentiation of Eq (5.58) with respect to $\alpha_{\beta-}$ and incorporating necessary conversion of potential difference (ψ_o - ψ_β) to charge according to Eq (5.53), we have at pH<<pH_{PZC}

$$\frac{\partial p^*Q_{Cl^-}}{\partial \alpha_{\beta^+}} = \frac{\partial}{\partial \alpha_{\beta^+}}\left(\log\frac{\gamma_{+-}}{\gamma_o}\right) - \frac{e}{2.3kT\,C_-}\left(\frac{\partial \sigma_o}{\partial \alpha_{\beta^+}}\right) \tag{6.77}$$

where, C_- is the capacitance of the plane of adsorption of anion and is considered to be independent of the concentration of complexated species if the ratio of the activity coefficients of the surface species is assumed to be unity and also independent of $\alpha_{\beta+}$. The slope value may be given by eN_s /(2.3 k T C_-). Therefore,

$$C_- = \frac{eN_s}{2.3kT\,Slope} \tag{6.78}$$

Knowing C_-, σ_0 and ψ_o we can calculate ψ_β from the relationship $C_- = \sigma_0 / (\psi_o - \psi_\beta)$. Similarly, from the slope value of the plot of p^*Q_{Na+} vs. $\alpha_{\beta-}$ one can obtain capacitance value for the adsorption of Na^+ as counter ion in the β plane at pH>>pH_{PZC}

$$\frac{\partial^* Q_{Na^+}}{\partial \alpha_{\beta^-}} = \frac{\partial}{\partial \alpha_{\beta^-}}\left(\log \frac{\gamma_{-+}}{\gamma_o}\right) - \frac{e}{2.3kT\,C_+}\left(\frac{\partial \sigma_o}{\partial \alpha_{\beta^-}}\right) \quad (6.79)$$

where, $$C_+ = \frac{eN_s}{2.3kT\ Slope} \quad (6.80)$$

The values of C_- and C_+ may show some variation due to difference in the ionic radii of anion and cation. Therefore, their planes of adsorption and consequently potential values may differ little. Based on this observation Bousse and Meindle[188] proposed four layer model where positive and negative counterions are placed in two different planes before OHP. Usually, this minor difference is overlooked and an average, common plane of adsorption is assumed whose potential and capacitance values are C_1 and ψ_β respectively .

6 .7 Estimation of ψ_d or ζ (Zeta Potential) and σ_d

As stated in chapter 5, the charge and potential values on the diffuse layer (slipping plane) are estimated from the electrokinetic measurements such as electroosmosis, electrophoresis, streaming potential, sedimentation potential etc of colloid particles. These methods are briefly discussed below giving the necessary relationships between ζ - potential and the rate of movement of either liquid or solid phase under the influence of applied electrical field or the potential generated due to the movement of relative phases.

6.7.1 Electroosmosis

When an electrical protential (EMF) is applied to a colloidal suspension confined in a capillary tube both the liquid and the solid particles move with a velocity equal but opposite to the other. In such a case, it is easier to measure the electroosmotic velocity of the liquid directly as the colloidal particles act as markers. However, in the absence of colloidal particles the liquid velocity is determined by measuring the volume of liquid transported through the capillary in unit time. It is well- known that when a liquid flows through a tube, a velocity gradient from the center to the wall of the tube occurs. At the same time, on the application of electrical field, the wall of the capillary gets charged (usually negative for glass or quartz tube) and a diffuse layer of oppositely charged or counter ions is formed. The liquid flows past the immobile double layer. However, any variation in the velocity near the wall is too small to discern through microscope, since the region extends only several times the thickness of the double layer ($1/\kappa$). In other words, the thickness of the double layer which holds a layer of immobile liquid is negligibly small compared to the radius of the tube. The liquid near the wall therefore appears to move with the same velocity at the center. We shall show how the distance between the two charged layers may be eliminated from the final relationship between zeta potential and the velocity of the electro osmotic flow [190].

As stated earlier, that the electrical double layer may be considered as an electrical condenser consisting of two charged parallel plates of distance d apart and each carrying charge σ_0 (C/cm^2) . If ζ is the potential between the two plates or layers and ε is the dielectric constant of the medium between the charged layers, then it can be shown from the principle of electrostatics

$$d = \frac{\varepsilon \varepsilon_o \zeta}{4\pi \sigma_o} \tag{6.81}$$

where, ε_0 is the permittivity in vacuum . The velocity gradient across the double layer is U_{eo}/d where U_{eo} is the electro osmotic flow rate . There are two factors which affect the value of U_{eo} , namely , the frictional force between the moving liquid layer and the wall of the capillary tube , and is $\eta U_{eo} / d$ where η is the viscosity of the liquid . The other factor is the electrical force causing electroosmosis to take place i.e., $E_{eo}\ \sigma_0$ where E_{eo} is the potential gradient generated across the capillary tube . At steady state both the opposing factors are equal and therefore

$$\eta U_{eo} / d = E_{eo} \sigma_o \tag{6.82}$$

Combining Eqs (6.81) and (6.82) it follows that

$$\zeta = \frac{4\pi \eta U_{eo}}{\varepsilon \varepsilon_o E_{eo}} \tag{6.83}$$

If r is the radius of the capillary , then the volume V of liquid flowing through per second is $\pi r^2 U_{eo}$. Substituting for Ueo in Eq (6.83) we have

$$\zeta = \frac{4\eta V}{r^2 E_{eo} \varepsilon \varepsilon_o} \tag{6.84}$$

When a porous plug or a membrane of area Am is used instead of a single capillary for electro osmotic flow, U_{eo} may be replaced by V/A_m in Eq (6.83) so as to give

$$\zeta = \frac{4\pi \eta V}{\varepsilon \varepsilon_o E_{eo} A_m} \tag{6.85}$$

The current strength I_e is related to E_{eo} by the expression

$$E_{eo} = I_e / (A_m \lambda_o) \tag{6.86}$$

where λ_o is the specific conductance of the liquid . From Eqs (6.85) and (6.86) we get

$$\zeta = \frac{4\pi V \eta \lambda_o}{\varepsilon \varepsilon_o I_e} \tag{6.87}$$

The equation is valid only if almost all of the current I_e is carried through the bulk liquid. If the concentration of electrolyte is very low a significant part of the current may be carried through the double layer near the wall where there is high concentration of charge density. In such a case Eq (6.86) may be written as [190]

$$E_{eo} = \frac{I_e}{A_m \lambda_o + 2\pi r \lambda_s} \tag{6.88}$$

where, λ_s is the specific surface conductance . The above equation along with [6.85] can be solved only numerically. Since the value of the correction factor is generally small, it is neglected in most cases .

6.7.2 Electrophoresis

Among all the electrokinetic measurements, electrophoresis or movement of colloid particles under the influence of applied electrical field is the most popular one and used widely for the characterization of both inorganic and bio colloids . Very dilute suspension of a colloidal suspension is confined within a capillary tube of either circular or rectangular cross section and a suitable electric field is applied to both ends of the capillary . However, the application of electrical field causes both medium and the particles to move. It is therefore essential to take both the phenomena into account to obtain a full picture of the system.

The relationship between electrophoretic movement and zeta potential is based on the Smoluchowski's treatment (see refs.191 and 192). If the liquid is regarded as fixed the solid particles must move with a velocity (U_{ep}) equal and opposite to that of the liquid (U_{eo}).

$$U_{ep} = -U_{eo} = (\varepsilon\varepsilon_o\zeta)/\eta \tag{6.89}$$

The relationship is valid if the thickness of the double layer ($1/\kappa$) around the particle is very small compared to its radius (a). This means that ratio of particle radius to the effective thickness of the electrical double layer (κa) is greater than 300. Under this condition , the forces imparted to the liquid by the applied field are transmitted to the particle as the liquid flows along its surface. On the other hand , when the double layer is thick compared to the particle radius ($\kappa a << 1$) i.e., when particles of small radii are suspended in dilute electrolyte solution , one can ignore the electrical forces acting on the double layer as they are not transmitted to the particle [190]. The two forces which act on the particles are, (1) the direct electrical force $F_e = \sigma_0 E_f$ where E_f is the magnitude of the electrical field and (2) the viscous drag of the liquid which for a spherical particle is given by the Stoke's equation. At constant velocity both the forces are equal and hence

$$\sigma_o E_f = 6\pi V a\eta \quad \text{so that}$$

$$U_{ep} = V/E_f = \sigma_o/6\pi a\eta \tag{6.90}$$

Now , from the principle of electrostatics , the potential ζ of a charged sphere is related to σ_0 by the expression $\sigma_o = 4\pi a\varepsilon\varepsilon_o\zeta$. Substituting in Eq (6.90) we have

$$U_{ep} = 2\varepsilon\varepsilon_o\zeta/3\eta \tag{6.91}$$

This is referred to as Huckel's equation proposed in the early 1920's. This is also sometimes known as the smoluchowsky equation in many test books and is applicable when $\kappa a >> 300$. When $\kappa a << 1$, Huckel proposed the elimination of the factor 2/3 from Eq (6.91). In practice, most of.the colloidal systems are in the range which fall in the range $300 << \kappa a >> 1$. The difference between Eqs (6.89) and (6.91) is the factor

2/3. The question is which of the two equations should be used. In the early 1930's Henry studied the effect of particle shape and size on the electrical field and proposed the following equation

$$U_{ep} = \left(\frac{2\varepsilon\varepsilon_o\zeta}{3\eta}\right) f(\kappa a) \quad (6.92)$$

where the function f(ka) varies from 1.0 to 1.5 as κa varies from zero to infinity. Henry's equation is valid only when ζ is small ($\zeta \approx 25$mV) . For higher values the conversion of mobility to potential becomes more problematic.

A variety of sophisticated instruments are available in the market for the determination of ζ potential which is calculated from the electrophoretic mobility of very fine solid particles suspended usually in 1:1 electrolyte solution of moderate ionic strength. One of them which uses standard procedure is the one supplied by Rank Bros.,Cambridge,U. K. and described in detail elsewhere [193]. The suspension is confined in a capillary tube of either cylindrical or rectangular cross section and having a total capacity less than 10 ml. The electrical field is applied at the two ends which are raised in the form of a "U" tube whose base is horizontal and constitute the central part of the capillary and immersed in a thermostat bath. However, it is much easier to observe and measure the movement of colloid particles in a rectangular or flat cell. The particles are illuminated by a light source at the back of the apparatus . In the earlier models, the movement of the particles was used to be measured with the help of a microscope ; but was later on replaced by closed circuit TV camera fitted with squared graticule of known size . This enables one to observe and measure the movement of more than one particle at a time . In more recent developments laser light is used to illuminate the particles and all the steps of measurement including calculation are performed in a computerized system . One such apparatus is made by Malvern Instruments Ltd., U. K.

A potential of about 60 -70 volt is applied through platinized platinum or palladium electrodes which are placed at 5 - 10 cm apart in the two vertical ends of the capillary tube .The speed is measured by noting the time taken by a particle to just cross one or two side by side graticules in a perfectly horizontal manner . The speed of at least 5 - 6 particles are noted and an average value is taken . The field strength is determined by measuring the effective distance between the electrodes , by comparing the conductivity of a salt solution of known strength in cell with the literature value . Field of the order of 10V / cm is considered normal . The speed is represented by the following relationship

$$U_{ep} = \lambda / (i\,R\,t) \tag{6.93}$$

where, U_{ep} = average mobility of the suspended particles, mm sec^{-1} / volt cm^{-1}.
i = applied dc current, amp.
R = specific resistance of the medium (ohm - cm).
t = average time taken by the particles to travel a certain or specified distance (usually the distance of the grid lines on the television monitor, sec
λ = cell constant which is specific for the instrument and other system parameters

The mobility values are then converted to potential (ζ) using equation (6.92) along with suitable correction factor available in the form of a chart for different particle sizes. However, care should be taken to convert the data into an uniform system of units. At 298 K the constant for the conversion of mobility to zeta potential is about 12.83. Equation (6.92) is applicable to an accuracy of about 1 % when the ratio of particle radius to the effective thickness of the double layer is greater than 300.

6.7.3 Streaming Current and Streaming Potential

When a liquid is forced under a given hydrostatic pressure through a capillary, the charges in the mobile part of the double layer close to the wall of the tube move towards one end. For negatively charged wall of glass or quartz tube charge in the mobile layer is positive and therefore a positive current flows in the direction of the liquid flow (see Fig.6.7). This is called the "streaming current "(I_s). The accumulation of excess positive charge near one end of the capillary tube sets up an electric field which causes a current, known as "conduction current" (I_c), to flow in the opposite direction. At steady state, the two currents are equal. The electric field caused by the migration of charge also results in a potential difference between the two ends of the capillary tube and is called streaming potential (E_s). The relationship between streaming current /potential and zeta potential may be derived in the simpler way as follows by aboiding complex mathematical derivations [190,191].

Let us consider the flow profile of a solution in a capillary as exemplified in Fig. 6.8. Let a be the radius of the capillary, r is the distance between the center of the capillary and the boundary of the moving liquid column, δ is the thickness of the immobile liquid layer attached to the wall of the capillary tube. The velocity of the liquid at a distance r from the center of the tube is given by the Poiseuille's equation

$$U_{(r)} = P\,(a^2 - r^2) / 4\eta l \tag{6.94}$$

where, P is the applied pressure and l is the length of the capillary. The double layer is assumed to be confined to a narrow distance δ near the wall of the capillary, so that the velocity of the liquid at r approaching the value of a ($r \to a$) is significant in determining the current, since bulk of the liquid carries no net charge. Therefore, substituting $r = a - \delta$ and when $r \to a$ Eq (6.94) may be written as

$$U_{(r)} = P(a + r)(a - r)/4\eta l = P a \delta / 2\eta l \tag{6.95}$$

The streaming current Is is given by

$$I_s = 2\pi a \sigma_o \{U_{(r)}\} = \frac{\pi a^2 \sigma_o P \delta}{\eta l} \tag{6.96}$$

If λ_o is the sp. conductance of the liquid used in the experiment, then the actual conductance θ of the liquid is given by

$$\theta = 1/R = (\pi a^2 \lambda_o)/l \quad \text{ohms}^{-1} \tag{6.97}$$

Streaming potential is , therefore , given by

$$E_s = R \times I_s = \frac{I_s l}{\pi a^2 \lambda_o} \tag{6.98}$$

Combining Eqs (6.96) and (6.98)

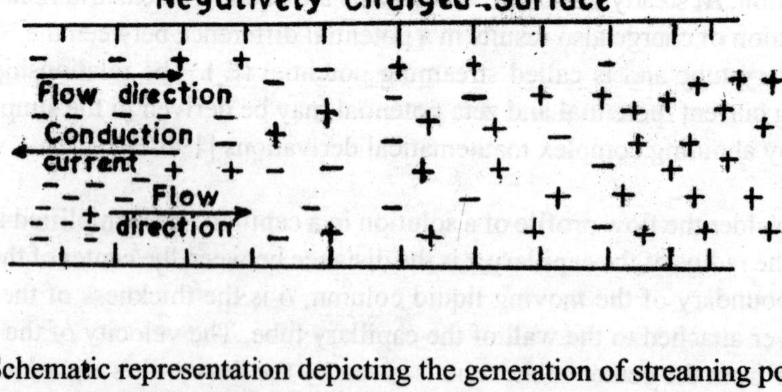

Fig. 6.7 Schematic representation depicting the generation of streaming potential and current when colloidal particles suspended in an electrolyte solution is allowed to flow under pressure through a glass capillary tube.

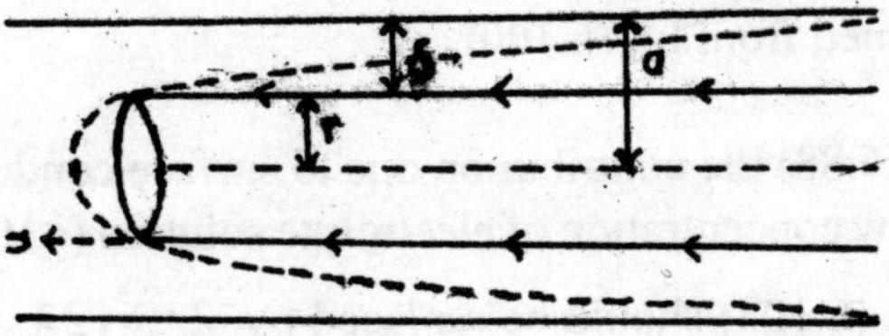

Fig. 6.8 Hydrodynamics of the flow of liquid through a capillary tube.

$$E_s = \frac{\sigma_o P \delta}{\eta \lambda_o} \tag{6.99}$$

Substituting δ with the basic electrostatic equation (6.81)

$$E_s = \frac{P\zeta\varepsilon\varepsilon_o}{4\pi\eta\lambda_o} \tag{6.100}$$

It may be noted here that V/I_e in Eq (6.87) is equal to E_s / P in Eq (6.100) suggesting that pressure induced streaming potential and current induced electro- osmotic flow rate are closely related. However, the equation deduced by Hunter [190] for streaming current by a more rigorous method is as follows:

$$I_s = \frac{\pi a^2 P\varepsilon\varepsilon_o\zeta}{l\eta} \tag{6.101}$$

As stated earlier that the streaming potential, E_s generated by current I_s also causes a conduction current in the reverse direction and is given by

$$I_c = \frac{\pi a^2 E_s \lambda_o}{l} \tag{6.102}$$

At steady state $I_s + I_c = 0$ and so

$$E_s = \frac{P\zeta\,\varepsilon\varepsilon_o}{\eta\,\lambda_o} \tag{6.103}$$

The value derived from equation (6.103) will be at least one order of magnitude higher than that obtained from Eq (6.100).

As in the case of Eq (6.88) the contribution due to surface conduction is significant, particularly at very low concentration of electrolyte solution (<10^{-3} M) , In such case, λ_o in Eqs (6.100) and (6.103) should be replaced by $\lambda_o + (2\lambda_s / a)$. However, surface conductance is not easily accessible by experiment and can be determined indirectly. It's contribution is taken into account by multiplying the experimental ζ - potential determined from streaming potential with the ratio R_{exp} / R_{calc} , where R_{exp} is the measured value of the resistance of the capillary and R_{calc} is the expected value , based on measurements carried at high salt concentration, when surface conductance is negligibly small [190].

6.7.3.1 *Measurement in a Capillary Tube*

If the measurement is made in a single capillary, then a reversible electrode is required at either end of the capillary [190]. Pressure to drive the liquid through the capillary is applied by using low pressure gas supply, and is measured with the help of either a manometer or a pressure transducer. The streaming potential is measured with the help of a high impedance (> 10^{11} - 10^{12} ohm) voltmeter such as pH meter so that current drawn from the system is close to zero. Similarly, if one wishes to determine streaming current, the electrodes have to be shorted out so that there is no alternating back path of the current. In practice, the current can be estimated by connecting a small known resistor (about 100 ohm) across the electrodes and measuring the resulting voltage drop with a micro voltmeter.

6.7.3.2 *Measurement in a Porous Plug*

Most of the studies on streaming potential have been made on porous plugs of granules of solid materials of particle size varying from tens of micrometer to millimeter. Both the Eqs (6.100) and (6.103) are also applicable to a plug consisting of such particles provided that flow rates are not too high. This can be verified from the linear plot of E_s versus P. The plug is confined between the two platinum gauge electrodes to measure conductivity of the liquid. To measure the streaming potential, two additional electrodes just outside the platinum ones are used, since the stresses in the platinum gauge interfere with high impedance voltage measurement [190].

Since the diffuse layer model does not envisage the existence of ζ - potential , the diffuse layer potential (ψ_d) is considered to be same as that of surface potential (ψ_o). However, the electrokinetic potential should be lower than the calculated value of ψ_o. However, several authors have found either higher values of zeta potential measured from electrophoresis or at best the values closely similar to those of ψ_o [193,194]. This is attributed to the difficulties associated with the interpretation of electrophoretic mobility and its conversion to ζ potential. There is a great deal of uncertainty in locating the shear plane which depends not only on the concentration of electrolyte solution but also on the applied electrical field. The shear plane does not necessarily lie on the outer edge of the layer of the adsorbed ions and indeed, is located quite a distance away from the surface because of bound water at the interface. Various authors [169-195] have suggested that the water close to the interface is highly structured and is therefore immobilized .The viscosity of water in this region is believed to be higher compared to bulk of the solution [171,191] and therefore influence the electrophoretic mobility and consequently ζ potential of the solid . This is known as the viscoelectric effect and this can be substantial in systems with high surface potential and in solutions of high ionic strength. Under this situation the shear plane moves further away from the surface [123,143]. As the concentration of solution increases, the charge density in the diffuse layer increases, leading to increased force of attraction between solid surface and the surrounding ions. On the other hand, thermal diffusional movement of the ions becomes restricted and the magnitude of the zeta potential decreases, which is observed experimentally [191].

Apart from the uncertainty in the location of shear plane in the Gouy - Chapman diffuse layer model, the very accuracy in the calculation of ζ potential from electrophoretic mobility is also questionable because of the difficulties involved in obtaining correct data for viscosity and permittivity of solution in the region close to the surface. Secondly, for higher potential, a correction is needed for 'relaxation effect' which is some form of slowing down effect caused by the deformation of double layer around a charged particle moving in an electrical field [143,152]. This relaxation effect is substantial at high ζ potential. Indeed, with increase in surface potential (ψ_o) zeta potential also increases but the latter gradually reaches a maximum value and remains constant or even tends to decrease. The mathematical relationships available for such correction are reasonably complex to apply in many cases. Thirdly, the relationships for the conversion of mobility to ζ- potential have been determined for particles of known geometries [152,153]. But, in practice, the particles are highly irregular in shape. Thus, calculation made on the assumption of spherical or cylindrical shaped particles will tend to introduce error in the value of ζ– potential.

In view of the problems associated with the calculation of ζ potential from electrophoretic mobility of oxide particles, the value is of little use as a parameter in the quantitative modeling of double layer. However , this parameter finds wide use for information about the sign of surface charge of soil, clays, minerals and suspended solid particles in aquatic environments etc. It is also used to determine / verify the point of zero charge where potentiometric titration method cannot be applied accurately. In such a case, zeta potential is measured in a medium of constant ionic strength but at different pH values. At a certain pH (isoelectric point) the solid particles may undergo charge reversal which can be determined by plotting ζ potential against pH. In the absence of specific adsorption of any ions other than the potential determining ones, both ψ_o and ζ will be zero at isoelectric point (iep) i.e.,pH_{PZC} and iep will be identical. In material science, electroosmosis and electrophoresis find use in the purification of many materials, deposition of rubber, leather and other coatings onto non-conducting materials. In the field of mineral processing electrophoresis is widely used to determine zeta potential and isoelectric point of ores and minerals. In biochemistry, electro kinetic measurement is almost routinely used for the characterization and estimation of proteins, enzymes, viruses etc.

Assuming that for all practical purposes, especially at low ionic strength $\psi_d \approx \zeta$, we may calculate charge at the plane of the diffuse layer or slipping plane using Eq. (5.10). Though σ_d may be calculated from the net surface charge i.e., - B[{SOH_2^+} - {SO^-}] provided the concentrations of these two charged species can be determined with reasonable accuracy, it can be estimated from more rigorous relationship between σ_d and ζ for flat surface or spherical particles and compare the results with those found experimentally.
Knowing the values of σ_d, ψ_d (or ζ) and ψ_β one can calculate capacitance value of OHP i.e., C_2 using Eq . (5.20) . But in most model fitting process C_2 is generally assumed to be 15 - 20μF / cm². The integral capacitance of the diffuse layer is obtained from Gouy-Chapman theory and is given by

$$C_d = \frac{e\sqrt{8RT\,c\varepsilon}}{2kT} \tag{6.104}$$

The above relationship suggests that C_d is dependent on the square root of electrolyte concentration and little on the value of σ_d. Koopal et al [156] derived the following interrelationship between the capacitance values of different planes of the electrical double layer.

$$1/C_t = 1/C_1 + (1 - C_r)(1/C_2 + 1/C_d) \tag{6.105}$$

where, C_r is the charge ratio (- σ_β / σ_0) , and all other symbols have their usual significances . At low ionic strength $\sigma_\beta \approx 0$ and therefore , C_r also tends to zero and Eq.(6.105) reduces to Eq .(5.18)

$$1/C_t = 1/C_{st} + 1/C_d \tag{5.18}$$

where; $1/C_{st} = 1/C_1 + 1/C_2$

In other words, at low ionic strength C_d is low compared to C_{st} and therefore C_t depends primarily on the C_d, but at higher ionic strength the charge ratio Cr assumes significant value and therefore needs to be incorporated in the calculation of total capacitance which depends mostly on C_{st}.

CHAPTER 7

SOIL AND ITS CONSTITUENTS AS ADSORBENTS

7.1 General Characteristics of Soil and its Inorganic Constituents

Soil is one of the most important constituents of our ecosystem. It is a highly heterogeneous entity occurring as earth's crust. It mainly consists of solid and liquid phases, but may also contain gaseous pockets generated by chemical / biological interactions. The liquid phase is either moisture absorbed in the solid or water trapped within the pores. The solid phase consists of both inorganic and organic compounds, the latter occurring primarily as a coating on the former. The adsorptive power or reactivity of soil depends on mutual interaction between inorganic and organic components. Thus, soil provides an environment for the sustenance of plant life by supplying the necessary micronutrients consisting of both inorganic and organic compounds of biological origin. The great adsorptive power of soil provides the necessary medium for the removal of toxic elements or compounds which tend to pollute the earth's surface.

Soil is a product of weathering of igneous rocks (known also as primary rocks) over millions of years in the presence of air, water and carbon dioxide. The chemical changes that occur during weathering led to the disintegration of dense solids into porous and light materials whose chemical compositions differ greatly from those of parent rocks. In most cases, the original crystalline structure collapsed to either new structure or to poorly crystalline structure or even to amorphous materials with extremely fine particulate texture. These are called secondary minerals. Secondary minerals of sedimentary origin may also get mixed with weathered rocks making it difficult to distinguish between the two types of mineral entities. Indeed, the boundary between sedimentary rock and soil material is often diffuse.

The most abundant primary minerals in soils are quartz (SiO_2), Feldspar (Na or $K(AlSi_3O_8)$), micas, pyroxenes, amphiboles, olivines, besides some accessory minerals of igneous and metamorphic origin which occur in very small quantities [196, 197] . In most soils primary minerals constitute sand and silt fractions.

The secondary minerals in soil include carbonate (calcite), sulphur bearing minerals such as gypsum and pyrites, layer silicates and hydrous oxides of Al , Fe , Mn and Ti. The hydrous oxides are the precipitation products from weathered solution rich in concentration of the concerned cations. They occur in finely divided state, generally with poor crystallinity, though crystalline oxides of these cations are also found in soils. Some poorly crystalline aluminosilicate compounds like allophane and imogolite formed by intense weathering of igneous rocks are also found in soils. However, layer silicates of size fraction less than 2μm constitute the so called clay fraction of the soils. Hence an understanding of the properties of inorganic components should begin with the structure of layer silicates.

7.2 Structure of Layer Silicates

The terms 'clay minerals' is interchangeably used to refer phyllosilicates or aluminosilicates having layer structure. When such minerals occur in soils in finely divided state (< 2μm) along with hydrous oxides of Al and Fe, they are known as 'clays'. In the subsequent discussion , we shall use the term 'layer silicates' to signify various types of phyllosilicates that occur in soils .The structure of layer silicates is described by the way in which [SiO_4] tetrahedral units are linked together in three dimensional arrangements. While discussing the structures of clay minerals, the terms 'sheet' and 'layer' are frequently encountered. According to Baily et al. [198] a sheet is a combination of the planes of atoms, and a 'layer' is a combination of sheets. In layer silicates, the corners of the base of a [SiO_4] tetrahedron are shared by three other silica tetrahedra forming a ring containing 6Si and 6O atoms which are again connected to the neighboring ring through shared O atoms. All the oxygen atoms shared by the tetrahedral rings are coplanar forming the basal plane of oxygen atoms (O_B). The oxygen atoms at the apices of [SiO_4] tetrahedra are oriented in the same direction i.e., normal to the basal plane. The apical oxygen atoms (O_A) are also coplanar and the distance between the two planes of oxygen atoms is called the tetrahedral sheet (Fig.7.1a). The compact ring structure formed by the [SiO_4] tetrahedral is shown in Fig. 7.1b.

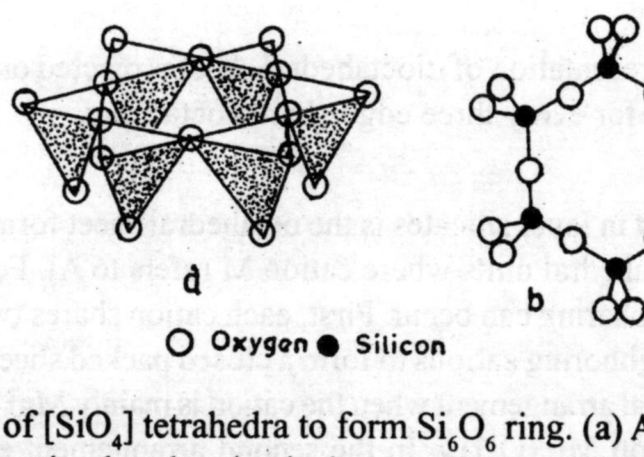

Fig. 7.1 Linking of [SiO_4] tetrahedra to form Si_6O_6 ring. (a) A polyhedral representation showing basal and apical oxygen atoms. (b) Ball and string representation.

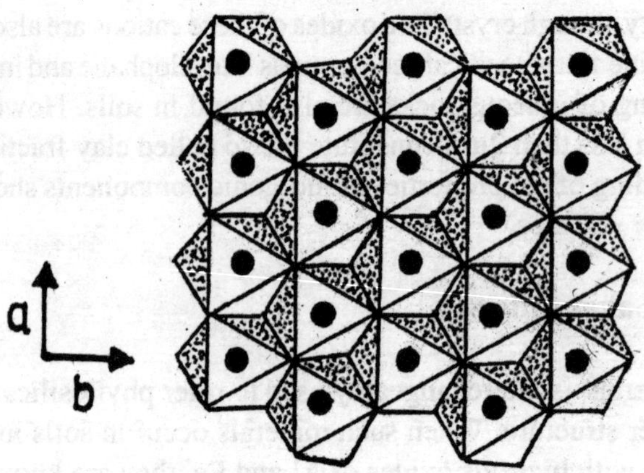

Fig.7.2 Polyhedral representation of edge-shared trioctahedral sheet projected on c-axis.

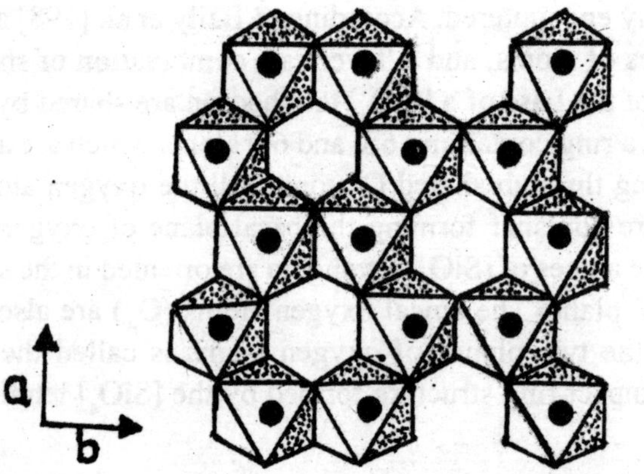

Fig. 7.3 Polyhedral representation of dioctahedral sheet projected on c-axis, showing one vacant site for every three edge-shared octadedra.

The other structural unit in layer silicates is the octahedral sheet formed by sharing of the edges of $[MO_6]$ octahedral units where cation M refers to Al, Fe, Mg etc . There are two ways by which sharing can occur. First, each cation shares two oxygen atoms with each of the six neighboring cations to form a closed packed sheet (Fig .7.2) . This is also called trioctahedral arrangement when the cation is mainly Mg^{2+} giving a brucite-like layer of composition $Mg_6(OH)_{12}$. In the second arrangement each cation in an octahedron shares two oxygen atoms with the three neighboring octahedra giving a ring of six octahedra with the middle one being empty (see Fig 7.3). This means that

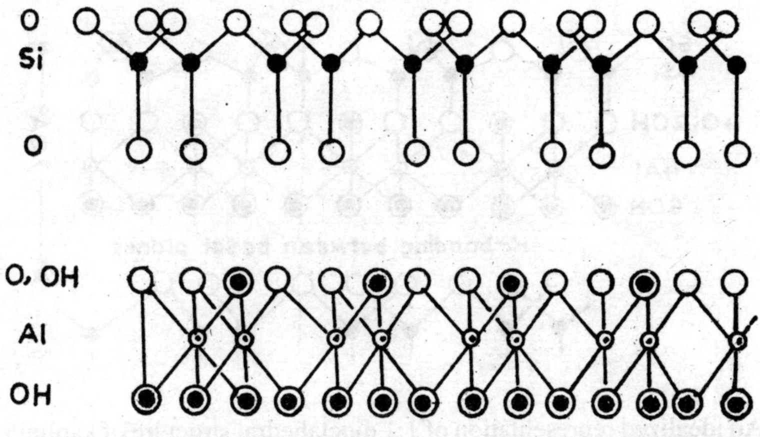

Fig 7.4 (a) Linear chain of $[SiO_4]$ tetrahedra showing basal and apical oxygen atoms. (b) Ball and string representation of $[Al(O,OH)_6]$ octahedra joined by sharing edges to form infinite chain.

for every three octahedra in the linear chain two are filled and the remaining one being empty, thus giving a gibbsite-like layer of composition $Al_4(OH)_{12}$.This is called dioctahedral arrangement when the cation is either Al^{3+} or Fe^{3+}. Figure 7.4a and 7.4b illustrate linear tetrahedral chains of silica and dioctahedral $[M(O,OH)_6]$ respectively, which are the essential constituents of dioctahedral layer silicates. The layer silicates are formed by combining i.e., superimposing the trtrahedral sheet(s) over the octahedral sheet(s) along c - axis in different ways. Indeed, the entire classification of clay minerals is based on the nature of such combinations as exemplified for some common clay minerals that occur in soils.

7.2.1 1:1 Dioctahedral Layer Silicate (Kaolinite)

In this group of clay minerals one tetrahedral sheet is superimposed in inverted manner on an octahedral sheet containing anion group of $(OH)_4O_2$. In the resulting 1:1 arrangement one plane consists entirely of OH and the other consists of one OH and two O per octahedron. The two O atoms are shared as apical O atoms of the tetrahedral sheet by octahedral Al as can be found in mineral kaolinite. The ideal structure formed by 1:1 superimposition is illustrated in Fig.7.5. Besides kaolinite, dickite and nacrite are the other minerals belong to this group and have an ideal full cell composition of $Al_4[Si_4O_{10}](OH)_8$. This corresponds with the theoretical composition of SiO_2 46.5%, Al_2O_3 39.5% and H_2O 13.96%.The formula suggests that there is no substitution of either Si^{4+}in the tetrahedral layer or Al^{3+} in the octahedral layer, which means that the net layer charge is zero. However, natural kaolinite carries little negative

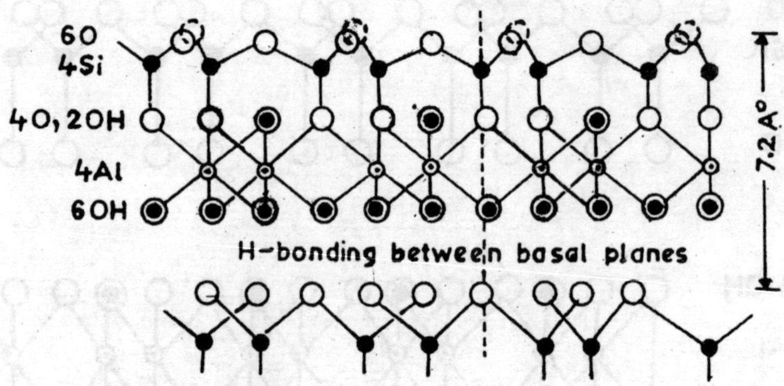

Fig. 7.5 An idealized representation of 1:1 dioctahedral structure of kaolinite formed by the inverted superimposition of silica tetrahedral chain over $[Al(O,OH)_6]$ octahedral chain. Dotted line idicates boundary for an unit cell.

charge, because of minor substitution of octahedral Al^{3+} with Fe^{2+} or Mg^{2+}. Dickite and nacrite differ from kaolinite mainly in the manner in which successive layers (1:1) superimpose (stacking), although the individual layer in the three minerals also differ slightly in tetrahedral twist and lateral repeat distance [199]. If the ideal kaolinite composition given above is considered as one packet, dickite has two such packets and nacrite has four such packets. In most kaolinite the ideal composition of Si/Al ratio of 1:1 is followed, although in the octahedral sheet some Fe^{3+} occasionally substitute for Al^{3+} [200]. Using the XRD method the kaolinite group of minerals is identified by its strong 1st order basal plane reflection (001) at 0.714 - 0.721 nm and medium to strong 2nd order basal plane reflection at 0.357 - 0.379 nm. In TEM it appears as hexagonal platelets. Surface area varies within a small range of 10 - 20 m^2/g. The cation exchange capacity varies between 1and10 meq /100g. Surface charge is highly dependent on pH, suggesting that isomorphous substitution is not the predominant source of charge. Indeed, the dissociation of silanol (SiOH) and aluminol (AlOH) groups at the edges of the clay particles contribute mainly to surface charge. We shall discuss this aspect in more detail later on in this chapter. When Al^{3+} in the octahedral layer is extensively replaced by either Mg^{2+} or Fe^{2+}, all the octahedral holes tend to be filled up giving 1:1 trioctahedral minerals like serpentine, crysotile etc. But, they seldom occur in soils.

7.2.2. 2:1 Layer Silicates

This group comprises a large number of clay minerals where an octahedral sheet is sand witched between two tetrahedral sheets. The octahedral unit is coordinated though shared or linked oxygen atoms to two tetrahedral units. If Al or Si predominantly occupy octahedral and tetrahedral sites respectively, the mineral is pyrophyllite ($Al_4Si_8O_{20}(OH)_4$).If the octahedral sites are fully occupied by Mg, the mineral is talc ($Mg_6Si_8O_{20}(OH)_4$). In the former only two third of the octahedral sites is occupied by Al and therefore the mineral group is called dioctahedral, while the latter all the octahedral sites are occupied by Mg and therefore the mineral is called trioctahedral. Between these two extreme mineral groups of ideal compositions there lies a large number of 2:1 layer silicates formed by cation substitutions in either tetrahedral or in octahedral sheet or in both. But in soils only three types 2:1 layer silicates namely micas, smectites, and mixtures of micas and smectite are generally found. We shall therefore confine our discussion to these three groups of clay minerals. It may be mentioned here that soil clays are usually poorly ordered and much smaller in size than the pure minerals found in their specific deposits.Intermixing of layers, coatings of finely divided hydrous oxides and overlapping with neighboring particles or sheets render their structures and consequently surface properties extremely complex.

7.2.2.1 Micas

Micas are characterized by interlayer charge imbalance created by the substitution of either tetrahedral Si with Al or octahedral Al with Mg or Fe(II) . This imbalance or deficiency of charge is compensated by an interlayer ion which is uniquely K^+ in the case of mica. The interlayer ion is firmly held between the adjacent layers and is an integral part of the structure (see Fig.7.6). In common mica or muscovite the deficit of layer charge arises entirely from the substitution in tetrahedral sheet. Consequently, the seat of charge deficiency in micacious clay is on the silica sheet and therefore lies on the surface, in contrast to montmorillonite where the charge deficiency is in the octahedral sheet and therefore lies in the center of the unit layer. Because of this difference, the unit layers of mica structure are relatively fixed in position so that large polar molecules cannot enter between them and cause expansion. However, in phengite , the other dioctahedral mica, the deficit arises from the substitution in both tetrahedral and octahedral sheets . Still other dioctahedral micas are of glauconite - celadonite series where charge deficiency arises from substitution in octahedral sheet. But they occur in marine sediment and their K^+ content is very low compared to the terrestrial mica [200].

The particle size of soil micas is less than 2mm in spherical diameter. They contain less K^+ and more water than coarse grained or macroscopic micas. The trend becomes more predominant with still finer (<0.3 μm) particles. In this size range the structural homogeneity among particles generally disappears in many clays. Ca, Mg or Al also occur as interlayer cations in hydrated state, besides normal occurrence of K [200]. Such clays are interstratified or mixed clays and their XRD patterns differ

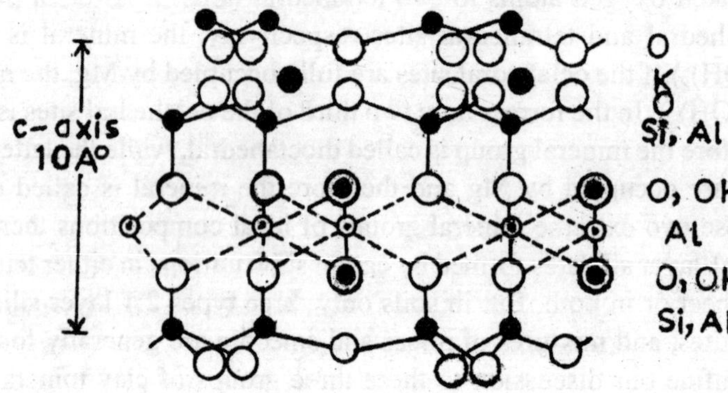

Fig. 7.6 An idealized 2:1 layer structure of mica projected on b-axis. The interlayer anionic charge deficit is balance by potassium. Soil illite has similar structure. Adapted after ref. 200

greatly from those of pure micas which exhibit characteristic (001) reflection at 1.0nm. It has been suggested that large surface area exposed by extremely fine clay particles makes the metal ions on the basal planes more vulnerable to hydration or for exchange with other metal ions in solution than those of interlayer K^+. Despite high charge (1.0 per unit anion i.e. $O_{10}(OH)_2$) micas show relatively low CEC (10 - 30meq / 100g).

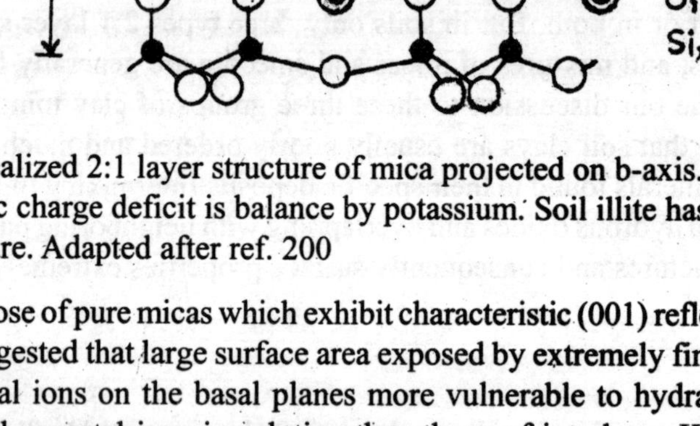

Fig.7.7 2:1 layer structure of vermiculite projected on b-axis. Hydration of interlayer Mg^{2+} results in the expansion of basal planes along c-axis. Adapted after ref. 200.

One consequence of these features is that it is difficult to define clay mica.' Illite ' is the most widely used group name but not of any specific clay mineral name, It is known to occur almost universally in lake , river and marine sediments and in soil in more than one form or other. It occurs in soil in interstratified form with smectite, contains more structural OH groups but less K^+ than normal coarse grained mica or muscovite, especially at the edges of adjacent tetrahedral sheets to form "frayed edges". It is characterized by non expandable basal plane reflection at about 1.0nm corresponding to the repeat distance of each 2:1layer of mica. The frayed edges are the outcome of intense weathering whereby mica or micaceous clay lose their interlayer cations i.e., K^+ and creating vacant ditrigonal cavities between two adjacent tetrahedral sheets. These negatively charged vacant sites have strong affinity for cations like Mg^{2+} and Ca^{2+} from surrounding environment. This accounts for the presence of Mg and Ca in almost all illite samples.

Illite occurs in many parts of the world as distinct deposits, the prominent among which are Muloorina (southern Australia) and Fithian (Illinois, USA). The chemical analyses of illite samples from these two sources, as reported in literature [223] are as follows (in weight percent): SiO_2 53.4 and 47.2, Al_2O_3 14.8 and 23.1 Fe_2O_3 15.8 and 14.8 , MgO 3.2 and 4.1, TiO_2 1.3 and 1.7 , CaO 1.7 and 0.9, Na_2O 0.9 and 2.1, K_2O 7.3 and 6.0 respectively.

Trioctahedral micas such as phlogopite , where more than 70% of the octahedral sites are occupied by Mg^{2+} and biotites with 20-60 % octahedral sites are occupied by either Mg^{2+} or Fe^{2+}, occur rarely in soils. This is because trioctahedral micas are readily weathered to dioctahedral vermiculites or minerals of similar structure by losing interlayer K^+. Micas are the major source of potassium in soils.

7.2.2.2 Vermiculites

As stated above vermiculites are formed by the weathering of both di - and trioctahedral micas , present in basic or ultrabasic rocks . Both di and trioctahedral vermiculites resembling 2:1 layer structure of parent mica, are found in soils [200]. Weathering or alteration of the precursor micas replaces the interlayer K^+ by exchangeable Mg^{2+} in hydrated state, thus rendering the basal plane expandable to 1.4 nm (see Fig 7.7). Typical or idealized formula of vermiculite is $[Mg(H_2O)]_n^{\ 2+}(Mg,Fe)_3(Si_{4-x}Al_x)O_{10}(OH)_2]$, where $[Mg(H_2O)]^{2+}$ serving as exchangeable cation. However, vermiculites inherit the charge imbalance created by substitution either in the octahedral or in the tetrahedral sheet or both. The net deficit which is about -0.7 per unit formula, results in high cation exchange capacity (CEC) value of 120 - 150 meq / 100g. The interlayer hydrated Mg^{2+} can be easily replaced by K^+ by suspending vermiculites in KCl solution and in such case the basal spacing is collapsed to 1.0 nm from 1.4 nm. Like smectites, vermiculites adsorb organic compounds like glycol or glycerol resulting in the expansion of swelling or basal plane spacing. But, unlike smectite the increase is limited to the maximum of 1.4 nm.

7.2.2.3 Smectites

Smectites represent a group of clay minerals belonging to 2:1 layer structure with variable basal plane spacings. Structurally, though smectites are similar to vermiculites, the former carry smaller charge (< - 0.65 per $O_{10}(OH)_2$). Consequently, smectites not only exhibit low CEC values, but also lower interlayer binding strength than vermiculites. Smectites, therefore, undergo expansion of basal plane with increase in the hydration of interlayer cations. This is known as swelling of smectites. While dehydrated smectites show basal plane spacing of 0.95 - 1.0nm, but on exposure to water vapour they quickly rehydrate and the basal plane separation between two unit layers may increase up to 2 nm or even more . Unlike micas the interlayer cations are accessible and can be exchanged by other cations from their saturated solutions.

The dioctahedral smectites have two extreme varieties as far as the substitutions in octahedral and tetrahedral sheets are concerned . These are:

Beidellite	$M_x^+(Al_2)(Si_{4-x}Al_x)O_{10}(OH)_2$
Montmorillonite	$M_x^+(Al_{2-x}Mg_x)(Si_4)O_{10}(OH)_2$

where, M_x^+ is the charge compensating monovalent cation, which in this case is Na^+ and is exchangeable with other cations. The number of tetrahedral Si and octahedral Al atoms in the above chemical formulas, if multiplied by 2, will give formulas per unit cell of their crystalline structures, as also presented in some books. For example, the formula of montmorillonite has also been represented as $M_x^+(Al_{4-x}Mg_x)(Si_8)O_{20}(OH)_4$. In the above examples x varies from 0.65 to 1.3. Therefore while beidellite is essentially free from Mg, montmorillonite is rich in Mg content (6-7% MgO). If Al is extensively replaced by Fe(III) in montmorillonite, the mineral is known as nontronite. This clay mineral occurs extensively in marine sediment, but very rarely even in iron - rich soils. In practice, especially in soils, the smectite minerals do not follow the ideal compositions as given above. They usually occur in compositions which are mixtures of beidellite and montmorillonite. The general formula of a soil montmorillonite is as follows: $M^+_{(x+y)}(Si_{8-y}Al_y)(Al_{4-x}Mg_x)O_{20}(OH)_4$. If y is about 0.2 and x is about 0.6, then the total charge deficiency is -0.8 which is generally found in montmorillonite. The structure of a typical montmorillonite is shown in Fig. 7.8.

The most common trioctahedral smectite where Mg is the main octahedral cation is saponite. This gives octahedral sheet a positive charge which partially balances the negative charge arising from the substitution in tetrahedral sheet by Al. The remaining imbalance is compensated by interlayer mono-valent cation.

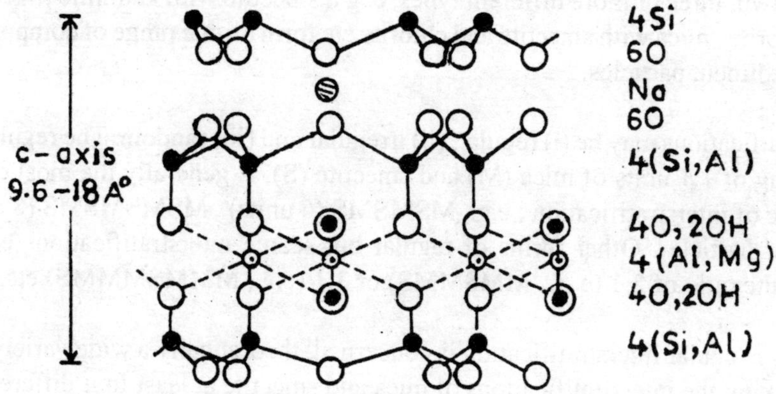

Fig.7.8 2:1 layer structure of montmorillonite projected on b-axis showing exchangeable cation between the layers. Adapted after ref. 200.

Smectites readily form complexes with a large number of organic compounds and the literature on this subject is indeed vast. Detailed discussion is beyond the scope of this chapter. Positively charged compounds like N-alkyl ammonium or pyridinium salts can displace interlayer cations in montmorillonite . The adsorption of uncharged polar organic compounds like acetone, glycol and glycerol on smectites leads to swelling i.e. expansion of interlayer spacing along c - axis by more than 2-3 times the normal value of about 1 nm. The high swelling property of smectite is used to differentiate it from vermiculite wherein the maximum expansion is limited to 1.3 -1.4 nm.Besides individual and discrete occurrences, bentonites are the best known sources of smectites. Bentonites, which are formed by the alteration of volcanic glasses, consist mostly of montmorillonite free silica, cristobolite and minor amounts of bidellite.

7.2.2.4 Interstratified or Mixed Layer Clays

Many soils and sediments have been found to contain clays which often undergo swelling like smectites. But, chemical analysis of such clays reveals the presence of potassium, suggesting that the clays are not true smectites. On the other hand, various physico - chemical measurements do not reveal the presence of discrete fine grained micas, instead suggests the occurrence of mixed assemblages micas and smectites. These are known as interstratified or mixed layer clay minerals [200]. This is the most complicated and perhaps the most interesting aspect in the field of clay mineralogy. Interstratifications may take place between different types of clay minerals wherein the unit layers usually have cations associated with them either in separate and hydrated state or as part of the hydroxide sheets. Thus, 2:1 layer plus cation may be referred to as unit structure or simply unit. Interstratified minerals may contain units of two, three or more different types, e. g., smectite with kaolinite, vermiculite with chlorite, mica with smectite and chlorite etc form a wide range of composition in soil or sediment particles.

Interstratifications may be (i) regular, (ii) irregular and (iii) random. The regular form, consisting of 1:1 units of mica (M) and smectite (S), is generally the most common sequence of interstratification, e.g., MSMSMS (6 units), MSMSMSMS (8 units) or more per particle. Other forms of regular but scarce interstratifications can have units in the ratio of 2:1 (e.g. MMSMMS) or 3:1 (e.g., MMMSMMMS) etc.

As far as irregular interstratification is concerned, theoretically a wide variety is possible. Taking the interstratifications of mica and smectite at least four different types of arrangement may be visualized. These are:(a) random or haphazard (e.g., MSSMSMS, MMMSM, SMMMSM MMSMSMMS etc.), (b) maximum order (e.g.,MSMSMSMSMM,MSMSMMMSMSMM),(c)segregated(e.g., MMMMMMSSSS, MMMMSS, SSSSSS, MMMMM) and (d) zoned (e.g. SSMMMMMMSS, SMMMMS, SMMMMMMSS). The phenomenon of interstratifications is studied by precision XRD method and high resolution transmission electron microscopy with selected area electron diffraction facility.

Since, soil is formed by the weathering of rocks of different origins, it is natural that different types of interstratified clay minerals would occur in it. Because of the complexity of the nature of occurrence it is hardly possible to study the structure and nature of mixed or inter stratified clays in soils. In addition to the regular trioctahedral inter stratification as weathering product of biotite, various irregular interstratifications occur in soils. Regular dioctahedral interstratifications include mica - beidellite or montmorillonite as discussed above, and chlorite - montmorillonite. Indeed, montmorillonite which was thought to be the main clay mineral in many soils of sedimentary origin have subsequently been found to be interstratified or mixture of mica and montmorillonite. Inter stratifications of kaolinite and montmorillonite have also been found in some sediments and soils.

7.2.3 2:1+1 Layer silicates (chlorite)

Chlorites, which belong to 2:1+1 layer silicates, occur extensively in soils. The minerals belonging to this group have a structure based on two layers of mica like units joined through an intermediate octahedral sheet of brucite ($Mg(OH)_2$) of about 0.4 nm thickness. Since each 2:1 mica like unit has a thickness of about 1.0 nm, the total thickness of 2:1:1 layer chlorite is about 1.4nm. Therefore the presence of 2:1 mica layer plus a brucite sheet is the defining characteristic feature of chlorite minerals.

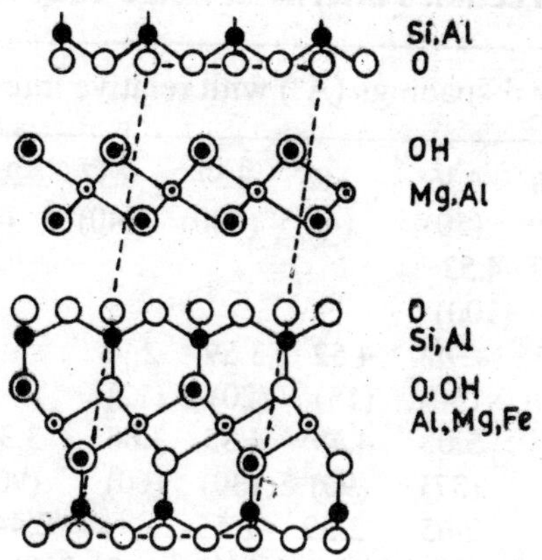

Fig.7.9 2:1 + 1 layer structure of chlorite projected on a-axis. The unit cell is shown by broken lines. Adapted after A. V. Milovsky and O. V. Konnov, Mineralogy, Mir Publishers, Moscow, 1985.

Like micas, a negative charge in 2:1 layer is balanced by a positive charge in the intermediate hydroxide sheet created by the substitution of some Mg^{2+} by Al^{3+} or Fe^{3+}. The resulting structure is harder, less hydrated and non - expandable compared to smectite. While the chlorite occurring within the parent rocks generally possess trioctahedral sheets , those occurring in soils have dioctahedral sheets. An idealized unit formula is $[AlMg_2(OH)_6]_x$ $[Mg_3(Si_{4-x}\ Al_x)O_{10}(OH)_2]$. The cation exchange capacity ranges from 10 to 40 meq /100g and surface area varies from 70 to150 m^2/g. Unlike other clay minerals, the structure of chlorite may not be represented by a singular format because of its six-fold symmetry. Consequently, there are six ways by which the two mica structural units that can be joined through an octahedral layer of brucite [200]. However, for illustration sake, a simplified form of the structure is shown in Fig. 7.9. Typical x-ray diffraction patterns of some important clay minerals that are present in many soils are shown in Table 7.1.

7.3 Structural Formula of Clay Minerals

The structural formula of clay minerals can be determined , generally by two methods, from the total chemical analysis that is all the constituent oxides including water [200]. If the type of clay mineral is precisely known from XRD data and its unit cell parameters are available then method 1 may be used. In this method the atomic mass per unit cell (M) can be estimated from the formula $M = \rho V N_A$ where ρ is the density in g / cm^3 , V is the unit cell volume in cm^3 and N_A is the Avagadro number (6.03 x 10^{23})

Table 7.1 X- Ray Diffraction Patterns of Some Clay Minerals

Mineral name	Major d-spacings (A°) with relative intensities in parentheses*									
Kaolinite	7.16 (>100)	4.46 (40)	4.36 (50)	3.85 (40)	3.57 (100)	3.37 (40)	2.56 (60)	2.49 (80)	2.37 (60)	1.49 (40)
Montmorillonite	15.2 (90)	4.97 (75)	4.53 (100)							
Montmorillonite (Glycerolated)	18.4 (90)	9.0 (20)	4.98 (50)	4.52 (15)	3.59 (20)	2.98 (12)				
Illite (Mica)	10.0 (100)	5.03 (37)	4.49 (90)	4.35 (30)	3.66 (60)	3.36 (90)	3.07 (50)	2.58 (50)	2.56 (90)	
Chlorite	14.0 (>100)	2.65 (116)	2.59 (80-300)	2.55 (40-148)	2.44 (26-210)	2.39 (100-200)	2.26 (60-90)	2.0 (98-170)		
Silica	4.26 (35)	3.34 (100)	1.817 (37)	1.84 (15)						

*The relative intensity values of chlorite vary widely depending upon the crystal orientations according to the six fold symmetry elements.

Here, the atomic mass M depends upon the number of formula units (z) of the concerned mineral. The steps involved are as follows [200]:

(i) Divide the percent composition of the individual oxides with their correspond ing formula weight to obtain formula percent .

(ii) Multiply the formula percent values with M/100 to obtain formula per unit cell mass.

(iii) Multiply formula per unit cell mass of each component oxide with the corre sponding number of cations in the oxide (e .g . 2 in Al_2O_3) and add up the values to obtain total number of cations per unit cell .

(iv) In the manner similar to step (iii) obtain total number of anions per unit cell.

The above method suffers from several shortcomings. First, many clays, especially soil clays, are frequently associated with impurities and therefore have disordered structures. The unit cell parameters determined for these clays do not represent the true values for the concerned clay minerals, even after subjecting them to various stages of purification. Secondly, extremely fine colloidal particles not only carry charges, but also occur in highly hydrated state. It is difficult to define and also to determine true density of such materials. Thirdly, and possibly the most important of all, is the discrepancy in the values of both structural and inter layer water contents which exhibit considerable variation in certain expanding clays like vermiculites and smectites depending upon the humidity of the environment in which the clay samples are dried. Therefore, the values of $H_2O(-)$ and $H_2O(+)$, that is water evolved between room temperature and 105° C and between 105° and 950°C respectively, are somewhat subjective . Therefore, in method 2 the structural formula is determined on the basis of anhydrous sample. However, unlike method 1 the composition of anion in the original clay is assumed to be constant in this method. For example, the anion composition of an unit layer of 2:1 dioctahedral clay is assumed to be $O_{20}(OH)_4$ for which the corresponding positive or cation charges are balanced accordingly. The essential steps in this method are as follows [200] :

(i) The chemical analysis may be carried out either on an ignited (950°-980°C) or on an air dried sample. If an air dried sample is used, the loss of total water content has to be compensated by the gain in the weight % of the other ox ides. In other words, the sum of the anhydrous oxides should be made up to 100.

(ii) Each oxide percent calculated on anhydrous basis is divided by its corresponding molecular weight to obtain 'moles' of oxides per 100g .

(iii) The moles of oxides per 100g as calculated in step (ii) are multiplied by the number of cations in the corresponding oxides to give the number of cations per 100g.

(iv) The values obtained in step(iii) are multiplied by the corresponding valence or oxidation state of cations in the oxides to give cation charges per 100g . These values when added together to give the total cation charge per 100g . Let this value be designated as Σcc.

(v) The cation charge for each oxide is multiplied by 44 and divided by Σcc to give the number of cations per $O_{20}(OH)_4$ i . e. 44 negative charge per unit layer structure.

(vi) The values obtained in step (v) for individual cation when multiplied by their corresponding oxidation state gives cation charge per anion unit i.e . $O_{20}(OH)_4$. The individual cation charge when added together becomes equal to total negative charge , that is 44 . This step is used only for the purpose of verifi cation of the correctness of calculation.

Table 7.2 Calculation of the Structural Formula of an Illite Sample

1	2	3	4	5	6	7	8	9
Oxides	Analysis percent	Dry anal. percent	Formula weight of oxides	Mole or formula per-cent(a.m.u)	Cations per 100 a.m.u.	Cation charge per 100 a.m.u.	Cations per $O_{20}(OH)_4$	Cation charge per $O_{20}(OH)_4$
SiO_2	51.53	54.36	60.09	0.905	0.905	3.620	6.692	26.768
Al_2O_3	30.30	31.97	101.94	0.314	0.628	1.884	4.644	13.932
Fe_2O_3	4.55	4.80	159.70	0.030	0.060	0.180	0.444	1.331
MgO	2.32	2.45	40.32	0.061	0.061	0.122	0.451	0.902
K_2O	4.90	5.17	94.20	0.055	0.110	0.110	0.813	0.813
Na_2O	1.00	1.06	61.98	0.017	0.034	0.034	0.251	0.251
$+H_2O$	3.70							
$-H_2O$	1.80							
Total		99.81				5.950	13.295	43.997

Column 5 = Column 3 / Column 4
Column 6 = Column 5 x No of cations in the oxide formula
Column 7 = Column 6 x oxidation state of cation in oxide
Column 8 = Column 6 x 44/ Σcc of column 7
Column 9 = Column 8 x oxidation state of cation to check that the sum = 44

From column 8 the empirical formula of the clay mineral may be written as follows:

$$Na_{0.251}K_{0.813}\ Mg_{0.451}\ Fe(III)_{0.444}\ Al_{4.644}Si_{6.692}$$

For a micaceous clay like illite, the cations may be rearranged in the tetrahedral and in the octahedral sites as follows:

Tetrahedral sites $Si_{6.692}\ Al_{1.308}$ Total = 8

Octahedral sites $Al_{3.336}\ Fe(III)_{0.444}\ Mg_{0.451}$ Total = 4.231

The total octahedral sites is little higher than the theoretical value of 4 suggesting that some Fe(III) as well as some Mg possibly occur as impurities. The structural formula of the illite sample is represented by

$$[K_{0.813}Na_{0.251}]^{XII}\ [Si_{6.692}\ Al_{1.308}]^{IV}\ [Al_{3.336}Fe(III)_{0.444}Mg_{0.451}]^{VI}\ O_{20}(OH)_4$$

Assuming that Na is the only exchangeable cation, the exchange capacity of the clay mineral is calculated as follows:

CEC = (1.055x1000)/31 = 34meq /100 g of ignited clay or 1x1000/31 = 32.25 meq/ 100g of air dried clay.

7.3.1. Assignment of Cations to Different Layers According to their Coordination Number

In method 1 it is necessary to know the number of unit layers to be accommodated in each unit cell. The unit layer is meant by unit structural layer, that is 2:1 for micas and smectites, 1:1 for kaolinite etc. In case the number is two, the number of cations and anions calculated in steps (iii) and (iv) are to be divided by two to give the number per unit layer corresponding to two octahedral and one tetrahedral layers. The cations are then assigned to these layers in the following manner applicable to both the methods.

1) Si atoms are placed in the tetrahedral sites. If the number of Si atoms are less than the total available sites i.e., 8, the balance is completed by Al or by Fe(III) in the case of iron rich clay mineral like nontronite.

2) The remaining Al Fe , Ti , Mn and Mg are all placed in the octahedral sites , which is 4 in case of dioctahedral and 6 in the case of trioctahedral clays.

3) Na , K and Ca are assigned to sites between the aluminosilicate layers .

The entire procedure has been amply illustrated by Brown et al [200] using a mica and a montmorillonite samples. However, for the benefit of those readers who may not have immediate access to the above reference, the procedure is illustrated in Table 7.2 using an illite sample of Rajasthan, India. The sample was treated with acidified (3M acetic acid) solution of oxalate, followed by washing and drying.

7.4. Inorganic Accessory Minerals in Soils

7.4.1. Allophane

Besides layer silicates (phyllosilicates) most of the soils is enriched in various oxides / hydrated oxides of Si, Al, Fe(III), Ti and Mn whose contents vary from region to region and type of soil . One such constituent is amorphous aluminosilicate gel which is broadly known as allophane.

Allophane was first found in soils derived from volcanic detritus, though it is present in most soils as an intermediate product of weathering [200]. Consequently, allophane has a high surface area (70 - 300 m^2/g) and high CEC (100 - 150 meq/100g) [199].

The composition of allophane varies widely from soils of one region to another. However, the most common constituents are hydrated Al_2O_3, SiO_2, Fe_2O_3 and Na_2O with minor amounts of Mg, Ca and K. It is not certain whether allophane is a mixture of the hydrated oxides of Al and Si or an amorphous aluminosilicate compound in which oxygen is shared between Al and Si. The range of chemical composition of allophane collected from weathered igneous rocks and treated to remove organic matter and free iron oxides is 32 - 40% SiO_2 , 35 - 40% Al_2O_3 , other oxides 2 - 3%, 5 - 6% Na_2O , 15 - 18% $H_2O(+)$ and 20 - 25% $H_2O(-)$. The identification of this mineral is difficult by any single method and therefore a number of methods such as DTA/TG , IR spectra etc are required [200 , 201].

7.4.2. Oxides and Hydrated Oxides of Iron

Oxides including hydrated oxides or hydrous oxides of iron occur almost universally in their various mineralogical forms in most soils. Though some oxides may occur as distinctly separate mineral grains, most of the hydrated oxides occur in the form of very fine, poorly crystallized surface deposits / coatings on soil clays [200]. Consequently, they influence not only the surface properties of soil clays, but also the fertility of soils. In view of this, study of the hydrated oxides of iron (and also of the other hydr(oxides) of Al, Ti, and Mn) forms part and parcel of the mineralog of soils and soil clays [59] . All the oxides and hydroxides are the products of intense weathering of metal ion (iron, aluminum etc) bearing silicate rocks. We shall now discuss important characteristics of those oxides and hydrous oxides of iron that are commonly found in soils.

7.4.2.1. Ferrihydrite / Hydrous Ferric Oxide(HFO) (Fe_2O_3 , nH_2O)

In soils or in other surface environments ferrihydrite is generally formed by the rapid oxidation and hydrolysis of Fe(II), that is when Fe(II) containing water appears on the surface [202]. The oxidation may be biotic and abiotic. When formed in the laboratory under highly oxidizing conditions at alkaline pH, the product is known as "hydrous ferric oxide" or "amorphous ferric hydroxide" whose interfacial properties have been discussed in the preceding chapters. Both natural and synthetic products are characterized by ill - defined composition, fine particle size(3 - 7 nm), poor crystallinity etc, which render their identification very difficult, especially when present less than 10% of total constituents. However, some broad reflections observed in XRD pattern of natural ferrihydrite suggest that it consists of planes of oxygen hexagonally closed packed as in the case of hematite. But, the stacking is so frequently interrupted by faults that no three dimensional regularity is observed [202]. The deficiency of Fe^{3+} in the octahedral sites is made up by additional protons associated with oxygen. However, it is interesting to observe that the IR spectra do not suggest the presence of distinct hydrogen bonded OH groups. It is therefore concluded that OH groups combine to form H_2O. Because of the wide variation in H_2O/Fe_2O_3 mole ratio in ferrihydrites, the composition is represented by $Fe_2O_3 \cdot nH_2O$.

Thermodynamically, ferrihydrite is the least stable of all the hydrous oxides of Fe(III) and spontaneously converted to either goethite or lepidocrocite though dissolution and recrystallization processes during which other elements such as Al may be incorporated in the goethite structure . However, the rate of transformation is rather slow and is inhibited by the presence of dissolved silica. As a result, both goethite and ferrihydrite occur together in soils. Because of its large area, ferrihydrite acts as a powerful adsorbent for many charged species of environmental importance, such as phosphates, arsenate, fluoride silicates and some dissolved organic anions at pH below its point of zero charge (PZC). It is also a powerful adsorbent for cations like Pb^{2+}, Hg^{2+}, Cd^{2+} etc.

7.4.2.2 *Goethite (α -FeOOH)*

This yellowish brown compound is one of the most stable form of Fe(III) oxides [59]. It occurs in almost every surface environment, especially in soils and sediments around the world, either singly or in combination with other hydrous oxides of iron. As stated above, goethite is formed by a slow process of recrystallization of redissolved Fe^{3+} solution. Therefore, the substitution of Fe^{3+} by Al^{3+} in goethite structure is very common [59, 202]. The extent of such substitution may range from about 5.6% to as high as 30%, resulting in considerable decrease in unit cell dimension of pure goethite due to smaller size of Al^{3+}. Indeed, goethite is isostructural with diaspore (α-AlOOH), groutite (α- MnOOH) and VOOH, all of which have orthorhombic unit cell system. It is identified by its characteristic XRD pattern and by its needle shaped (acicular) or sub rounded crystals under electron microscope. Structurally, the double chain of edge shared oxygen octahedra is linked to another double chain of oxygen octahedra by sharing corner oxygen atoms (see Figs. 7.10a,b). Since the chains are infinite in one direction, the repeat distance along this direction (c-axis) becomes stabilized at approximately 0.302 nm. The other dimensions of the orthorhombic cell systems are a = 0.4608 nm and b = 0.9956 nm for acicular goethite of density 4.26 g cm^{-3}. However, in many soils and sediments, instead of acicular crystals, goethite occurs in non-dimensional or irregularly shaped crystals, thus making it difficult to identify from microscopic observation.

7.4.2.3 *Lepidocrocite (γ - FeOOH)*

Although less frequent than goethite , lepidocrocite , a polymorph of goethite occurs widely in minor amounts in soils and sediments of humid temperate regions such as

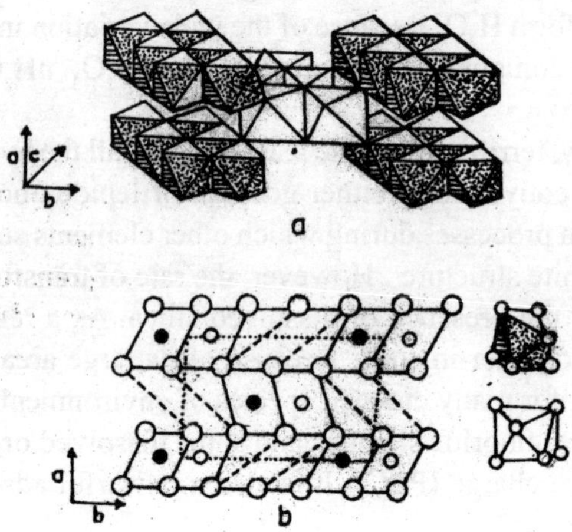

Fig. 7.10 (a) Polyhedral representation of the chain structure of goethite projected on 001} plane.

(b) The same polyhedral structure is represented in schematic form showing the vacant sites and orthorhombic unit cell dimensions by dotted outline. Hydrogen bondings between oxygen and hydroxyl groups across the vacant octahedral are shown by broken lines. Adapted after ref. 200.

seasonally water logged soils [202] . It is often associated with goethite and sometimes with ferrihydrite, but seldom with hematite. It is frequently seen as bright orange colored mottles in certain soils other than in the tropical region. But lepidocrocite and maghemite (γ - Fe_2O_3) are produced from unstable ferrous ferric hydroxide called "green rust" formed by the oxidation of ferrous iron. In all these cases lepidocrocite appears as lathe shaped crystals. Unlike goethite, substitution of Al for Fe(III) is rarely observed in lepidocrocite.

Structurally, pairs of edge shared O/OH octahedra join with each other also by sharing edges in forming chains of octahedra in zig - zag fashion running parallel to c - axis [59]. Many such chains are joined in the same manner forming an infinite two dimensional sheet of one octahedron thick. A layer perpendicular to b - axis is formed when two such sheets are joined by sharing oxygen atoms through hydrogen bonding in such away that each oxygen common to two sheets is shared by four octahedral (see Fig .7.11) .Like goethite, the unit cell system is orthorhombic having dimensions a = 0.388 nm , b = 1.25 nm and c = 0.307 nm . The density of well crystalline form is 4.09 g cm-3. It is also an active adsorbent for both anions and cations depending upon pH of the medium.

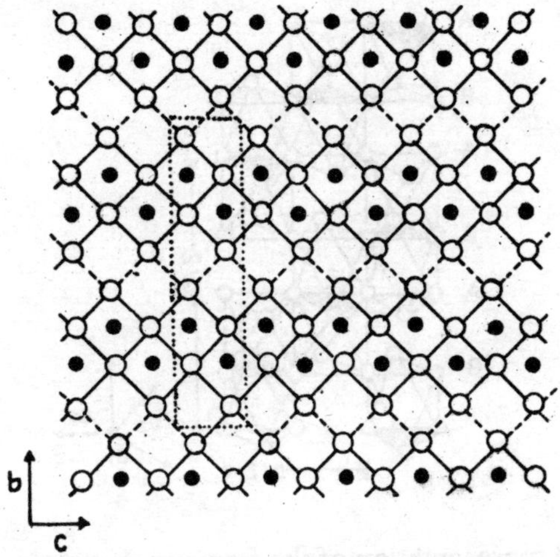

Fig. 7.11 Layer structure of lepidocrocite projected on {100} plane. Broken lines indicate the linking of individual layers by hydrogen bonding. Dotted outline indicates orthorhombic unit cell. Adapted after ref. 200.

7.4.2.4. Akaganeite (β - FeOOH)

Though this polymorphic form of FeOOH is very rarely found in soil, it is generally found in marine concretionary deposits, in certain hot springs with high chloride contents, in yellow brown surface deposits of white island volcano, New Zealand [202]. The conditions in which this particular oxyhydroxide of iron is found to occur are high Cl^- concentration ($\approx$ 0.1 M), low pH (3 - 4) and high warm temperature (~ 60°C). It can also be synthesized in the laboratory under similar conditions, e.g., rapid hydrolysis of $FeCl_3$ from homogeneous solution containing urea at 60° -70°C. The particles are somatoids, rods or wheat shaped. The colour is yellow brown and the unit cell system is tetragonal with dimension a_o = 1.048 nm and c_o = 0.302 nm. Its density is about 3.269 g cm^{-3}.

7.4.2.5. Hematite (α - Fe_2O_3)

Along with goethite, hematite is thermodynamically most stable form of iron oxide in oxidizing environment like air. Therefore, it occurs widely in most of the soils, especially in red soils where its presence suppresses the yellowish brown colour of goethite. Indeed, the redness of certain soils depends upon their hematite content. The structure of hematite (see Fig. 7.12) consists of hexagonal closed packed layers of oxygen stacked perpendicular to c - axis (in the order ABAB....) with Fe^{3+} occupying 2/3 of the octahedral sites. Neighboring [FeO_6] octahedra are linked by sharing one

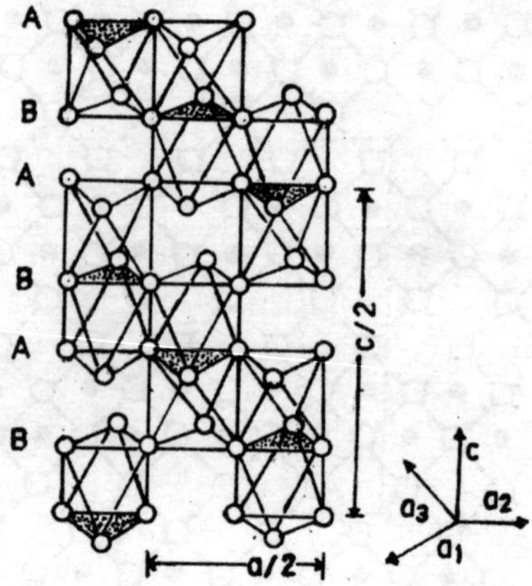

Fig. 7.12 Polyhedral representation of the framework structure of hematite. For every three octahedra there is one vacant octahedron. Half of the hexagonal unit cell dimension is indicated in the figure. Adapted after ref. 202.

face (along c - axis) and an edge (along a - axis) thereby causing slight distortion in their shapes. Octahedral distortion results in the displacement of Fe^{3+} ions, with the consequence that unlike oxygen Fe^{3+} occurs at different or zig - zag shaped level along c-axis. The parameters of the hexagonal unit cell are: a_o = 0.504nm and c_o = 1.377nm. The absence of H bonds together with octahedral distortion renders the crystalline structure of hematite very compact and therefore exhibits high density (5.26 g cm^{-3}). Soil hematite however does not occur in pure state. Apart from goethite and ferrihydrite, other oxides such as Al_2O_3, TiO_2 and SiO_2 (quartz) are present in intimate association with hematite [203].

7.4.3 Other Minor Oxides of Iron , Aluminum and Titanium [203]

The other oxides of iron that are found in certain soils in small quantities are maghemite (γ - Fe_2O_3) magnetite (Fe_3O_4) . Both the oxides are magnetic and both have similar crystalline structure , namely cubic closed pack oxygen . While maghemite is considered to be formed by the dehydration of lepidocrocite caused by bush or forest fires in the tropics and subtropics region , magnetite is either lithogenic (inherited from parent rock) or biogenic (biooxidation of Fe^{2+}) origins. Oxyhydroxide of Al such as diaspore (α - AlOOH) and boehmite (γ-AlOOH) or simple hydroxide such as gibbsite ($Al(OH)_3$) are rarely found in soils. Similarly, the three polymorphic forms of TiO_2, such as rutile, bruckite and anatase occur in very trace quantities as detrital of parent rocks, mostly igneous and metamorphic origins . In some soils, anatase occurs in small quantity as an alteration product of ilmenite ($FeTiO_3$)

7.4.4. Manganese oxides

Though occur in small quantities in soils, the oxides of manganese have profound effect on the fertility and productivity of soils. Many trace elements which act as micronutrients for the growth of plants occur in association with manganese oxides. Like iron, manganese in its different oxidation states (+2, +3, +4) forms a wide range of compounds with oxygen either in the absence or in the presence of water, depending on their genetic history. Because of close similarity in their structural properties, oxides and hydr(oxides) of iron and manganese occur side by side in nature and have profound effect on certain environmental issues relating to earth's surface. While the oxide and hydroxide of Mn(II) are not very stable under normal atmospheric condition , Mn(III) forms several varieties of stable oxidic minerals like hausmannite(Mn_3O_4), jacobsite($MnFe_2O_4$), bixbyte{$(Mn,Fe)_2O_3$}, braunite ($Mn^{2+}Mn_6^{3+}O_8[SiO_4]$), manganite ($\gamma$-MnOOH) etc.

By far, the most important of all manganese oxides are those formed by Mn(IV) . They occur widely in nature not only as variety of crystalline minerals , but also in finely divided state in soils and sediments of natural water systems like river , lake and deep sea . The main building block of all types Mn(IV) oxide is the edge shared [MnO6] octahedra linked in various ways giving rise to at least three predominant types of structure namely, chain (single or double), ring or framework and layer . Interestingly, the oxides and oxyhydroxides of iron (III) can also be broadly classified in the same manner. This topic has been reviewed by several authors [54, 204-206].

7.4.4.1 Chain Structure

The most stable crystalline forms of Mn(IV) oxide belong to this category of structure. The primary building unit of these structures is edge-shared octahedra which may form either single or double chain extending along c-axis. When it forms single chain the structure becomes analogous to that of rutile (TiO_2) and is observed in mineral pyrolusite or synthetic β - MnO_2. Each chain is cross-linked to the neighboring chain through corner sharing of oxygen atoms of the adjacent octahedral to give a tetragonal symmetry with unit cell dimension of a_o= 04.38nm and c_o=0.287nm. This highly dense (~ 5.0 g cm^{-3}), well-crystallized mineral very rarely occur in soil.

When the double chains of edge shared octahedra are cross-linked to the neighboring double chain by sharing corners of the adjacent octahedra, the structure so generated is of the mineral ramsdellite. The alternating double chains are linked together by sharing opposite edges, thus producing continuous pyrolusite type chain along c-axis to give orthorhombic symmetry with unit cell dimension of a_o = 0.453nm, b_o=0.927nm, c_o=0.287nm and 4 formula units MnO_2 per unit cell [204-206] . Ramsdellite is also well-crystalline, dense (~ 4.73 g cm^{-3}) mineral and does not occur in soil . However, the purpose of providing the above information is to understand the differences in the structures of various polymorphic forms of the other oxides and oxyhydroxides on manganese as discussed below.

The tetragonal and orthorhombic symmetry of single and double chain structures as depicted above are most ideal. In reality, the former undergoes a rotation of about 10° and the latter of about 27° from ideal orthogonality i.e., perpendicular to the chain axis due to cation repulsion [204]. While oxygen atoms in the single chain structure of pyrolusite or β-MnO_2 are rarely replaced by OH, in the double chain structure such replacements are quite frequent as observed in the orthorhombic structure of groutite (α-MnOOH), diaspore (α-AlOOH), goethite (α-FeOOH) etc. All these minerals which can also be prepared in the laboratory, have almost similar unit cell parameters ($a_o \sim 4.6$, $b_o \sim 10.2$, $c_o \sim 2.8$). If any of the oxygen atoms is replaced by OH, the hydrogen bonding of the two neighboring O^{2-} and OH^- ions further facilitate the twisting or rotation of the $[Mn(O,OH)_6]$ octahedra in the cross- linking chain so that the distance between the two hydrogen bonded oxygen atoms is much shorter than the other O-O distances due to rotation [204]. However, the interatomic distances within the octahedral layer are not altered due to such rotation. The γ-MnO_2, which exhibits ion-exchange properties , also possesses similar double chain structure where some of the O^{2-} are replaced by OH^- and the corresponding Mn^{4+} by Mn^{3+} to maintain charge balance . This results in the random replacement of some double chain with single chain in the octahedral layer growth along c-axis.

7.4.4.2 Framework or Ring Structure

The structures of this group of Mn(IV) oxide minerals is based on the double chain of $[MnO^6]$ octahedra running as usual along c-axis . Two such chains are cross-linked through two other double chains by sharing the adjacent corner oxygen atoms giving rise to three dimensional framework with pseudo-tetragonal symmetry , having unit cell parameters of $a_o = 0.996$nm and $c_o = 0.286$nm. The resulting framework structure has large cavity or tunnel which can accommodate H_2O as well as monovalent cation (Na^+ or K^+) or divalent cation (such as Ba^{2+} or Pb^{2+}) as found in the structures of minerals cryptomelane and hollandite respectively. Each large cation is surrounded by eight oxygen atoms situated at the corners of a slightly distorted cube. Figure 7.13 illustrates hollandite type structure which frequently occurs in soil. This mineral exhibits good cation exchange properties . In order to maintain charge balance within the cavity, some Mn^{4+} near the cavity is replaced by Mn^{3+}. This is reflected in the average Mn - O distance of 0.198nm which is significantly larger than 0.189nm observed in the chain structure [206].

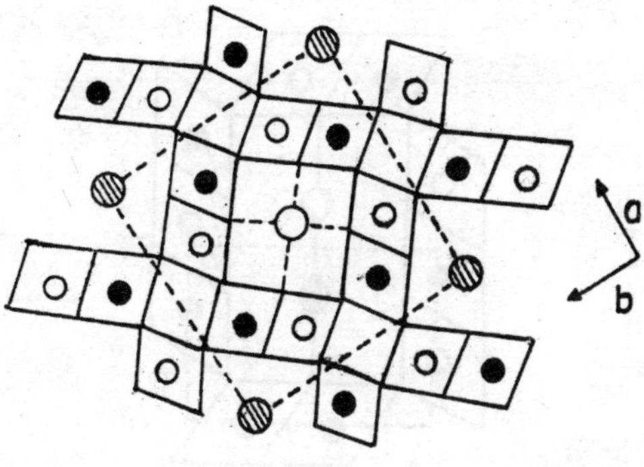

Fig.7.13 Edge shared double chains of $[MnO_6]$ octahedra join corners forming frame work structure of α-MnO_2 with large tunnel which is occupied by either Pb^{2+} or K^+ (represented by larger circles) for charge balance to give hollandite or cryptomelane type structures respectively. Smaller circles represent Mn. Open circles represent atoms at 0 level and filled or shaded circles represent atoms at c/2 level. Dotted lines show unit tetragonal cell. Adapted after ref. 204.

Akaganeite or synthetic β-FeOOH is isostructural with Ba^{2+} or K^+ containing hollandite of tetragonal symmetry. Its structure accommodates H_2O molecule and anions like Cl^-, F^-, NO^{3-} etc . as stated in section 7.4.2.4 . The spindle or wheat shaped morphology or akaganeite, observed typically under transmission electron microscope(TEM), is actually the built-up of parallel needles or rod-like sub-crystals packed into an orthogonal array [206]. The rods are typically $0.6nm^2$ in cross section and about 600nm in length. Powerful microscopic examination reveals that the rods are themselves hollow with internal diameter of 30nm.

The structure of psilomelane consists in cross linking of two triple chains of $[MnO_6]$ octahedral through two double chains by sharing their adjacent corner oxygen atoms. This results in three dimensional framework structure with a series of tunnel or tube running in the direction of b-axis. The mineral has monoclinic symmetry with unit cell dimension of a_o=0.958nm, b_o=0.287nm, c_o=1.385nm, β=92°30'. Thus, b_o dimension of psilomelane corresponds to c_o dimension of hollandite. The tunnels are occupied by Ba^{2+} (and sometimes K^+) and H_2O, generally in molar ratio of 1:2 [204,206]. The occurrence of this mineral in soil is not as frequent as hollandite, but it ooccurs in sediments of deep-sea and even in fresh water lakes. However, there is no Fe(III) oxide of similar structure. The structure is illustrated in Fig.7.14.

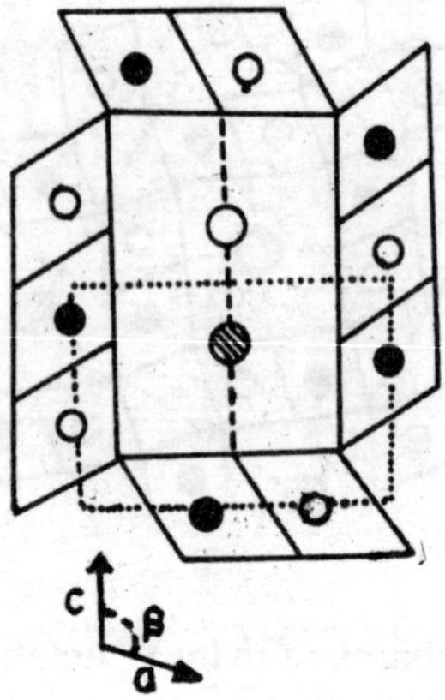

Fig 7.14 Edge shared triple chains of $[MnO_6]$ octahedra join at corners with similar octahedral double chain forming another framework structure of α-MnO_2 with large tunnel occupied generally by hydrated Ba^{2+} ion (psilomelane). Open circles represent atoms at 0 level and filled or shaded circles represent atoms at b/2 level. Dotted lines show 1/4th of the unit monoclinic cell. Adapted after ref. 204.

7.4.4.3 Layer Structure

Apart from hollandite, the other commonly occurring Mn(IV) oxides in soils belong to this group. This is because deep weathering of manganiferrous rocks in humid condition results in the formation of very fine, poorly crystalline oxides of both iron and manganese. The main characteristic features of the oxides belonging to this group are that both Mn(IV) and Fe(III) form layers of edge shared octahedral between which H_2O, OH^-, Na^+, K^+ and sometimes divalent ions reside. The long range ordering or octahedral growth along c-axis is restricted, and there exists extensive cation vacancies in the layers; all of which render these hydrated oxides highly porous with large surface area and have numerous adsorption sites.

The two most common Mn(IV) oxides with layer structure that occur in soil are lithiophorite and birnessite. In lithiophorite layers of edge shared $[MnO_6]$octahedra, alternate with layers of $[(Li,Al)(OH)_6]$ octahedra. The stacking sequence along c-axis is -O-Mn-O-OH-(Al,Li)-OH-O-Mn-O-and the distance between two consecutive (MnO_6) layers is about 0.95nm [204,206]. Interestingly, in lithiophorite octahedral vacancies either in $[MnO_6]$ or in $[(Al, Li)(OH)_6]$ layer are generally absent, though substitution of Mn^{2+} for Mn^{4+} do occur to maintain charge balance.

The structure of natural birnessite could not be determined because of short range of ordering, but certain information regarding its structure has been obtained from electron diffraction of synthetic sodium containing birnessite ($Na_4Mn_{14}O_{27}$. $9H_2O$) or its sodium-free analog (Mn_7O_{13}. $5H_2O$) [54, 55]. Broadly, H_2O and OH- group occur between [MnO_6] octahedral layers. The distance between two consecutive layers is about 0.72nm along c-axis. One out of every six octahedral site is vacant and Mn^{2+} or Mn^{3+} ions are believed to lie above and below these vacancies. The lower valence manganese ions themselves are coordinated to oxygen atoms in both [MnO_6] layer and (H_2O, OH) sheet in the upper and the lower layers respectively. Sodium possibly lies between the layers to satisfy charge balance, but the nature or its bonding is not clear. However, it is certain that the interlayer cations are exchangeable with other cations. Thus, the packing sequence proposed by Giovanoli et al [55] is possibly - Mn^{4+}-O-(Na^+,Mn^{3+})-(H_2O, OH)-(Na^+,Mn^{3+})-O-Mn^{4+}. Besides basal plane reflections at 0.72nm and 0.356- 0.36nm ; x-ray diffraction exhibit broad reflections at 0.24nm and 0.142nm corresponding to diffraction from (1010) and (1120) which are characteristics for all hexagonally close packed Mn(IV) oxides.

An extensive and stormy debate was raised in the late 1960s and in the 1970s whether poorly crystalline δ-MnO_2 with two weak and broad lines (0.240nm and 0.142nm) belongs to the same phase as naturally occurring birnessite or its synthetic varieties. Many distinguished authors participated in the debate for and against this controversy as well as on the nomenclature problem of the disordered phases of Mn(IV) oxide with layer structure. The topic is beyond the scope and objective of the present monograph. Interested readers may, however, consult the relevant literature [54,55,206,429]. It is now generally accepted that nomenclatures like sodium manganese(II, III) manganate(IV) ($Na_2Mn_{14}O_{27}$. $9H_2O$) and manganese(III) manganate(IV) (Mn_7O_{13}. $5H_2O$) need to be dropped and Mn(IV) oxides with layer structures including δ-MnO_2 should be clubbed under the birnessite variety. The present author, however, favors to retain the nomenclature δ-MnO_2 for near amorphous or broad and weak two lines phase to distinguish it from relatively better crystalline natural or synthetic birnessite with either orthorhombic super cell (Fig.7.15A ,B) or hexagonal closed pack (HCP) unit cell with characteristic reflections at 0.721nm at 0.356nm. Structurally, δ-MnO_2 is closely related to hydrous ferric oxide (FeOOH, xH_2O) which exhibits weak but broad bands at 0.255nm and 0.47nm due to larger ionic radius of Fe^{3+}(r = 0.065nm) . Because of the close similarity in the structures of these two hydrous oxides of manganese and iron exhibit epitaxial intergrowth in the sediments of deep sea [206]. Possibly, near amorphous or poorly crystalline ferrihydrite (Fe_2O_3, nH_2O) has also similar layer structure where cation vacancies occur in a random manner within the hexagonally closed pack oxygen as illustrated in Fig. 7.15C.

In buserite there occurs one extra layer of H_2O molecules between lower valence

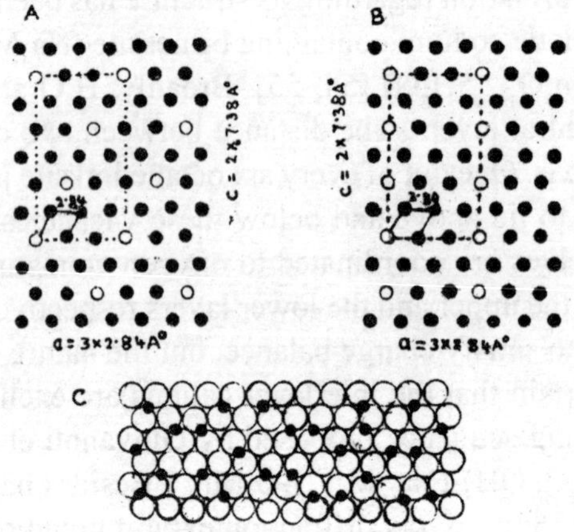

Fig.7.15 (A) Schematic structure of synthetic birnessite ($Na_4Mn_{14}O_{27}.9H_2O$)at zero level and (B) at about 0.72nm along c-axis. Orthorhombic super cell is indi cated by dotted outline.(C) Highly disordered structure of near amorphous hydrated ferric oxide(HFO) is closely similar to that of δ-MnO_2 with two broad lines at 0.240nm and 0.142nm. Adapted after ref. 55.

manganese and oxygen ions, thereby extending the thickness of the two consecutive octahedral layers to about 1.0nm. However, this hydrated oxide undergoes dehydration even on exposure to air, resulting in the structural collapse to attain birnessite structure [54]. X-ray diffraction patterns of some relevant oxides of iron and manganese are shown in Table 7.3. It may be mentioned here that x-ray diffraction is not the only or even an effective method for the identification of inorganic minerals (including clays), as many of them occur not only in small quantities (< 5%), but also in complex inter dispersed state in soils. It needs elaborate separation of different size fractions down to minus 1mm and also selective dissolution to separate specific minerals. Other physical methods employed in soil research are IR spectroscopy, electron microscopy, selected area electron diffraction etc.

7.4.5. Silicon Dioxide (SiO_2)

Silicon dioxide or silica is one of the major constituents in soils and occur in both crystalline and non crystalline forms. The most common crystalline form is quartz, derived from volcanic rocks. It generally occurs in the pure form. It has a compact frame work structure consisting of neighboring [SiO_4] tetrahedra joined by sharing common oxygen atoms via Si-O-Si links, giving a compact structure. The non crystalline form is frequently referred to as opaline silica which is largely of biological origin. It occurs in both plants and animals which after dying and subsequently decaying release silica in the soils and sediments.

Table 7.3: X-ray Diffraction patterns of Some Oxide Minerals Present in Soils.

Mineral name	Major d-spacings (Ao) with relative intensities in parentheses							
Goethite		4.18 (100)	2.53 (30)	2.49 (16)	2.45 (25)	2.19 (20)	1.72 (20)	1.56 (16)
Lepidocrocite	6.26 (100)	3.29 (90)	2.47 (80)	1.937 (70)	1.732 (40)	1.524 (40)	1.367 (30)	1.075 (40)
Hematite		3.66 (25)	2.69 (100)	2.51 (50)	2.20 (30)	1.838 (40)	1.69 (60)	1.484 (35)
Hollandite	6.98 (50)	3.47 (80)	3.13 (100)	2.40 (90)	2.15 (80)	1.83 (60)	1.55 (70)	
Cryptomelane	6.90 (90)	4.90 (80)	3.10 (80)	2.39 (100)	2.15 (60)	1.83 (60)	1.54 (60)	
Psilomelane	6.96 (55)	3.48 (50)	2.408 (100)	2.366 (50)	2.188 (85)	1.824 (50)		
Birnessite (Natural)	7.27 (S)	3.60 (W)		2.44 (M)			1.424 (M)	

S = strong, M= medium, W = weak

7.4.6 *Zeolites*

Although hydrated aluminosilicate minerals under this group seldom occur in common soil, these compounds occur widely in nature and can also be synthesized according to tailor made needs such as molecular sieves, adsorbents etc. The crystalline structure of zeolites is characterized by the combination of [AlO_4] and [SiO_4] tetrahedras forming three-dimensional framework consisting of four fold links interconnected into six-, eight - or twelve membered rings (see Fig. 7.16a,b). The resulting reticular structure has channels and inner voids to provide access of the adsorbing species The size of the voids and channels in the framework of zeolite structure depends on the chemical composition; larger voids being typical of highly siliceous zeolites [207]. The voids run in different directions from the surfaces through the crystals and are not of uniform cross - section, but consist of wider cavities interrupted by periodic constrictions called 'windows' [207,208]. The way the different kinds of channels are interconnected and the sizes of the windows determine the ease with which cations (or any gaseous adsorbed molecule) can enter into, move through and out of the crystals [208]. The Al^{3+} is small enough to remain fixed in the central position of the tetrahedron, but isomorphous replacement of Si^{4+} with Al^{3+} produces negative charge in the

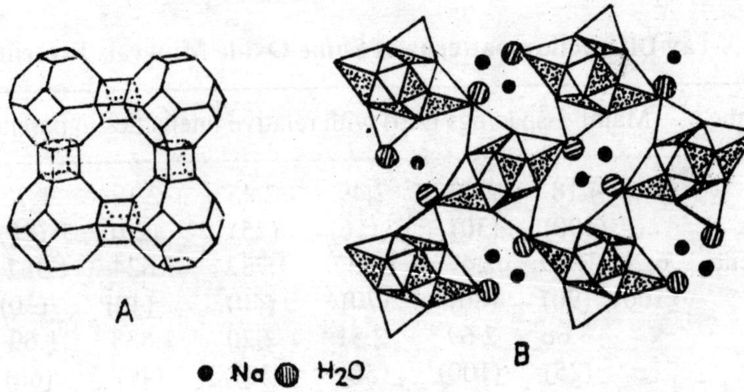

Fig.7.16 (A) General structural features of zeolites containing open channels and large "windows". (B) Structure of natrolite showing the position of Na^+ and H_2O in the hexagonal cavity formed by eight member tetrahedral ring. Adapted after A. V. Milovsky and O. V. Konnov, Mineralogy, Mir Publishers, Moscow,1985.

lattice. The net negative charge is balanced more frequently by the exchangeable cations such as Na^+ and Ca^{2+} and less frequently by K^+, Ba^{2+} and Ca^{2+}. In the inter-connecting voids the cations are coordinated to water forming aquo complexes of large radii. These cations are highly mobile and are apt to exchange with other divalent cations like Pb^{2+}, Cd^{2+} , Zn^{2+} and Mn^{2+} [208]. The composition of zeolites can be expressed by the formula $M_xE_yO_{2y}$. nH_2O, where M is the exchangeable cation and E is Si and Al in various proportions.

7.5 Surface Properties of Soil and its Inorganic Constituents

7.5.1 Origin of Charge at Soil / Electrolyte Interface

To understand the interaction between soil and environment it is necessary to understand the surface properties of soil and its constituents. However, the genesis of most of the soils is such that their constituents are themselves formed from the environment to which the soil particles are expo- sed. This shows that like air and water, soil is an integral part of our ecosystem which provides us with food, forest, oxygen, rain and host of many other benefits. The soil constituents vary from simple oxides and hydr(oxides) to large and complex organic molecules, all of which are either precipitated or adsorbed on aluminosilicate minerals . Therefore, the surface properties of soils are the sum total of all its constituents, unlike those of pure individual substrates as discussed in the preceding chapters. It is therefore obvious that interfacial properties will vary from one soil to another. Though sufficient information has been collected over the last five to six decades on this subject, it is sometimes still difficult to explain every aspect of observation from the established theories of surface chemistry. However, an attempt has been made to explain from colloid chemical point of view the results / observations that have emerged from careful experimentations by many researchers.

The methods used for the characterization of the surfaces of finely divided solids have been briefly discussed in chapter1. Similar procedures are applicable to soil colloids also. The precautions needed to be taken for the measurement of surface area, porosity, and particle size distribution of finely divided porous solid by using low temperature gas (nitrogen or argon) adsorption method, also need to be taken for soils. However, a few points may be added here considering the wide variety of materials that are likely to be handled. First, attempts have been made to calculate the area of certain 2:1 clay minerals like micas, smectites etc. from the areas of the upper and lower basal planes . This is achieved by dividing the product of a and b dimensions of unit cell with molecular mass $[Si_8Al_4O_{20}(OH)_4]$ per unit cell as follows

$$[(0.515 \times 0.892 \times 10^{-18}) \times (6.03 \times 10^{23}) \times 2] / 728 = 760 \ m^2 \ g^{-1}$$

The numerator is to be multiplied by 2 to account for top and bottom surfaces of unit cell. Though, in practice, this value is never obtained from actual experiment, but this value is the theoretical area that is available for adsorption of water or glycol or for cation exchange reactions. Indeed, this value of calculated area is used as a 'marker' for the whole group of 2:1 minerals [69]. For example, the surface area of pure illite is about 76 $m^2 \ g^{-1}$. This is because each crystal of illite consists of 10 sets of 2:1 layers of mica, Therefore, the measured area is 1/10 th of the theoretical area. In fact, higher surface charge in mica renders the interlayer bonding stronger and therefore shows lower surface area and also lower CEC capacity.

In the above theoretical calculation of surface area, the area occupied by the edges of clay particles has been tacitly neglected. While, this assumption may be partially valid for expanding clays like montmorillonite, vermiculite etc., considerable error is introduced for non - expanding clays like kaolinite, halloysite etc. Further, the calculated area does not take into account other soil constituents like silica, hydrous oxides of iron and aluminum and amorphous aluminosilicate like allophane.

While studying the surface chemistry of soil clays, generally two types of surface are encountered, namely (i) the silicate or siloxane type and (ii) the oxide or hydr(oxide) type . Based on this broad classification, we may discuss the surface properties of soil constituents.

7.5.1.1. *The Silicate or Siloxane Surface*

The tetrahedral sheet of silica in almost all clays has two planes of oxygen atoms, namely the basal plane where all the oxygen atoms are linked to form Si_6O_6 rings and the apical oxygen atoms which are shared by the octahedral aluminum hydr(oxide). Since all the oxygen atoms in the silica basal plane are coordinately linked with each other, the net charge density in this plane is zero. This is known as siloxane surface which is usually hydrophobic in nature as in the case of minerals like pyrophyllite and talc. However, in practice, the surface carries a resultant charge because of isomorphous substitution of Si by Al and the deficit in charge is balanced by interlayer K^+ as in the case of mica .When exposed to external weathering, the K^+ ions can be exchanged by suitable cations like Mg^{2+} and Ca^{2+} without disturbing the charge balance . For micas the charge density is about 2.2 nm^{-2} which means the area per unit charge is about 0.45nm^2. This corresponds to half the basal area of the unit cell i.e., the product of a and b dimensions of a common mica (0.5 x 0.9 nm^2). As stated earlier, the charge in 2:1 layer silicates may also arise from isomorphous substitution of Al by Mg in the octahedral sheet.

Soil micas, however, undergo alteration / hydration through weathering, resulting in a product with higher Al content and more hydration i.e. , incorporation of more OH groups, some of which are also attached to interlayer cation [140] . This is possible only when some K^+ ions are replaced by Mg^{2+} or Ca^{2+} as interlayer cations as found in smectites and particularly in vermiculites where the replacement of K^+ by Mg^{2+} is complete. Therefore, alteration of micas leads to frequent inter-stratification between micas and smectites or between micas and vermiculites. Since isomorphous replacement of Al is less in such clays, the interlayer charge deficit is less and therefore they expand and may even cleave at the interlayer separation. As far as these clays are concerned, it has been observed that the distribution of structural and dissociable charges is about 90% and 10% respectively. Consequently, most of the studies carried out so far were related to structural charge i.e., cation exchange properties. The charges arising from isomorphous substitution are independent of pH, cation and electrolyte concentration. Such charges are therefore called permanent charges. Careful potentiometric titration of Georgia kaolinite carried out by Schofield [209] in 1.M NH_4Cl and by Greenland [210] in 1M CsCl demonstrates the pH independent charge of these clays.However, proton adsorption behavior i.e., dissociable charge in 2:1 clays is gaining more interest as adsorption of both anions and cations under certain conditions depends more on the dissociable charge on the clays.

From Potentiometric titration of many clay minerals and soils if has been observed that the surface charge depends not only on pH but also on background electrolyte solution. This has been attributed to the protonation of metal-oxygen especially Al - O bonds at the edges, besides protonation of hydrous oxides present in the soils about which we shall discuss in the next section. But, it may be pointed out here that even silicate surface may undergo protonation and contribute to pH dependent charge. This important phenomenon is discussed below in detail.

Protons from acids or generated *in situ* during hydrolysis of adsorbed cations penetrate a few atomic sheets deep into the surface of most silicates with the release of structural cations. This is usually a relatively slow and irreversible reaction taking place at ambient temperature and pressure [211]. The reaction has however been found to be reversible at higher temperature and pressure [212]. Proton adsorption creates high acidity at soil / electrolyte interface and Si-O-Si and Si-O- M (where, M = Al or Fe(III) , Mg , Ca etc.) groups dissociate forming small polymeric species or even simple monomers that are either adsorbed on the surface or diffuse into bulk solution [213-215]. In trioctahedral clays protons exchange takes place with both octahedral Mg or Fe(II) and tetrahedral Al. Indeed silicic acid, formed by the protonation of the silicate surface, acts as proton sink from solution. It may therefore be concluded that both basal plane and apical oxygen atoms of silica can be protonated at very low and high pH and in the presence of strong electrolyte solution. Though well crystallized quartz is inert to this reaction, amorphous silica may undergo protonation reaction. The main implication of the protonation of silicate surface is related to soil acidity. The pH dependent hydrolysis of Al ion releases H^+ in solution. The excess acidity is generally neutralized by adding requisite amount of lime. This is a very important subject to the agricultural scientists.

The kaolinite surface is possibly the best example where charges due to siloxane, silanol (Si -OH) and aluminol (Al-OH) have been studied in detail [216]. The hexagonal platelets of kaolinite are believed to have only small negative charge on two opposite faces of the basal plane. While the top exposed face is considered to represent siloxane surface corresponding to the basal oxygen atoms of the tetrahedral silicate sheet, the bottom exposed surface is considered to represent Al-OH surface of octahedral gibbsite sheet. While the former face carries small permanent negative charge due to isomorphous substitution, the latter face carries pH dependent negative charge which however does not change to positive value even at low pH. This resistance to reach +ve value is attributed to the adsorption of silicate ion by the Al-OH group of the gibbsite sheet during the formation of the mineral from silicate rich weathering solution [217-218].

Since isomorphous substitution in the tetrahedral sheet of kaolinite is seldom observed, the small amount of permanent negative charge on the basal plane may possibly arise due to the presence of very small (1-3 %) quantity of smectite as impurity which may escape detection by XRD method [69]. However, the increase in the -ve charge on the edges with increase in pH is generally found to be more than that estimated from charged sites at the edges. The source of the additional negative charge is attributed to the gibbsite face of the bottom basal plane. Very small +ve charge at low pH is attributed to the amphoteric behavior of Al - OH groups at the edges. This has been demonstrated by the selective adsorption of oppositely charged colloid particles onto kaolinite at different pH values. The selective adsorption has been monitored or examined directly by transmission electron microscopy or TEM [219]. The negatively charged AgI particles get adsorbed on positively charged edges at pH 4-5. At pH ~ 3 extremely fine positively charged hydrous iron oxide tends to be adsorbed on the negatively charged basal plane of dickite, but not on edges having similar i.e., +ve charge [69, 216 , 220]. The difference in the nature of charge at the basal plane and at the edges of kaolinite particles, and for that matter on other clays, can also be demonstrated from the 'self flocculation ' of clays with change in pH . This means that +vely charged edges get attracted to -vely charged basal planes resulting in the so called 'card house ' type flocs .

7.5.1.2 The Hydroxylated Metal Oxide Surface

The surface properties of hydrous oxides in soils, especially in aqueous electrolyte medium , are essentially similar to their individual properties described in the preceding chapters. It may however be noted that the hydrous oxides in soil are rather a highly heterogeneous mixture, unlike pure individual oxides as used in the basic studies of solid - electrolyte interface. Co-precipitation, surface deposition / coatings of one cnstituent on another, presence of impurity ions etc. are the common features of ydrous oxides in soil. Therefore, pH_{PZC} values, surface ionization / protonation contants, surface complexation constants etc of soil constituents will naturally differ from the corresponding values in their pure states. Besides, the presence of a host of organic and inorganic ligands and their association with oxide cations cause considerable deviations from ideal oxide - electrolyte interface. Since all these complexities are inherent and cannot be avoided or altered, many researchers consider soil as a composite single entity and assume that its surface behaves in the same way as an individual oxide. On this basis, various models, both empirical and mechanistic, have been applied to the adsorption of cations and anions on soil. Recently, Bradle [221] has made an excellent review on this subject. We shall discuss more elaborately on this subject on this aspect in a separate chapter on adsorption.

7.5.2 Pretreatment / Purification of Naturally Occurring Clays or Soil Clays for the Study of their Surface Properties

In order to determine certain basic interfacial parameters like pH_{PZC}, constants for surface protonation / deprotonation and association or exchange with cations etc, it is of foremost interest to purify naturally occurring clay minerals including soil clays so that the parameters obtained from potentiometric or acid - base titration are, as far as possible, representative of the concerned clay mineral. Purification or rather pretreatment of naturally occurring / soil clays involves three main steps. These are: (a) dispersion in water, (b) decomposition of carbonates, (c) removal of impurity coatings and (d) extraction of pure clays by repeated sedimentation and decantation or dialysis of the treated clays . Among these , the most critical steps are (c) and (d) which involve the risk of possible change in the surface structure of clays . Decarbonation is normally affected by treating the material with dilute HCl at pH 3.0-3.5 or with weak acid like acetic acid at pH 4-5 depending upon the carbonate minerals present. Aluminum or iron hydr(oxide)s coatings can be removed by leaching with chelating buffers such as oxalates [59] or citrate-dithionite-bicarbonate [222] depending upon the nature of such coating. But, sometimes it is difficult to completely get rid of strongly adhered oxide coatings even after such treatment. Organic coatings are removed by treating with hydrogen peroxide.

As stated earlier, the above treatments for purification may bring about surface dissolution or even alteration in the clay structure. For example, it has been observed that pH_{PZC} of an acid treated illite sample is about 4.5, but shifts to about 8.0 when washed thoroughly with water [223] . In view of this, Baeyens and Bradbury [224] disapproved the use of any extensive purification procedure unless the objective of the investigation is to generate basic data on the concerned clay mineral . Where the objective is to characterize raw materials simple washing with classification down to clay size (< 0.2 mm) may be sufficient.

Some authors [225 - 227] have suggested that treating the washed clay sample with appropriate salt solution (generally NaCl or $NaNO_3$) can bring about homogeneity in the exchangeable cation and therefore helps to stabilize the 2:1 type clay structure. It has already been stated that H clays produced by excessive washing or even dialysis of pretreated clays results in the dissolution and structural breakdown of clays. Homoionization prevents this possibility. The excess salinity is removed by repeated settling and decantation until a reasonable limit (about 0.1mM) is reached. The washed clay is then dried either at ambient temperature or at higher temperature (~ 60°C). Clay sample is also dried at low temperature under vacuum (freeze drying).

7.5.3 Separation of Clay Fraction Using Size Classification by Sedimentation Method

The procedure followed by the author to separate clay fraction is as follows. The pretreated and dried sample is thoroughly ground in an agate mortar and pestle, homogenized and wet sieved to size finer than 35 - 40 μm (- 400 BSS) using 0.1M

NaCl solution as suspending liquid. The clay fraction containing particles finer than 2 µm is then separated by sedimentation in a 250 ml measuring cylinder using the well-known stokes law which is given by

$$R = \frac{2}{9}\left(\frac{\rho_p - \rho_s}{\eta}\right) g r^2 \qquad (7.1)$$

where, R = sedimentation rate (cm sec^{-1})
ρ_p = density of particles , g cm^{-3}
ρ_s = density of fluid , g cm^{-3}
g = acceleration due to gravity , cm sec^{-2}
η = viscosity of the suspending fluid , g sec cm^{-1}
r = equivalent radius of the suspending particles , cm

In a system where ρ_p , ρ_s , g and η are constants at a specified temperature the above equation may be expressed in the simplified form

$$R = K \cdot r^2 \qquad (7.2)$$

where, K is a constant. Therefore, R is directly proportional to square of the particle radius. The sedimentation velocity at 20°C of clay particle of 1 µm radius is about 347 x 10^{-6} cm sec^{-1}. Similarly, for 10 µm radius particle the rate will be 347 x 10^{-4} cm /sec as R is directly proportional to r^2. Knowing the time of sedimentation t (sec) elapsed to settle to a certain distance say 10 cm (as is usually done) under the action of gravity the particle size can be obtained from the following equation

$$t = 10 / Kr^2 = 1 / 3470 \times 1 / r^2 \qquad (7.3)$$

Thus, for clay particle having an effective radius of 1 µm , time required to settle to a depth of 10 cm at 25°C is about 7 h 5 min and for 30°C the value is about 6h 9 min. Various methods, such as decantation or pipetting may be used to withdraw clay suspension from a depth of 10 cm [228] . However, both the methods are some what tedious and takes considerable amount of time to collect sufficient quantity of sample. The clay suspension is centrifuged, the supernatant liquid is decanted off. The classified concentrated clay suspension is used either as it is or dried after filtering though 0.45 mm membrane filter.

The Stokes law is applicable to uniform dilute suspension where each particle settle without any hindrance or collision from the neighboring particles. The temperature should be maintained constant so that no convection current is set up. The density of soil particle or raw material is assumed to be constant in many instances, but it is better to determine density before carrying out size classification experiment. For clay rich soil the density value is around 2.65 g cm^{-3}.

7.5.4 *Early investigation on base titration of H - clay*

Potentiometric and conductometric titrations have long been used for characterization as well as for the determination of surface ionization and complexation constants of clay colloids in an electrolyte medium for use in formulating an appropriate model of electrical double layer. To render the surface acidic with exchangeable H^+, the clay mineral is converted to H^-form by two methods. The first one adopted by Marshall et al [229, 230] in the 1940s is dialysis or electrodialysis. It has been observed that H-clays prepared by this method are not only unstable and weakly acidic, but also in the form different from the original clay. This is because H^+ replaced cations from the interlayer position meant for balancing charge deficiency. This triggers the release of Al and or Mg in suitable hydrolysed form to occupy the position held by H^+ of extremely small size. The second method involves treating the clay with ion exchange resins in acid forms so that exchangeable cations are replaced with H^+. The hydrogen clays produced by this method are found to be more acidic and stable over a longer period than those prepared by the dialysis method. That the H-clay produced by the first method results in hydrolysed Al species which occupy the exchange sites has received indirect support from the work of Low [231]. The author has shown that base titration curves of H^- clay prepared by dialysis and having different proportions of surface H^+ and hydrolysed Al species are closely similar to the base titration of a mixture of $AlCl_3$ and HCl in corresponding molar ratios. Despite being altered due to dialysis, Marshall et al [229,230] have made extensive studies on different types of clays including beidellite, illite and montmorillonite by measuring conductivity, pH, pM, electrokinetic data during base titration. The extensive studies helped the subsequent investigators [232] to derive constants for surface ionization and metal ion association.

Garrels and Chist [232] have suggested the presence of two types of surface sites; one is 'stong'(designated as SOH) which reacts with base at pH <5.5 and the other is 'weak' (designated as TOH) which reacts with base at pH >6.0 - 7.0. However, the authors considered only the exchange reactions of cations without considering the ionization of surface sites.

$$SOH + Na^+ \longleftrightarrow SO^- Na^+ + H^+ \qquad {}^*K_{SONa} = \frac{\{SO^- Na^+\}\{H^+\}}{\{SOH\}\{Na^+\}} \tag{7.4}$$

$$TOH + Na^{+} \Longleftrightarrow TO^{-}Na^{+} + H^{+} \qquad {}^{*}K_{TONa} = \frac{\{TO^{-}Na^{+}\}\{H^{+}\}}{\{TOH\}\{Na^{+}\}} \tag{7.5}$$

It may be noted from the above equations that the authors did not differentiate between surface and bulk concentration of H^+ in the above two equations. Na^+ ion and surface site balance were represented by the authors as follows:

$$[Na]_T = \text{NaOH added} = [Na^{+}] + [SO^{-}Na^{+}] + [TO^{-}Na^{+}] \tag{7.6}$$

$$[SOH]_T = [SOH] + [SO^{-}Na^{+}] + [SO^{-}] \tag{7.7}$$

$$[TOH]_T = [TOH] + [TO^{-}Na^{+}] + [TO^{-}] \tag{7.8}$$

The overall electroneutrality was proposed to be maintained as follows

$$[H^{+}] + [Na^{+}] = [OH^{-}] + [SO^{-}] + [TO^{-}] \tag{7.9}$$

All the data required for the solution of the above mass and charge balance equations are available from the experiments of Marshall et al (229, 230). Using the data Garrels and Christ [232] calculated the constants according to Eqs (7.4) and (7.5). The values obtained at pH above 6 were attributed to ${}^{*}K_{TONa}$ and below it to ${}^{*}K_{SONa}$. Despite many questionable assumptions and procedural inaccuracies in the calculation, the authors obtained good agreement between calculated and experimental data for the base titration of the same the clay i.e., Putnam clay, by appropriate choice of the adjustable parameters.

James and Parks [60] made an attempt to remove some of the shortcomings of Garrels and Christ [232] by considering ionization of surface sites and differentiating between surface and bulk $[H^+]$ through the use of Eq (3.10). The entire exercise was aimed at formulating EDL based on triple layer model using the relevant constants for the two types of surface sites i.e., strong and weak. The exchange of H^+ with Na^+ (added in the form of NaOH solution) was considered as the process of surface complexation as in the case of oxides. All the mass and charge balance equations are similar to those given in chapter 5 and are also presented below.

Ionization of surface:

$$SOH = SO^{-} + H_s^{+} \qquad K_{a1SOH}^{Int} = \frac{\{SO^{-}\}\{H_s^{+}\}}{\{SOH\}} \tag{7.10}$$

$$TOH = TO^- + H_s^+ \qquad K_{a1TOH}^{Int} = \frac{\{TO^-\}\{H_s^+\}}{\{TOH\}} \tag{7.11}$$

Ion - exchange reactions and their equilibrium constants are similar to those expressed in Eqs (7.4) and (7.5) except that $\{H^+\}$ is expressed as a surface species $\{H_s^+\}$. Since neither positively charged surface nor any clear cut PZC was observed in the base titration of H - clay , the surface charge is therefore expressed by

$$\sigma_0 = -B\,([SO^-] + [SO\text{-}Na^+] + [TO^-] + [TO^-Na^+]) \tag{7.12}$$

The charge on IHP or Stern layer is thus given by

$$\sigma_\beta = B([SO^-Na^+] + [TO^-Na^+]) \tag{7.13}$$

and the diffuse layer charge ($-\sigma_d$) is given by Eq (5.10) . Consequently, the overall charge balance is expressed by Eq (5.14) and the potential - capacitance relationships also follow Eqs (5 .19) and (5 20) . By invoking the Boltzmann law i.e., Eq (5.23) the concentration of individual surface species (mole / kg) is then expressed as follows :

$$\{SO^-\} = K_{a1SOH}^{Int}\{SOH\}\{H^+\}^{-1}\exp(e\psi_o / kT) \tag{7.14}$$

$$\{SO^-Na^+\} = {}^*K_{SONa}^{int}\{SOH\}\{H^+\}^{-1}\{Na^+\}\exp[e(\psi_o - \psi_\beta)/kT] \tag{7.15}$$

$$\{TO^-\} = K_{a1TOH}^{int}\{TOH\}\{H^+\}^{-1}\exp(e\psi_o / kT) \tag{7.16}$$

$$\{TO^-Na+\} = {}^*K_{TONa}^{int}\{TOH\}\{H^+\}^{-1}\{Na^+\}\exp[e(\psi_o - \psi_\beta)/kT] \tag{7.17}$$

The total quantities of the two types of surface sites i. e., strong and weak are given by

$$\{SOH\}_{tot} = \{SO^-\} + \{SOH\} + \{SO^-Na^+\} \tag{7.18}$$

and $$\{TOH\}_{tot} = \{TO^-\} + \{TOH\} + \{TO^-Na^+\} \tag{7.19}$$

The relevant parameters were either assigned or calculated numerically or analytically. Generally all the equations can be solved using computer program FITEQL (chapter 6). The values of some of the parameters obtained by James and Parks [60] for the purpose of fitting the experimental data of Marshall et al for Putnam clay (beidellite) are as follows:

$pK_{a1SOH}^{Int} = 6.0$, $pK_{a1TOH}^{Int} = 7.4$, $p^{*}K_{SONa}^{int} = 3.0$, $p^{*}K_{TONa}^{int} = 7.4$, $C_1 = 2.5\ F / m^2$

and surface area = 650 m^2/g.

Though the agreement between experiment and theory was poor the authors obtained at least qualitative agreement in the nature and shape of the titration curves , irrespective of the solid concentration which varied from 1.0 to 10 % (w/v) . The results also supported the presence of free or unbound SO^- and TO^- sites , because if the constraints or parameters would have been calculated on the basis of only the exchange reactions i.e., Eqs.(7.4) and (7.5) without considering the ionization reactions, the titration curves were independent of solid concentration. But, in practice, the curves showed distinct shift or variation with solid concentration. Further, if ionization reactions were neglected the zeta potential values would be zero, irrespective of the extent of exchange reactions. But, in reality, this was also found to be not true.

Unfortunately, for H - clays prepared by treatment with ion exchange resins, the data available from base titrations are insufficient compared to those collected by Marshall et al [229 , 230] . The lack of complete data hindered the detailed analysis and fitting of EDL model as in the case of dialyzed or electrodialyzed clays. Subsequently, Schindler et al [233] also proposed two kinds of surface sites to explain the adsorption behavior of cations on soil. One kind consists of weakly acidic site (° XH) which can undergo ion exchange with both diavalent (M^{2+}) and monovalent (M^+) cations, while the other kind is the usual amphoteric surface hydroxyl groups (SOH) which form strong inner sphere surface complex with M^{2+} (SOM^+ and $(SO)_2M$) and outer sphere complex with Na^+.

7.5.5 Recent Attempts to Determine Point of Zero Charges of Clay Minerals

Since the base titration studies of H - clays by Marshall et al [229, 230] , various authors have carried out acid - base titration of soil clays and clay minerals. In most cases, the clay minerals were first converted to Na form. The experimental procedure (both rapid and batch methods) followed is similar to that used for the potentiometric titration of pure oxides and hydroxides (see chapter 4). The procedure for calculating net H^+ adsorbed or desorbed (i.e., OH^- adsorbed) as function of pH and electrolyte concentration in chapter 4 has also been applied to clays [234]. Beene et al [223] have studied the interfacial properties of homoionic varieties of Muloorina and Fithian illites using potentiometric titration in chloride solutions of their corresponding homoions. This means that Na-, K- and Cs- illites were titrated batch wise respectively in different concentrations of NaCl, KCl and CsCl solutions. Prior to titration, pure clay samples were separated by dispersion and size classification in aqueous medium, treated with

0.1M HCl to remove coated iron oxide, washed free from electrolytes by extensive dialysis and finally dried at 80°C. One of the important observations made by Beene et al is the distinct variation in the nature of the titration curves (pH vs surface charge) of the three homoionic varieties of Muloorina illite in different concentrations of their of their corresponding chloride solutions. In all cases, the pH at which the titration curve intersects the line at zero surface charge, PZNPC shifts to higher value with decrease in the concentration of electrolyte from 0.1 to 0.001M. Though such shifts are closely similar for NaCl and KCl solutions, for CsCl the shift takes place at higher pH values. However, the nature of the titration curves in NaCl solution can be differentiated from the other two homoionic forms by the occurrence of sharp inflection in the former at PZNPC values. The titration curves, on the other hand, are rather smooth for K- and Cs- illites. Typical examples of the titration curves obtained by Beene et al [223] for different homoionic forms of Muloorina illite in 0.01M salt solutions are shown in Fig. 7.17.

While the experimental range of surface charge for K-illite (~ - 0.08 C/m^2) and Cs-illite (~ -(0.07-0.08) C/m^2) is closely similar, the highest value for Na-illite is little lower (~ -(0.05 - 0.06)C/m^2). The sift of PZNPC, notably for K- and Cs-illites, was attributed to the low hydration energies of K^+ and Cs^+ and their strong affinity to occupy vacant sites at the edges, especially frayed edges compared to hydrated Na^+. Such specific adsorption is not accompanied by the release of H^+ and therefore shift in PZNPC value is higher. In the case of Cs^+ specific adsorption is so strong that even in low CsCl concentration (0.0001M) a considerable shift in PZNPC takes place.

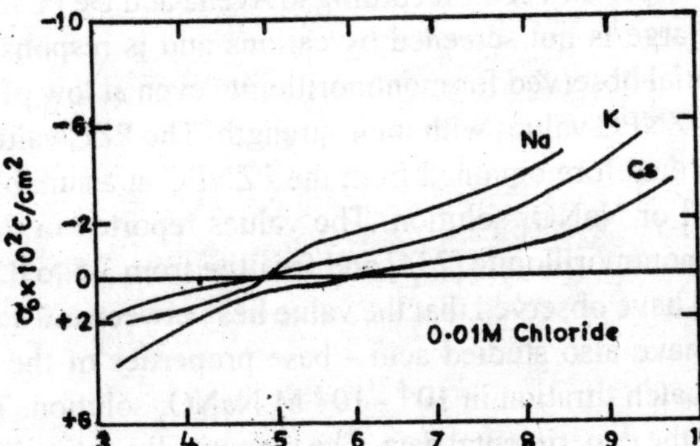

Fig.7.17 Variation of surface charge of homoionic form of Muloorina illite as determined by potentiometric titration carried out in 10-2M homocation chloride solutions(namely NaCl, KCl and CsCl) as function of pH. Redrawn after data from ref. 223.

An interesting observation made by Beene et al [223] is that for an "ill prepared" high alumina containing Fithian illite, the titration curve of homoionic clay in NaCl solution of different concentrations tend to converge to a common point at pH around 3.0. At the same time individual curves closely approach one another at pH above 6.5. Such phenomena, however, has not been observed even for ill prepared low alumina Muloorina illite.

The salient point of these studies is that unlike hydr(oxides) no common point intersection of the titration curves is obtained at different ionic strengths of electrolyte solution. Indeed, the titration curves drifts to acid side with increase in ionic strength as observed for montmorillonite [224, 235 - 237], illite [223,238,243] and attapulgite [239]. But this behavior has not been observed for kaolinite [240] or for synthetic allophane [201]. Consequently , PZNPC i.e., the intersection of adsorption / desorption curves of H^+ per unit weight of clay with zero line of adsorption occurs at different pH values for montmorillonite and illite depending upon ionic strength of the medium.

The shift is believed to be due to permanent charge arising from lattice substitution. Further, for montmorillonite and illite good linear relationship between pH at which titration curves show zero charge and logarithm of molar concentration of electrolyte has been observed many authors [234,235, 240]. This suggests that permanent negative charge greatly influence the dissociation of SiOH or AlOH groups at the edges which, in turn, depends on pH as well as ionic strength for charge development. The slope values of the linear relationships are significantly higher for illites (av. ~ 0.73) with about 1.0-1.1 substitutions per unit cell than for montmorillonite (av. ~ 0.49) with about 0.6 substitution per unit cell. According to Avena and De Pauli [235], a fraction of the structural charge is not screened by cations and is responsible not only for negative zeta potential observed for montmorillonite even at low pH values, but also for the shift in the PZNPC values with ionic strength. The PZC values of montmorillonite and illite are therefore obtained from the PZNPC at a suitable ionic strength; usually 0.1 M NaCl or $NaNO_3$ solution .The values reported in the literature vary from 5.5 to 8.5 for montmorillonite [234] and for illite from 3.5 to 8.5 [223,225,241]. Gu and Evans [243] have observed that the value lies between 6.0 and 7.5 for Fithian illite. The authors have also studied acid - base properties of the sodium form of Fithian illite using batch titration in 10^{-1} - 10^{-3} M $NaNO_3$ solution. The authors used two site model to fit the experimental data. The numerically optimized density values of permanent negative (XH) and variable charge (SOH) sites were found to be 0.087 and 0.152 mmole/g. Using the constant capacitance model pK_{X-Na}^{int}, pK_{a1}^{int}, and pK_{a2}^{int} values obtained by FITEQL optimization at 10^{-2} M $NaNO_3$ are 1.57, 7.98 8.65 respectively. On the other hand, for iron oxide coated amorphous aluminum silicate (synthetic allophane) the pH_{PZC} value is found to occur at about 4.8 [201]. This value is higher than that for amorphous silica (~ 3.0), but lower than that for synthetic amorphous aluminum silicate (~ 6.2). The authors, however, did not envisage two types of surface, but assumed amphoteric single site model like hydr(oxide) surface. Thus, the authors used triple layer model to derive not only the acid / base dissociation constants (pK_{a1}^{int}=3.46 and pK_{a2}^{int}=7.16), but also the surface complexation constants with anions and cations (pK_{aX^-}=5.12 and pK_{aM^+}=5.58) .

The present author [242] has carried out acid titration of a montmorillonite(West Bengal, India) and an illite(Rajasthan, India) sample in their Na forms. The samples were partially purified by treating with acidified oxalate solution and washed by repeated settling and decantation. The samples were then converted to Na form by suspending in 0.5M NaCl solution. Most of the excess salt was removed by repeated settling and decantation after diluting the suspension with deionised water. The washed clay suspension was filtered though 0.45 mm membrane and dried in air at 32°C until constant weight. Figures 7.18 and 7.19 show the net adsorption of protons on montmorillonite and illite samples respectively as function of pH and ionic strength of the medium. The figures show that the adsorption of proton decreases slowly from pH 7 to 4, especially at higher concentration (> 10^{-2} M) of electrolyte, and then rapidly decreases to almost zero adsorption at pH around 3.0. It has been frequently observed that there occurs a rapid exchange between H^+ in solution and exchangeable Na+ in the clay samples in the pH range of 3.0 - 4.5.

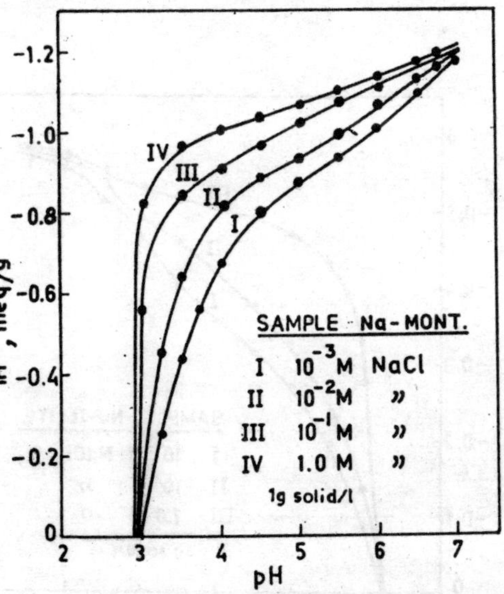

Fig.7.18 Base titration of a sodium montmorillonite sample in different concentrations of NaCl solution.

$$Na - clay + H_2O \longleftrightarrow H - clay + NaOH$$

As the released NaOH is gradually neutralized by the addition of acid, the reaction tends to shift towards right with increase in pH above 4.5. The resultant penetration of H^+ into the lattice tends to release octahedral cations in the solution [214, 243, 244]. While a part of the released cations occupy interlayer exchange sites in the clays, others go into the solution. In either case the released cations undergo hydrolysis with rising pH forming not only a variety of soluble hydrolytic species, but may also form neutral or insoluble hydroxides. The soluble hydrolytic species possess stronger affinity for adsorption than the corresponding free cations. The extensive adsorption of non - exchangeable hydrolytic species slows down the adsorption of protons with increase in pH and at pH around 7.0 practically no sites are available for adsorption of protons either on the basal surface or at the edges irrespective of the concentration of electrolyte solution. Indeed the surface saturation takes place at pH around 7.0 [230]. Therefore, while the first near convergence of the proton adsorption curves at pH about 3.0 represents the minimum negative or near zero charge condition arising from Si - OH groups, the second near convergence represents maximum negative charge condition arising from both Si - OH and Al - OH groups at the edges. It may be noted from the figures that there is no common point of intersection of the titration curves either at the minimum or at the maximum negative charge conditions. This is in accordance with that observed by other investigators and discussed in the preceding paragraphs.

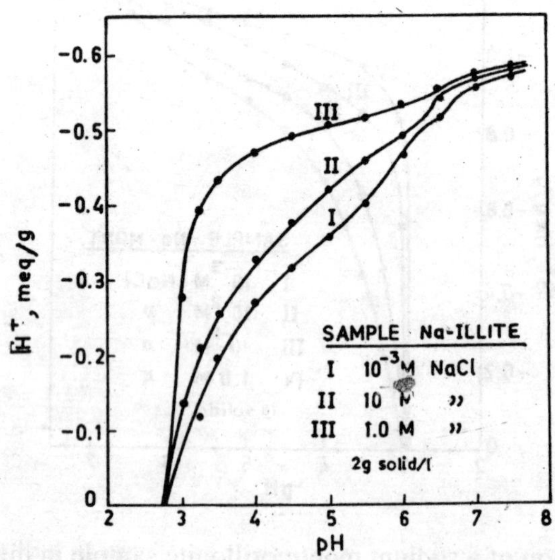

Fig.7.19 Base titration of a sodium of illite sample in different concentrations of NaCl solution.

CHAPTER 8

MODELING OF ADSORPTION FROM AQUEOUS MEDIUM

Modeling is an important step in the process of understanding a scientific phenomeon. It makes an attempt to conceptualize the cause and effect of any phenomenon, be it natural or created artificially. Such a model is generally conceived in the micro state by incorporating the different parameters involved in order to provide a mathematical or relational picture of the phenomenon. Some of the parameters may be experimentally accessible and the others are adjustable so as to fit the model with experimental data. The success of a model mainly depends on its mathematical simplicity and therefore requires lesser number of adjustable parameters for fitting experimental data. Another important criterion for a successful model is its predictive capability beyond the normal range of experiment. A good model is able to do this with satisfactory accuracy.

In the field of adsorption of inorganic ions from aqueous medium, modeling is generally attempted from two different approaches . In the first or empirical approach, attempt is made to fit the experimental data using empirical model equations without taking into consideration the mechanism involved in the adsorption process. In the second or semi-empirical approach, attempts have been made to accommodate the mechanism of physico-chemical interaction between adsorbate ion and solid adsorbent. In this case, the properties of solid / liquid interface assume great importance and requisite parameters are derived under equilibrium conditions. Since adsorption of inorganic ion on an inorganic solid is a dynamic process, especially in natural aquatic environment, modeling can also be approached empirically from kinetic point of view. It is, therefore, very pertinent to consider kinetic aspects of adsorption before discussing modeling under equilibrium condition.

8.1 Modeling of Adsorption Under Dynamic Condition : Kinetics of Adsorption

From the kinetic point of view the adsorption of inorganic ions onto a solid surface consists of three following steps:

1. Mass transfer of ions from the bulk solution to the particle surface i.e., across a

film of liquid.
2. Adsorption of ion onto the surface sites.
3. Intra-particle diffusion of ion through pores or diffusion occurring uniformly / homogeneously around the particles.

Since step 2 is considered to be very fast, either step1 or step 3 is thought to be the rate controlling step. But, in most of the literatures adsorption is considered as a two step process, wherein the external mass transfer i.e., step 1 is not considered to be rate limiting [214, 245-247]. Therefore, in such a case the first step is the rapid adsorption of ion onto the surface followed by the second step of slow uptake of adsorbate ion which may take days or even weeks or months [248]. The slow step has been attributed to various processes such as ion pair formation, surface precipitation, formation of solid solution between adsorbate and adsorbent besides slow diffusion through micropores [249]. The contribution of this step assumes more significance with increase in ionic strength of the medium or increase in molar ratio of adsorbate / adsorbent [27]. Indeed, the rate- limiting process can change with increase in adsorption density.

External mass transfer depends upon the concentration gradient across the liquid boundary film. In case of low concentration of adsorbate, the difference is so small that it does not effectively control the rate of adsorption. Likewise, in a well-stirred system the concentration gradient is also reduced to minimum. As far as the rate of intra-particle diffusion is concerned, the rate of adsorption depends upon the state of aggregation and consequently the porosity of adsorbent particles. For micro and mesoporous solid, the rate is proportional to the concentration of adsorbate ion in solution, inversely proportional to the square of the average radius of particles / agglomerates and pores constituting the structure of solid [249] . The intra particle diffusion is generally characterized by its dependence on the square root of time $t^{1/2}$. This is generally verified from the linear plot of adsorption / desorption vs. $t^{1/2}$. The slope values of the linear plots give the rate constants for intra-particle diffusion. This is however valid for flat surface or rather semi infinite surface. Helffereich [251] used the following equations correlating half time ($t_{0.5}$) with either film diffusion or pore diffusion coefficients:

$$t_{0.5} = 0.23\,(r\,l / D_f)\,(C_A / C_B) \text{ (film diffusion)} \tag{8.1}$$

$$t_{0.5} = 0.03\, r^2 / D_p \text{ (pore diffusion)} \tag{8.2}$$

D_f and D_p are the coefficients of film and pore diffusion; r, l, C_A / C_B are the radius of the adsorbent particle, thickness of the liquid film and ratio of concentration between surface and bulk (loading factor of the adsorbent) at equilibrium respectively. It has been reported [252] that when film diffusion is a rate limiting step, the value of D_f should be in the range of 10^{-6} - 10^{-8} cm² / sec and for pore diffusion D_p should be in the range of 10^{-11}- 10^{-13} cm² / sec. However, as stated, in a well agitated system film diffusion is most unlikely to be rate - limiting. Adsorption in micro porous solid involves inter-diffusion of cations and protons, the latter diffusing more easily than the formers. Consequently, the rate of cation sorption is controlled by the acidity of the OH groups in the micro pore region [250, 253]. This has been demonstrated for the adsorption of Na^+ onto non- porous silica (quoted in ref. 249).

Monitoring of the initial rate of rapid adsorption is confronted with several problems such as time expended for proper mixing , mode of stirring , measuring adsorption in the presence of solid, etc. Attempts have been made to use ion selective electrodes [254, 255]. However, for many cations and anions, the response is slow and erratic and therefore cannot be reliable . Since the adsorption of cations releases H^+ ions in solution, attempt has been made to monitor drop in pH in the initial stage [256]. At low ionic strength of the medium, change in conductivity has also been used to monitor the initial rate [257].

However, among all the techniques, the pressure jump relaxation technique has been widely used by the Japanese researchers to measure rate of adsorption in the initial fraction of a second with reasonable accuracy [258, 259]. The principle of pressure jump method is based on the dependence of equilibrium constant on the pressure according to the following thermodynamic relationship.

$$\left(\frac{\partial \ln K}{\partial P}\right) = -\frac{\Delta V}{RT} \tag{8.3}$$

where ΔV is the standard molar volume change in the reaction, P, T, R, are the pressure, the temperature and the universal gas constant respectively . Any variation in the system pressure results in the shifting of equilibrium; the return of the system to the original (i.e., the relaxation) is related to the rates of the elementary reaction steps. The relaxation time constant can therefore be used to elucidate the mechanism of reaction.

The modes of construction of the apparatus used for measuring relaxation time constant by different workers are essentially similar in principle. A typical variety has been explained by Hayes and Leckie [40]. Since this is a highly specialized nature of apparatus and used scarcely in adsorption studies, the details are not given here. The main components include suitable autoclave, pressure pump, sample and reference cells, bayonet socket and vacuum pump. The cells were capped with plastic membrane which effectively transmits pressure. A piece of brass shimstock is clamped onto the wall of the autoclave with bayonet socket. When the pressure in the autoclave is raised to sufficiently high, the shimstock bursts and the pressure returns to ambient value spontaneously in less than 100 microsecond. The chemical condition at ambient pressure is detected by change in conductivity which is related to the extent of reaction. For the interpretation of relaxation curves and the method of calculation of time constant for adsorption of cations on hydrous oxides, the readers may consult the above references.

8.1.1 Empirical modeling of rate process

When adsorption is not so fast as to warrant the use of very sophisticated method like pressure- jump method and is measurable in seconds or minutes, empirical kinetic models have been used to analyze the rate data . Generally, the pseudo first or second order models have been used. As early as 1898 Lagergren suggested a pseudo first order rate equation which is given by

$$dq / dt = k_1 (q_e - q) \tag{8.4}$$

where, q and q_e are the amounts adsorbed at time t and at equilibrium respectively , k_1 is the first order rate constant. Integrating Eq (8.4) for the boundary condition t = 0 to t = t and q = 0 to q = q gives

$$\ln \{(q_e - q) / q_e\} = -k_1 t \tag{8.5}$$

This linearized equation has been extensively used by many researchers, some of which are quoted in refs. [260 - 263]. Another frequently used model is based on pseudo second order rate law and is given by

$$dq / dt = k_2 (q_e - q)^2 \tag{8.6}$$

where, k_2 is the second order rate constant . Integrating between the limits of boundary conditions as above we have

$$1/(q_e - q) = 1/q_e + k_2 t \tag{8.7}$$

The above equation may be rearranged in the linear form

$$t/q = 1/(k_2 q_e^2) + t/q_e \tag{8.8}$$

The plot of t/q versus t gives straight line with a slope $1/q_e$ and intercept $1/(k_2 q_e^2)$ from which q_e and k_2 may be obtained . The pseudo second order equation has also been extensively used by different workers [264 -267]. Equations (8.5) and (8.8) have also been derived by Azizian [268] using specific conditions of adsorption. It has been observed [269] that at higher initial concentration of solute adsorption data follow pseudo first order reaction, whereas at lower or moderate concentration the kinetics follow pseudo second order reaction.

Tien and Huang [270] have derived the second order rate equation for the adsorption of Cu^{2+} on activated sludge releasing two moles of H^+ i.e., following 2:1 stoichiometry. This is a typical example of adsorption and desorption reactions, wherein the rates have been measured in minutes.

$$M + 2S \underset{k_d}{\overset{k_a}{\longleftrightarrow}} MS_2 \tag{8.9}$$

where, S and MS_2 are the surface species , M is the free metal ion , k_a and k_d are the rate constants for adsorption and desorption respectively. Accordingly, the adsorption rate can be expressed in the following form:

$$dq/dt = k_a [M] (q_m - q)^2 \tag{8.10}$$

where, [M] is the concentration of free metal ion in solution , q_m is the maximum adsorption capacity which is equivalent to q_e in Eqs (8.5) and (8.8), expressed generally in moles/g of adsorbent on dry basis . From mass balance relationship it is known that

$$[M] = [M]_o - \phi q \tag{8.11}$$

where, $[M]_o$ is the total metal ion concentration (mole/l) and ϕ is the volume concentration of solid (g / l). Integrating Eq. (8.10) under the similar boundary condition as in the case of Eq. (8.6) we have

$$\frac{t}{q} = \frac{1}{k_a [M] q_m^{\,2}} + \frac{t}{q_m} \tag{8.12}$$

or $$\frac{t}{q}=\frac{1}{k_a\{[M_o]-\phi q\}q_m^2}+\frac{t}{q_m} \tag{8.13}$$

Similarly, the rate of desorption of heavy metals from sludge is proportional to the adsorption density q and is expressed as follows:

$$-dq/dt = k_d q \tag{8.14}$$

Integrating Eq. (8.14) with initial condition t = 0 and $q = q_m$ yields

$$\ln(q/q_m) = -k_d t \tag{8.15}$$

By plotting $\ln(q/q_m)$ against t, k_d can be determined from the slope. During the desorption process q gradually decreases. Therefore, the amount desorbed $q_d=q_m-q$. Substituting in Eq. (8.15) q_d can be expressed as follows:

$$q_d = q_m\{1-\exp(k_d t)\} \tag{8.16}$$

This implies that $q_d = q_m$ at infinite time.

8.1.2 The Elovich Equation

Besides mass transfer (diffusion) and pseudo first or second order controlled kinetic models of adsorption, attempt has also been made to analyze kinetic data using the Elovich equation which was originally derived for kinetics of chemisorption at gas solid interface [271]. The equation in its simplest form is given by

$$dq/dt = a\exp(-bq) \tag{8.17}$$

where,

a = constant relating to the initial velocity of reaction

b = constant relating to activation energy

The equation was derived for uniform or near uniform surface, the essential characteristic of which is that the energy of activation for adsorption increases linearly with surface coverage. The non-uniform surface is considered to be divided into several patches of uniform elements wherein activation energy increases with surface coverage linearly. Integrating Eq. (8.17) with the initial condition q = 0 at t = 0 we have

$$q = \frac{1}{b}\ln\left\{\frac{(t+k)}{k}\right\}$$

where,

k = 1/ab so that

q = 1/b ln(t + 1/ab) - 1/b ln(1/ab) (8.18)

In a gas- solid system the kinetics of the adsorption of a gas generally exhibits pre-Elovichian stage. This leads to the boundary condition of $q = q_o$ at t = 0 giving the integrated form of equation (8.17) as follows

$$q = \frac{1}{b}\ln\left\{\frac{(t+k')}{k'}\right\} + q_o \tag{8.19}$$

where, $k' = e^{bq_o} / ab$

To interpret the interaction between adsorbed species Allen and Scaife [272] proposed the following generalized equation

$$dq/dt = a_\gamma \exp\{-b\,(q - \gamma q_o)\} \tag{8.20}$$

which on integration within the boundary conditions as above gives

$$q = \frac{\ln\{(t+k'')/k''\}}{b} + q_o \tag{8.21}$$

where, $k'' = \frac{1}{a_\gamma b}\exp\{-(\gamma - 1)bq_o\}$ and γ is a constant to account for the interaction between the adsorbed species. The treatment of kinetic data using Eqs. (8.18) - (8.21) for a chemisorption process may show discontinuities if more than one type of surface reactions occur. However, in most cases a plot of ln{(t + k) / k} against q exhibits excellent continuity over a wide range of adsorption or time.

The Elovich equation, in its different forms, has not been used much in the study of the kinetics of chemisorption of inorganic ions from aqueous solution, though it has a great potential of providing valuable information on surface reactions of complex nature . However, for this purpose generation of accurate rate data at short intervals of time is necessary. While the constant 'a' can provide activation energy value in the initial stage of adsorption and desorption processes, constant 'b' can provide overall activation energy value of the process, if linear Elovich plot is obtained over a wide period of time. An interesting example of the application on the adsorption of silicate on gibbsite at pH 9.2 has been cited by Hingston [273]. It has been observed [273] that activation energy values determined from the slopes of pseudo first order plots for the initial stage of adsorption at different temperatures agree well with those determined from Elovich plots.

8.2 Modeling Under Equilibrium Conditions

Basically, the equilibrium state is the condition at which the rates of adsorption and desorption are so slow that no change in the concentration of adsorbate is observed even after a long interval of time (say, days). Theoretically, equilibrium is defined when the chemical potential of the liquid phase is equal to the chemical potential of the solid phase at constant temperature. Adsorption under such condition can be modeled from two different approaches. The empirical model approach aims at empiric description of experimental data, while the semi-empirical or the mechanistic model is based on the basic aspects concerning the mechanism of adsorption. In the empirical approach, the model equation is generally derived on the basis of experimental data or observation so as to satisfy the model equation. In such a case, the model equation should be as simple as possible with minimum number of adjustable parameters relating to experimental variables. Nevertheless, empirical models are useful if the objective of the investigation is to present empirical description of the adsorption system.

In the mechanistic or semi-empirical approach the model equation is derived *a priori* based on certain basic / fundamental aspects of adsorbate - adsorbent interaction in aqueous solution of electrolyte such as charge and mass balance of different components etc. Consequently, the number of adjustable parameters are higher than those envisaged in empirical models, as the mathematical expressions include or encompass larger number of experimental variables. Such models are therefore more valid than the empirical models [221]. However, attempts are now being made to extend or cover additional aspects such as competitive adsorption, surface heterogeneity of solid phase etc. According to Bradle [221], the main difference between the two approaches is that mechanistic models include electrostatic term, while the empirical models do not.

8.2.1 Empirical Models

Empirical models are generally based on distribution or partition of adsorbate between the solid or liquid phases under equilibrium condition and at constant temperature. The distribution coefficient (k_d) is defined as $k_d = q_j / c_j$, where q_j is the species j adsorbed onto the solid surface (moles per unit mass or area) and c_j is the concentration (moles per unit volume) of species j present in the aqueous phase in equilibrium with the solid phase. For the adsorption of inorganic ions the distribution coefficient is highly dependent on pH, composition and concentration of background electrolyte,

concentration of competing ions, concentration of species j, concentration of dissolved carbonate etc. It is also assumed that all the dissolved species of j have equal affinity for the adsorbent surface and vice versa. Considering the effect of all these factors the variation in the value of k_d can be several orders of magnitude [274]. Therefore, k_d has limited value in predicting the adsorption behavior of both adsorbent surface and dissolved species j unless the conditions of the system are specified.

In the earlier literature partition coefficient has been widely used, especially in modeling adsorption related phenomena in various geochemical systems [275-278]. Indeed, k_d values estimated from laboratory experiment have been applied in modeling of natural water systems though with limited success [276,279, 280]. This is primarily because of the extremely crucial role played by release of H^+ or OH^- during the adsorption of inorganic ions . However Johnson [281] has claimed to obtain satisfactory agreement between laboratory experiment and field trial for the adsorption of Cu^{2+} and Zn^{2+} onto suspended particles in a river water contaminated with acid mine drainage .To make such models more realistic attempts have been made to include site density (N_s) and proton release coefficient (χ) as adjustable parameters besides k_d. Since N_s can be determined experimentally with reasonable accuracy, the other two need to be evaluated or adjusted. Besides being interrelated, these two parameters are highly dependent on pH and solution conditions. Of the two methods used to determine k_d and χ , Kurbatov plot is the oldest and the most widely used one [256]. The general equation for adsorption of metal ion M^{n+} onto an uncharged adsorbent surface (SOH) may be expressed as

$$M^{n+} + \chi(SOH) \Longleftrightarrow M(SOH)_\chi^{(n-\chi)+} + \chi H^+ \qquad (8.23)$$

Neglecting activity coefficient, the mass law expression of the above reaction is given by

$$K_e = \frac{\left[H^+\right]^\chi \left[M(SOH)_\chi^{(n-\chi)+}\right]}{\left[M^{n+}\right]\left[SOH\right]^\chi} \qquad (8.24)$$

Taking logarithm and rearranging

$$\log \frac{\left[M(SOH)_{\chi}^{(n-\chi)+}\right]}{\left[M^{n+}\right]} = \log K_e + \chi \log[SOH] + \chi\, pH \tag{8.25}$$

or ,

$$\log \frac{\left[M^{n+}\right]_{sorbed}}{\left[M^{n+}\right]_{solution}} = Z + \chi\, pH \tag{8.26}$$

or , $\log k_d = Z + \chi\, pH$ (8.27)

where, $Z = \log K_e + \chi \log [SOH]$

Eq. (8.27) clearly shows that k_d is directly related to pH so that the slope value of linear plot of log k_d against pH gives the proton release coefficient χ. However, it has been generally observed that linear relationship is followed only over a limited range of pH so that χ determined by the above method has limited applicability. Further, the values of χ obtained by the Kurbatov method are not only fractional, but also show unusually low values [282 -284]. Clearly, the macroscopic values have little significance in describing microscopic reaction stoichiometry [280,285].The experimentally observed macroscopic value is the net result of adsorption reactions, plus the adjustment of surface ionization and ion - pair formation under a new equilibrium condition. Therefore, difference between experimentally determined and predicted values of proton release coefficient has been observed by several authors and this aspect has been discussed in more details by Dzombak and Morel [27].

The other method, which is claimed to be a better one, has been suggested by Perona and Leckie [286]. This is based on the graphical subtraction of isotherms (in log - log plots of adsorption vs. concentration) constructed at different pH values with close intervals not exceeding 0.5. From these isotherms, concentration values are determined for a constant adsorption density. For each data set one can calculate χ from the following relationship

$$\chi = (\Delta \log c_j / \Delta pH)_{qj} \tag{8.28}$$

Various aspects of the application of the method has been discussed by Honeyman and Leckie [285]. The advantage of the method is that it takes into account the effect of adsorption density on the proton release coefficient. However , the present author did not observe any improvement in the values of χ which , like the Kurbatov method, also showed unusually low values for the adsorption of cations on δ^{-} MnO_2 [284].

8.2.1.1 *Adsorption Isotherms*

The relationship depicting the distribution of adsorbate between solid and liquid phases at constant temperature is also called adsorption isotherm. The first isotherm for adsorption of gas on homogeneous planer surface was put forward by Langmuir in 1916. Langmuir's theory was based on several premises which are as follows:

1) All the adsorption sites have equal affinity for adsorbate.

2) The bond with which adsorb ate molecules are attached to the adsorbent surface may be either physical or chemical, but should be sufficiently strong enough to prevent lateral movement of molecules along the surface. This is known as localized adsorption.

3) There should not be any mutual interaction between the adsorbate molecules.

4) At equilibrium the rate of adsorption is equal to the rate of desorption. In other words, the number of atoms or molecules adsorbing onto per unit surface per unit time is equal to those leaving the unit surface per unit time.

If the adsorbing gas is in equilibrium with the adsorbent and if the adsorbent surface is homogeneous as mentioned above, the concentration of adsorbed gas is same all along the surface. The concentration of adsorbed gas may be expressed either in moles/gm of adsorbent (a) or in moles / unit volume (c) or in moles / unit area (α). Let θ be the fractions of the total surface covered by the adsorbed gas and θ_o be the fraction of the surface vacant for adsorption then $\theta + \theta_o = 1$ and

$$\theta = c / c_m = a / a_m = \alpha / \alpha_m \tag{8.29}$$

where c_m, a_m and α_m are the values of different forms of expressions of surface concentrations covered completely with a monolayer of gas (maximum surface concentration). Similarly, the corresponding fraction of vacant sites is

$$\theta_o = c_o / c_m = a_o / a_m = \alpha / \alpha_m \tag{8.30}$$

c_o , a_o and α_o are the corresponding values of different expressions for concentration of vacant sites . Since the sum of occupied and vacant sites on the surface is equal to total number of sites

$$\alpha + \alpha_o = \alpha_m \tag{8.31}$$

or , in accordance with the definition

$$\theta + \theta_o = 1 \tag{8.32}$$

The equilibrium constant for adsorption of an ideal gas is given by

$$K = \frac{[adsorbed\ complex]}{[bulk\ concn. of\ gas],[concn. of\ vacant\ sites]} = \frac{\alpha}{p\alpha_o} = \frac{\theta}{p\theta_o} \tag{8.33}$$

Combining Eqs. (8.31) ,(8.32) and (8.33)

$$K = \frac{\alpha}{p(\alpha_m - \alpha)} = \frac{\theta}{p(1-\theta)} \tag{8.34}$$

From Eq . (8.34) we get

$$\theta = \frac{K\,p}{1+K\,p} \tag{8.35}$$

and

$$p = \frac{\theta}{K(1-\theta)} \tag{8.36}$$

Similarly,

$$\alpha = \frac{\alpha_m K\,p}{1+K\,p} \tag{8.37}$$

or

$$a = \frac{a_m\,K\,p}{1+K\,p} \tag{8.38}$$

All the above expressions are the different forms of Langmuir isotherm. At very low concentration Kp<<1 , the above equations reduces to well known Henry's equation which states that at low pressure (or concentration) amount adsorbed is proportional to the pressure i.e., $\alpha \rightarrow \alpha_m$ K p or $a_m \rightarrow a_m$ K.p. At high pressure or concentration K.p >>1, the isotherm reach maximum coverage of surface i.e., $\alpha \rightarrow \alpha_m$ or $a \rightarrow a_m$ or $\theta \rightarrow 1$.

For adsorption of inorganic species from solution phase on oxides or other inorganic substrates, the isotherm may be written as

$$q_j = b\left(\frac{K c_j}{1+K c_j}\right) \tag{8.39}$$

where q_j is the amount of species j adsorbed , c_j is the equilibrium concentration of species j in solution and K and b are the constants . The value of b is the upper limit of adsorption i.e., a_m for adsorption of gas in Eq. (8.38) . Equation (8.39) can be reduced to the following linear forms

$$\frac{c_j}{q_j} = \frac{1}{bK} + \frac{c_j}{b} \tag{8.40}$$

$$\frac{q_j}{c_j} = Kb - Kq_j \tag{8.41}$$

$$\frac{1}{q_j} = \frac{1}{b} + \frac{1}{bK}\frac{1}{c_j} \tag{8.42}$$

From the above linearized forms of the Langmuir isotherm both b and K can be obtained by plotting either c_j / q_j vs. c_j or q_j / c_j vs. q_j or $1/ q_j$ vs. $1/c_j$.

Another well known adsorption isotherm used frequently by surface chemists is the Freundlich Isotherm which may be expressed in either of the following forms

$$q_i = m\,(c_j)^n \tag{8.43}$$

$$q_i = m\,(c_j)^{1/n} \tag{8.44}$$

where , m and n are adjustable constants ; the value of n varies between 0 and 1 . The parameters are estimated by plotting log q_j against log c_j when the resulting straight line gives log m as intercept and n as slope. The units of q_j and c_j may be expressed either in mg or moles per unit mass and per unit volume respectively. However, some authors like Kinniburgh [287] have expressed Freundlich isotherm as follows:

$$q_i = (m\, c_j)^n e \tag{8.45}$$

where e is the error in constant and absolute term. By taking logarithm on both sides and plotting log q_j against log (c_j), the relevant constants can be obtained.

Although Langmuir and Freundlich isotherms were proposed more than ninety years ago, they still remain the most widely used isotherm equations. Besides inherent simplicity their success is attributed also to their ability to fit a wide variety of adsorption data. The two adjustable parameters required for fitting each of the two isotherms are determined by the weighted linear regression method. This however, tends

to ignore, in particular, the distribution of error implied, which can lead to biased estimates of the parameters. However, the values of the parameters vary with the linear form of the Langmuir isotherm used for regression analysis. This is because, each of the three linear transformation changes the original error distribution rather for better or for worse [288]. The best transformation is not necessarily the one which gives the highest correlation coefficient, but rather the one in which the resulting error distribution closely matches the "true" error distribution [287]. Some authors [288] have favored the plot of q_j / c_j vs. q_j (Eq. 8.41) over the other two forms, because this form tends to exaggerate deviations from the fitted equation and therefore provide caution against overoptimistic interpretation of the goodness of the fit. Subsequently, the different forms of the Langmuir isotherm were compared with independent nonlinear least square [NLLS] method of data fitting [289]. It was observed by Kinniburgh [287] that NLLS gave the results which were most unbiased with minimum variance.

Most of the authors dealing with the adsorption of inorganics on well defined or well characterized adsorbents have not shown as such the linear plots of any of the three forms of the Langmuir isotherm. However, the applicability of the Langmuir isotherm has been demonstrated for the adsorption of heavy metals on soil or soil clays [290-293]. Since the adsorption isotherms are extremely sensitive to the variation in pH, the construction of isotherm not only requires constant temperature but also at constant pH. Some hydrous oxide like δ - MnO_2 goes on releasing H^+ in solution even for days during the adsorption of heavy metals onto its surface [294]. Kanungo and Parida [294] therefore carried out adsorption isotherm of Cu^{2+} on different crystalline forms of MnO_2 in the presence of acetate buffer at pH 4.4. Langmuir plots for the isotherms exhibit linear behavior over a wide range of concentration for δ-MnO_2 as adsorbent, but exhibit some deviations for other polymorphic forms, both at low and high equi-

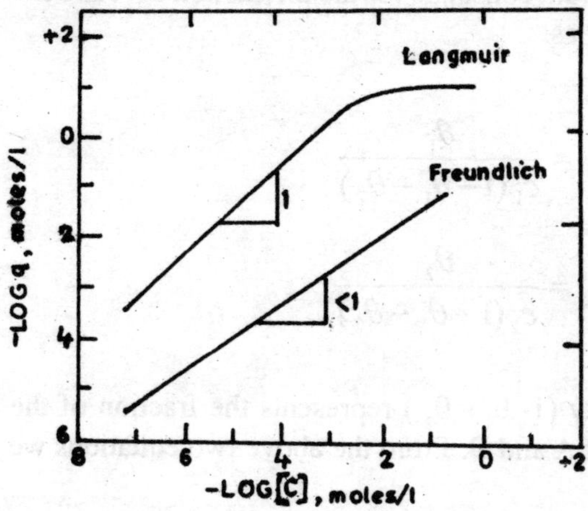

Fig. 8.1 Schematic representation of the log - log plot of concentration of free ion vs. amount adsorbed showing the difference between the Langmuirian and the non-Langmuirian (Freundlich) types of adsorption isotherms. The figures given for x- and y-axes are meant to illustrate typical ranges and not actually used in the plots.

librium concentrations. This has been attributed to the variation in the affinity of the sites i.e., surface heterogeneity on the faces of the different crystalline structures [295,296]. Now a days, many authors do not use any of the above three forms of Langmuir isotherm as such. Instead, Freundlich type plot is used to distinguish between Langmuirian and non - Langmuirian isotherm. For an ideal Langmuirian isotherm the slope value (n) is close to unity over a wide range of concentration and then falls sharply to a value close to zero i.e., parallel to x - axis . Adsorption at this point gives the value of maximum adsorption as envisaged in the Langmuir equation. Figure 8.1 illustrates Freundlich type plots showing deviations from the Langmuir isotherm. Rubin and Mercer [297] have made an excellent review on the application of Langmuir isotherm to the adsorption of both inorganic and organic ions on inorganic oxides and activated carbon. Though it is relatively an old literature, the essence of discussion is valid even today.

The adsorption of two different adsorbates on an adsorbent can also be fitted to Langmuir equation after modifying it to include their competitive effects. The modified equation is called competitive Langmuir equation which may be derived as follows.

Let c_1 and c_2 be the concentrations of two dissolved ionic species A and B and let q_1 and q_2 be the fractions of adsorbent surface covered by them respectively. Like Eq.

(8.34) the equilibrium constants for the interaction of A and B with adsorbent surface may be expressed as

$$K_1 = \frac{\theta_1}{c_1(1-\theta_1-\theta_2)} \tag{8.46}$$

$$K_2 = \frac{\theta_2}{c_2(1-\theta_1-\theta_2)} \tag{8.47}$$

where, the quantity $(1-\theta_1-\theta_2)$ represents the fraction of the surface lying vacant after adsorption of A and B. From the above two equations we get

$$\theta_1/\theta_2 = K_1 c_1 / K_2 c_2 \tag{8.48}$$

which suggests that fractional coverage is proportional to the product of equilibrium constant and concentration of individual component species. Combining Eq.(8.48) with Eq.(8.46) or with Eq. (8.47) we obtain

$$\theta_1 = \frac{K_1 c_1}{1+K_1 c_1+K_2 c_2} \tag{8.49}$$

$$\theta_2 = \frac{K_2 c_2}{1+K_1 c_1+K_2 c_2} \tag{8.50}$$

Since, by definition $\theta_1 = q_1/q_m = q_1/b$ and similarly, $\theta_2 = q_2/b$, the above two equations may be rewritten as follows

$$q_1 = \frac{bK_1 c_1}{1+K_1 c_1+K_2 c_2} \tag{8.51}$$

$$q_2 = \frac{bK_2 c_2}{1+K_1 c_1+K_2 c_2} \tag{8.52}$$

A simple and typical example is the competition for surface sites between H^+ and metal ion when adsorption is carried out at different pH values in the presence of different concentration of indifferent electrolyte solution. At low pH or at high electrolyte concentration adsorption of metal ion is low because of competition from H^+, but at higher pH or at lower electrolyte concentration adsorption of metal ion is predominant compared to H^+. Generalizing Eqs. (8.51) and (8.52) for the adsorption of ith adsorbate in a system containing j number of species, we have

$$q_i = \frac{b K_i c_i}{1 + \sum_{j=1}^{j} K_j c_j} \tag{8.53}$$

Besides competitive adsorption, another source of deviation of Langmuir isotherm from its ideal behavior is the heterogeneity of adsorption sites. This means that the interaction between adsorbing ion and surface site cannot be described by a single affinity parameter. There are two different ways by which the effect of surface heterogeneity can be included into modified surface single-site Langmuir isotherm [221]. First, let us consider a discrete number (say S) of different types of sites which are characterized by different concentrations and affinity constant for the adsorbate. Adsorption is expressed as the sum of adsorption on S types of sites, each one following Langmuir isotherm. The resultant multisite isotherm may be expressed as follows

$$q_i = \sum_{j=1}^{S} \frac{b K_i c}{1 + K_i c} \tag{8.54}$$

with 2S adjustable parameters and j referring to each adsorption site [297]. In the second approach the surface is considered as an assembly of different groups of sites, each characterized by its own distinct affinity constant which is distributed continuously according to "site affinity distribution function (SADF). The overall isotherm can then be derived by integrating each group of site (or local isotherm) along SADF. If θ_t (c) is the overall isotherm and θ (K, c) is the "local" or single site isotherm, the former can be obtained as follows

$$\theta_t(c) = \int \theta(K,c) f(K) dK \tag{8.55}$$

where f(K) is the SADF and f(K)dK is the fraction of sites covered within the range of K and K + dK and c is the concentration of the adsorbate.

Application of Eq .(8.55) for predictive purposes requires full knowledge of both the functions of θ (K , c) and f(K) .Taking θ (K , c) as Langmuir equation for multi component system, factor K which is related to heat of adsorption [$K = K_o \exp -(\Delta H_{ads} / RT)$], Eq . (8.55) can be solved analytically for three types of distribution function f (K) resulting in the following three isotherm equations [299-301].

Langmuir - Freundlich (LF):

$$\theta_t(c) = \frac{(K c)^m}{1+(K c)^m} \text{ or } q = \frac{(K c)^m b}{1+(K c)^m} \tag{8.56}$$

Generalized Freundlich (GF):

$$\theta_t(c) = \left(\frac{K c}{1+c}\right)^m \quad or \quad q = b\left(\frac{K c}{1+c}\right)^m \tag{8.57}$$

To'th:

$$\theta_t(c) = \frac{K c}{\left[1+(K c)^m\right]^{1/m}} \quad or \quad q = \frac{b K c}{\left[1+(K c)^m\right]^{1/m}} \tag{8.58}$$

These equations are characterized by three adjustable parameters, b, K and m. While the significances of b and K remain as usual , m represents the heterogeneity index ranging from 0 to 1, corresponding to flat and to very sharp distribution . For m = 1 all composite isotherms will revert to single site Langmuir isotherm. Van Riemsdijk et al [114] have used the equations to describe σ_0 - pH curve for mono component system. They have also attempted to modify these equations under certain conditions to derive σ_0 - pH curves for multi component system using both 1 pK and 2 pK models . Since this subject is confined in specific theoretical areas , interested readers may consult the original articles .

Kinniburg [287] has presented several more multisite adsorption isotherms which have been claimed to be applicable not only over a wide range of concentration , but also for a wide variety of adsorption data . These are as follows:

Redlich - Peterson (RP)

$$q = \frac{b K c}{1+(K c)^m} \tag{8.59}$$

Multisite Langmuir (LK) Already given in Eq .(8.54)

Temkin (TK)

$$q = a \ln(K_{ad} c) \tag{8.60}$$

where, a is constant related to the affinity of the surface towards adsorbate, K_{ad} is a constant related to heat of adsorption.

Dubinin - Radushkevich (DR) :

$$\log q = -m [(\log K c)]^2 + \log b \tag{8.61}$$

The equation has been extensively used by many investigators to analyze the adsorption behavior of both cations and anions (see Chapters 10 and 11). The DR equation in its modified form may be written as

$$\ln q_{ads} = \ln q_m - \beta\varepsilon^2 \tag{8.62}$$

where, q_{ads} is the amount adsorbed (mg/g), q_m is the maximum adsorption capacity (mg/g), β is the activity coefficient related to the mean sorption energy $(mole/kJ)^2$, ε is the Polanyi potential which is given by

$$\varepsilon = RT \ln (1 + 1/C_e) \tag{8.63}$$

where, C_e is the equilibrium concentration (mg/l). The plot of $\ln q_{ads}$ vs. ε^2 gives straight line whose slope value gives β and intercept gives q_m. The sorption potential is independent of temperature, but depends upon the energy of interaction between sorbent and sorbate and is given by

$$E = 1/(-2\beta)^{0.5} \tag{8.64}$$

All the above isotherm equations have been critically examined by Kinniburg [287] from experimental adsorption data and the findings are summarized below.

Both RP and LF isotherms are essentially Freundlich isotherm, but with different asymptotic properties which mean that both give linear log q vs. log c plots. While LF isotherm approaches an adsorption maximum at high concentration, RP isotherm approaches linear behavior at low concentration. For To'th isotherm log q vs. log c gives a continuous smooth curve throughout the intermediate range of concentrations. The DR isotherm is a parabola in log q - log c plot that often fit data well in the intermediate range of concentration but its asymptotic properties are unsatisfactory compared to the other isotherms. However, they are largely removed in the modified DR isotherm which closely resemble To'th isotherm in terms of fitting ability. LF, To'th and RP isothems are converted to Langmuir isotherm when the heterogeneity parameter m is set to unity, whereas the DR isotherm is converted to Langmuirian at m=0.25. Temkin isotherm is also applicable to heterogeneous surface on which adsorption energy decreases as the surface coverage increases. This is also a very widely used isotherm.

All the above isotherm equations were derived for mono component adsorption. However, these can also be used to describe binary (or competitive) adsorption. When all the adsorption sites remain occupied all the time, this can be done quite simply by replacing the concentration terms in thé equations by a concentration ratio of the binary components.

8.2.2. Mechanistic or Semi Empirical Models

As stated earlier, the main difference between empirical models and mechanistic models is that the formers do not take into account the electrostatic interactions between free ions in solution and the charged solid surface. The scope of the present discussion is, in reality, concerned with this type of interaction. The affinity parameters determined from empirical models do not consider the modifications of the surface that result from the interaction between ions and electrified surface. The term "mechanistic model" therefore refers to all those models which describe adsorption by taking electrical interaction between adsorb ate and adsorbent surface into consideration.

In Chapter 3 we have discussed the origin of electrical charge at inorganic adsorbent / water interface and in Chapter 5 various models of the electrical double layer envisaged at the interface have been discussed also. The adsorbents include inorganic oxides, hydroxides, clays and other alumino - silicate compounds. The adsorption models vary with the placement of cations or anions in different potential of the double layer, depending upon the nature and the amount of charge carried by the ions. Adsorption generally takes place through two different mechanisms: (i) Ion - exchange which is predominantly electrostatic in nature and (ii) surface complexation which involves chemical interaction involving charge transfer with the surface. Several mechanistic models such as constant capacitance model (CCM), diffuse double layer model (DDLM), triple layer model (TLM), four layer model (FLM), CD - MUSIC model etc have been proposed. Together with complex formation with adsorbent surface, these electrostatic models present a powerful approach to describe the macroscopic behavior of adsorption on charged surfaces of inorganic hydr(oxides), soils, clays etc. In the surface complexation approach, adsorption of ions is viewed as analogous to the formation of coordination complexes in homogeneous solution, ions bind to the functional groups present at the surface and these reactions can be described by mass law equations. Like the nomenclature used in coordination chemistry "inner sphere complexes" are formed when adsorption takes place by direct coordination of adsorbing ions with the surface functional groups; and "outer sphere complexes" are formed when adsorption occurs little away from the surface by indirect bonding such as hydrogen bonding. However, surface complexation reactions are distinguished from reactions between soluble species in that surface charge and hence electrostatic energy of interaction varies with the extent of adsorption and with solution conditions such as pH and ionic strength. Such variation in surface charge due to changes in the electrostatic environment is taken into account by applying the coulombic correction factor derived from EDL, to surface complexation mass law constants. The apparent binding constants of ions with surface functional groups as envisaged by the simple mass law equations are empirical parameters, but are related to thermodynamic constants when rational activity constants of the surface species are duly incorporated in the mass law equations.

In the light of the above basic tenets of surface complexation, we shall discuss below the various mechanistic models mentioned in the preceding paragraph. The models are distinguished by differences in their respective molecular hypotheses. Each model represents a particular interfacial structure that results in the consideration of various kinds of surface reactions and eletrostatic correction factors attached to mass law constants. All the models reduce to a few sets of simultaneous equations that can be solved numerically using certain adjustable parameters so as to fit the experimental adsorption data. The sets of equations include:

(1) Mass law equations for surface reactions under consideration.
(2) Mole or mass balance equations for the surface OH groups.
(3) Equation for the computation of surface charge.
(4) Equation that describe the relationship between charge and potential of the electrical double layer.

8.2.2.1 Constant Capacitance Model

The basic features of the model constituting the electrical double layer have been discussed in Chapter 5. The model was developed by Schindler et al and Stumm et al in the 1970s [39,111,302]. The molecular hypotheses of the model are as follows:

(1) The charges resulting from amphoteric dissociation of surface OH groups reside on the surface plane. The two apparent equilibrium constants (K_{a1} and

K_{a2}), which describe the surface dissociation reactions, are valid only for a particular ionic strength [249].

(2) Only one plane in the interfacial region is considered i.e., surface plane where H^+, OH^- or any other ions adsorbed specifically as inner sphere complexes are present in this plane. For electrical neutrality, ions of background electrolyte form diffuse layer of counter ions at a distance which is generally equal to the average radius of the ions.

(3) The constant ionic medium reference state determines the activity coefficients of the aqueous species in the equilibrium constant and no surface complexes are formed with ions of the background solution.

(4) The standard or reference state refers to the conditions where the surface spe cies carry zero charge.

(5) Potential drops linearly with distance from the surface. The ratio between charge and potential gives capacitance value which is dependent on ionic strength.

As far as the application of CCM to the adsorption of inorganic ions on well characterized adsorbent is concerned, two kinds of binding sites have been proposed [233]. One kind consists of weakly acidic sites (SH) which can take part in ion exchange reaction with M^{2+} or Na^+. The other kind of site is the usual surface hydroxyl groups (SOH) which form complexes like SOM^+ and $(SO)_2M$ and bind Na^+ as outersphere complexes. The CCM is considered as the limiting case of basic Stern model for high ionic strength solutions ($I \geq 0.1$ M), although it is very often applied to adsorption from solutions of low ionic strength [298]. The molecular picture of the model is same as that illustrated in Fig. 5.5, except that surface layer may contain either adsorbed cation or adsorbed anion as the case may be.

The CCM is the simplest of all the surface complexation models with theoretically one adjustable parameter i.e., capacitance value. However, for certain adsorbents whose interfacial properties are not clearly defined such as soils or its constituents, minerals etc. it may be necessary to adjust three parameters namely, K_{a1}, K_{a2} and C for each ionic strength. It can only be used for the description of specifically adsorbed ions and is unable to describe changes in the adsorption of ions with changes in the ionic strength of the medium. The four sets of basic equations (as stated in the preceding section) that are necessary to formulate constant capacitance model are given below as typical example for the adsorption of either anion or cation. This has already been presented partially in section 6.4.2.

Mass law equations

Amphoteric surface dissociation

$$SOH_2^+ \longleftrightarrow SOH^o + H_s^+ \quad ; \quad K_{a1}^{int} \qquad (3.1)$$

$$SOH^o \longleftrightarrow SO^- + H_s^+ \quad ; \quad K_{a2}^{int} \qquad (3.2)$$

Cation adsorption

$$SOH^o + M_s^{2+} \longleftrightarrow SOM^+ + H_s^+ \quad ; \quad K_M^{int} \qquad (8.65)$$

Anion adsorption

$$SOH^o + A_s^{2-} + H_s^+ \longleftrightarrow SA^- + H_2O \quad ; \quad K_A^{int} \qquad (8.66)$$

Coulombic corrections

$$[H_s^+] = [H^+] \exp(-e\,\psi_0 / kT) \qquad (3.10)$$

$$[M_s^{2+}] = [M^{2+}] \exp(-2e\,\psi_0 / kT) \qquad (8.67)$$

$$[A_s^{2-}] = [A^{2-}] \exp(2e\,\psi_0 / kT) \qquad (8.68)$$

Mole balance equations

$$[SOH]_T = [N_s] = [SOH^o] + [SOH_2^+] + [SO^-] + [SOM^+] + [SA^-] \qquad (8.69)$$

$$[M^{2+}]_T = [M^{2+}] + [SOM^+] \qquad (8.70)$$

$$[A^{2-}]_T = [A^{2-}] + [SA^-] \qquad (8.71)$$

$$[H^+]_T = [H^+] - [OH^-] + [SOH_2^+] - [SO^-] - [SOM^+] + [SA^-] \qquad (8.72)$$

$$= [H^+] - [OH^-] + \sigma_p A S / F \qquad (8.73)$$

where, σ_p is the net charge due to surface proton, A is the surface area (m^2/g), S is the concentration of solid (g/l) and F is the faraday constant. The surface species assumes positive charge when a proton is added to it and assumes negative charge when a proton is removed from it.

Net surface charge

Net surface charge = charge due to total adsorption of cation and anion + net charge due to surface H^+

$$\sigma_o = F/AS\ \{2[SOM^+] - 2[SA^-] + \sigma_p\}$$

$$= F/AS\ \{[SOH_2^+] + [SOM^+] - [SO^-] - [SA^-]\} \quad (8.74)$$

Charge - potential relationship

$$\sigma_o = C\,\psi_o \quad (8..75)$$

The possible ion exchange reaction between weak site (XH) and M^{2+} is not included in the complexation reaction as the equilibrium reaction does not involve contribution from electrical double layer. Its contribution to the overall adsorption is also negligibly small. Ion exchange adsorption will, however, be discussed elaborately in chapter 9. The CCM has been widely applied in the modeling of the adsorption of cations [39,111,303] and anions [304-306] on hydrous oxides.

8.2.2.2. Diffuse Double layer Model (DDLM)

The model was first introduced by Stumm et al [113,125,308] and developed by Dzombak and Morel [39,245] for application over a wide range of adsorbate / adsorbent ratio. The molecular hypotheses of the model are as follows:

(1) There are two planes in the interfacial region: (i) The potential determining ions like H^+ and OH^- as well as all specifically adsorbed ions are included in the surface plane and (ii) all counter ions required to preserve electrical neutrality are placed in the diffuse layer plane.

(2) The Gouy - Chapman theory is applicable to derive charge - potential relation ship in the diffuse layer. The electrical potential at the beginning of the diffuse layer is equal to the surface potential. The molecular picture of the model is same as that presented in Fig. 5.3.

(3) The indifferent or background electrolyte ions do not form any complex with the surface.However, unlike CCM this model is applicable at variable ionic strength.

(4) The infinite dilution is used as Reference State for aqueous species; a zero surface charge reference state is used for surface species.

Experimental data required for model fitting are N_s, surface area K_{a1}^{int} and K_{a2}^{int}. However, the last two parameters may be considered as adjustable constants that are applicable to all ionic strength. Mass and mole balance equations given under CCM are similar, except that instead of Eq .(8.75) Eq. (5.10) is used in DDLM.

Constant pH isotherm (log - log plot of adsorption vs. concentration) of cation adsorption suggests that at lower pH and at low equilibrium concentration of cation, adsorption is proportional to concentration (slope = 1) and after a certain concentration the slope value sharply reduces to almost zero. This is a typical Langmuirian behavior where the initial linear portion represents strong site which first gets saturated by adsorption leaving weak sites for further adsorption. As pH value increases the range of initial linear region decreases as strong or high affinity sites get saturated quickly at lower equilibrium concentration, leaving wider range of concentration for weak sites. The schematic representation of such plots for adsorption of cations on hydrous oxides is shown in Fig. 8.2. An opposite behavior may be noted for the adsorption of anions and is depicted in Fig. 8.3. The two dotted lines connecting the inceptions of two slope values (shoulder) and the final plateau regions in the isotherms indicate densities of strong (type 1) and weak (type 2) sites respectively . However, in the case of anion adsorption the demarcation of switch over from strong to weak sites is somewhat blurred especially with decrease in pH when adsorption of anions increases sharply. Therefore, for the adsorption of anion generally one surface site is considered in the DDLM. However, generally more than one surface species are required to fit anion adsorption with respect to the variation of pH. This is because many anions of interest are di or triprotic acids whose protonation / deprotonation changes with pH. This aspect will be dealt more elaborately in connection with the adsorption of anions on inorganic adsorbents in chapter 11.

In the light of the above discussion on the adsorption isotherms of cations, it is apparent that at least at least two types of surface sites are required to model cation adsorption-

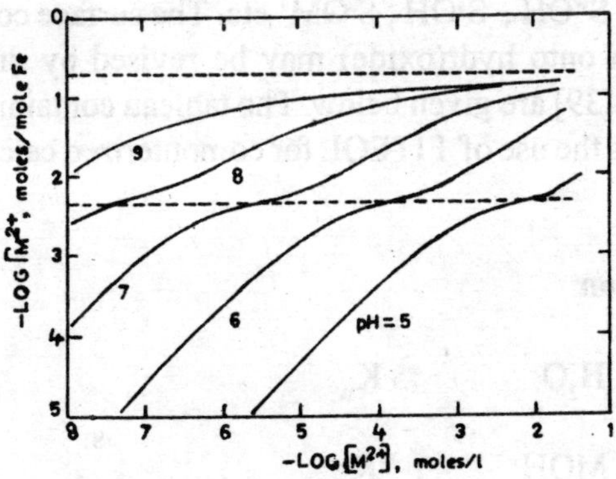

Fig. 8.2 Constant pH isotherms of the adsorption of a hypothetical divalent cation (M^{2+}) on hydrous ferric oxide. The upper and lower dashed indicate total (0.174 moles/mole Fe) and strong (0.004 moles/mole Fe) binding sites, respectively. The difference between the two values gives the density of weak binding sites. Calculated and redrawn after data given in ref. 27.

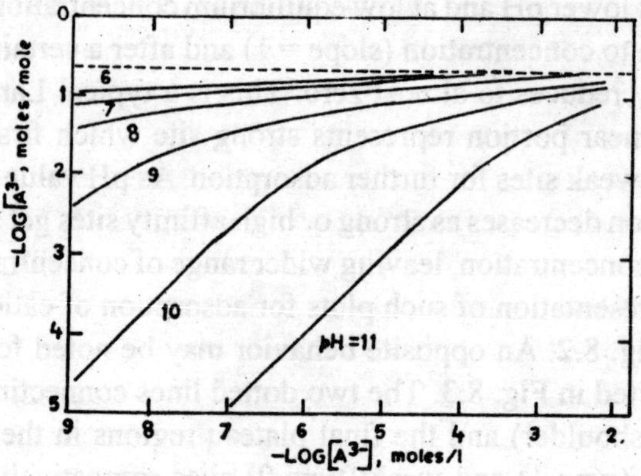

Fig. 8.3 Constant pH isotherms of the adsorption of a hypothetical anion (A^{3-}) on HFO. The dashed line indicates the density of total surface sites (0.174 moles/mole Fe). For anion adsorption no distinction is made strong and weak sites. Calculated and redrawn after the parameters given in ref.27.

tion. However, usually one kind of surface stoichiometry, namely SOM^+ (equivalent to $SOMOH_2^+$) is required. The hydroxylated species is considered at high pH and high M^{2+}/ sorbent ratio when surface precipitation is the ultimate fate of adsorption [39]. Both the surface types are assumed to have the same surface protonation and deprotonation constants i.e., K_{a1}^{int} and K_{a2}^{int} values. The concentration of weak or type 2 sites is typically much higher than strong or type 1 sites. To distinguish between strong and weak sites, superscripts 's' and 'w' are used respectively for each surface species, e.g., S^wOH, S^sOH, S^sOM^+ etc. The surface complexation reactions for cation adsorption onto hydr(oxide) may be revised by differentiating the two types of surface sites [39] are given below. The tableau containing the stoichiometric coefficients facilitates the use of FITEQL for computerized calculation is also shown in Table 8.1.

Mass law equation

$$OH^- + H^+ \longleftrightarrow H_2O \quad ; \quad K_W \tag{6.37}$$

$$M^{2+} + OH^- \longleftrightarrow MOH^+ \quad ; \quad K_{1H} \tag{6.38}$$

$$M^{2+} + 2\,OH^- \longleftrightarrow M(OH)_2 \quad ; \quad K_{2H} \tag{6.39}$$

Amphoteric surface dissociation

$$S^sOH_2^+ \longleftrightarrow S^sOH^o + H_s^+ \quad ; \quad K_{a1}^{int} \tag{8.76}$$

$$S^wOH_2^+ \longleftrightarrow S^wOH^o + H_s^+ \quad ; \quad K_{a1}^{int} \tag{8.77}$$

$$S^sOH^o \longleftrightarrow S^sO^- + H_s^+ \quad ; \quad K_{a2}^{int} \tag{8.78}$$

$$S^wOH^o \longleftrightarrow S^wO^- + H_s^+ \quad ; \quad K_{a2}^{int} \tag{8.79}$$

Cation adsorption

$$S^sOH^o + M_s^{2+} \longleftrightarrow S^sOM^+ + H_s^+ \quad ; \quad K_M^{int} \tag{8.80}$$

$$S^wOH^o + M_s^{2+} \longleftrightarrow S^wOM^+ + H_s^+ \quad ; \quad K_M^{int} \tag{8.81}$$

Mole Balance Equations

$$[H]_T=[H^+]-[OH^-]-[MOH^+]-2[M(OH)_2]+[S^sOH^+_2]+[S^wOH^+_2]-[S^sO^-]-[S^wO^-]-[SsOM^+]-[S^wOM^+] \tag{8.82}$$

$$[S^sOH]_T = [S^sOH^o] + [S^sOH_2^+] + [S^sO^-] + [S^sOM^+] \tag{8.83}$$

$$[S^wOH]_T = [S^wOH^o] + [S^wOH_2^+] + [S^wO^-] + [S^wOM^+] \tag{8.84}$$

$$[M^{2+}]_T = [M^{2+}] + [S^sOM^+] + [S^wOM^+] + [MOH^+] + [M(OH)_2] \tag{8.85}$$

Net surfac charge

$$\sigma_d = F/AS \{ [S^sOH_2^+] + [S^wOH_2^+] + [S^sOM^+] + [S^wOM^+] - [S^sO^-] - [S^wO^-] \} \tag{8.86}$$

Coulombic Corrections

$$[H_s^+] = [H^+] \exp(- e\,\psi_0 / kT) \tag{3.10}$$

$$[M_s^{2+}] = [M^{2+}] \exp(- 2e\,\psi_0 / kT) \tag{8.67}$$

Charge - Potential Relationship

$$\sigma_0 = 0.1174\, c^{1/2} \sinh(\psi_0 \times 19.46) \tag{5.10}$$

As far as the adsorption of anion is concerned many authors [304,309] have proposed mainly two types of adsorbing species derived from a hypothetical trivalent (A^{3-}) anion . Accordingly, anion adsorption may therefore be described as follows :

$$SOH^o + A_s^{3-} + H_s^+ \leftrightarrow SA^{2-} + H_2O \quad ; \quad K_{1A}^{int} \qquad (8.87)$$

$$SOH^o + A_s^{3-} + 2H_s^+ \leftrightarrow SHA + H_2O \quad ; \quad K_{2A}^{int} \qquad (8.88)$$

where A_s^{3-} is the hypothetical trivalent anion closed to the surface. The kinds of surface species required to describe adsorption depend upon pH and other experimental conditions. This is because, such anions of triprotic weak acids go from being fully protonated (H_3A) to being fully deprotonated (A^{3-}) in the pH range 3-12. Changes in the degree of protonation with pH run parallel to the nature of species to be adsorbed. The reaction schemes are given below. The tableau for anion surface complexation is presented in Table 8.2.

Table 8.1 Tableau for surface complexation for cation

Species Components	H^+	M^{2+}	S^sOH^o	S^wOH	$P = \exp(-F\psi_0/RT)$
H^+	1				
OH^-	-1				
M^{2+}		1			
MOH^+	-1	1			
MOH_2	-2	1			
S^sOH			1		
S^wOH				1	
$S^sOH_2^+$	1		1		1
$S^wOH_2^+$	1			1	1
S^sO^-	-1		1		-1
S^wO^-	-1			1	-1
S^sOM^+	-1	1	1		1
S^wOM^+	-1	1		1	1
	TOTH	TOTM	TOT(S^sOH)	TOT(S^wOH)	TOTP ($=\sigma_0$)

Mass Law Equations

$$OH^- + H^+ \leftrightarrow H_2O \quad ; \quad K_W \qquad (6.37)$$

$$H_3A \leftrightarrow 3H^+ + A^{3-} \quad ; \quad K_{1P} \qquad (8.89)$$

$$H_2A^- \leftrightarrow 2H^+ + A^{3-} \quad ; \quad K_{2P} \qquad (8.90)$$

$$HA^{2-} \leftrightarrow H^+ + A^{3-} \quad ; \quad K_{3P} \qquad (8.91)$$

$$SOH_2 \longleftrightarrow SOH^o + H_s^+ \quad ; \quad K_{a1}^{int} \tag{3.1}$$

$$SOH^o \longleftrightarrow SO^- + H_s^+ \quad ; \quad K_{a2}^{int} \tag{3.2}$$

$$SOH^o + A_s^{3-} + H_s^+ \longleftrightarrow SA^{2-} + H_2O \quad ; \quad K_{1A}^{int} \tag{8.87}$$

$$SOH^o + A_s^{3-} + 2H_s^+ \longleftrightarrow SHA^- + H_2O \quad ; \quad K_{2A}^{int} \tag{8.88}$$

Coulombic Corrections

$$[H_s^+] = [H^+] \exp(-e\psi_o / kT) \tag{3.10}$$

$$[A_s^{3-}] = [A^{3-}] \exp(3e\psi_o / kT) \tag{8.92}$$

Mole Balance Equations

$$[SOH]_T = [SOH_2^+] + [SOH^o] + [SO^-] + [SHA^-] + [SA^{2-}] \tag{8.93}$$

$$[H]_T = [H^+]-[OH^-]+3[H_3A]+2[H_2A^-]+[HA^{2-}]+\{SOH_2^+]-[SO^-]+2[SHA^-]+[SA^{2-}] \tag{8.94}$$

$$[A]_T = [H_3A] + [H_2A^-] + [HA^{2-}] + [A^{3-}] + [SHA^-] + [SA^{2-}] \tag{8.95}$$

Net Surface Charge

$$\sigma_0 = F/AS \{ [SOH_2^+] - [SO^-] - [SHA^-] - 2[SA^{2-}] \} \tag{8.96}$$

Charge - Potential Relationship

$$\sigma_0 = 0.1174\, c^{1/2} \sinh(\psi_o \times 19.46) \tag{5.10}$$

Indeed, Dzombak and Morel [27] have developed the simple Gouy - Chapman theory into a versatile model which can describe adsorption of both cations and anions in the presence of electrolyte solutions of different concentrations; can account for coulombic factor, surface heterogeneity and even surface precipitation at high sorbate / sorbent ratio.

8.2.2.3. Triple Layer Model

This is the extension of site - binding model [148,150] which, in turn, is the extension of the Stern model of the electrical double layer at the solid - aqueous solution interface.

Both the models have been discussed section 5.6. That there exists interaction between surface sites and ions of electrolyte solution has been supported by a large body of literature [176-178,253,308,309] .The adsorbing ions are separated from the surface oxygen by a hydration sheath of at least one H_2O molecule thickness so as to be accommodated at the β - plane [150,151] .The triple layer model of adsorption is therefore the logical extension of the site - binding model by including adsorption of cations and anions at the IHP or β - plane in addition to the ions of electrolyte solution. However, the triple layer model also envisages adsorption at the surface plane (o - plane) for strongly adsorbing ions. Taking the terminology from coordination chemistry these two types of adsorption have been termed as "outer sphere" and "inner sphere" complexes respectively. In the case of adsorption as inner sphere complex the adsorbing ion loses a portion (that facing the surface plane) of hydration sheath, whereas in the case of outer sphere complex ion retains its primary hydration sheath. The counter part of the dissociated ions on either o - or β - plane are placed in a plane close to IHP called d - plane or outer Helmholtz plane (OHP) from which the diffuse part of the double layer begins . For electroneutrality the sum of the net charges in the three planes is zero i.e., $\sigma_0 + \sigma_\beta + \sigma_d = 0$. This explains the increase in so with increase in ionic strength as a simple consequence of the increase in counter ion bindings . The schematic representation of the TLM model is illustrated in Fig. 8.4.

Model analogs of surface complexes have two parts : stoichiometry and closeness of approach of adsorbing ion to surface sites . A closer distance of approach of adsorbing ion represents a stronger interaction with the surface (equivalent to shorter bond length) as in the case of adsorption at o - plane. On the basis of similar argument adsorption at β - plane is weak due to relatively longer distance of approach [160].

Table 8.2 Tableau for anion surface complexation

species	components	H^+	A^{3-}	SOH^o	$P = \exp(-F\psi_o/RT)$
H^+		1			
OH^-		-1			
H_3A		3	1		
H_2A^-		2	1		
HA^{2-}		1	1		
A^{3-}			1		
SOH^o				1	
SOH^{2-}		1		1	1
SO^-		-1		1	-1
SA^{2-}		1	1	1	-2
SHA		2	1	1	-1
		TOTH	TOTA	TOTSOH°	TOTP

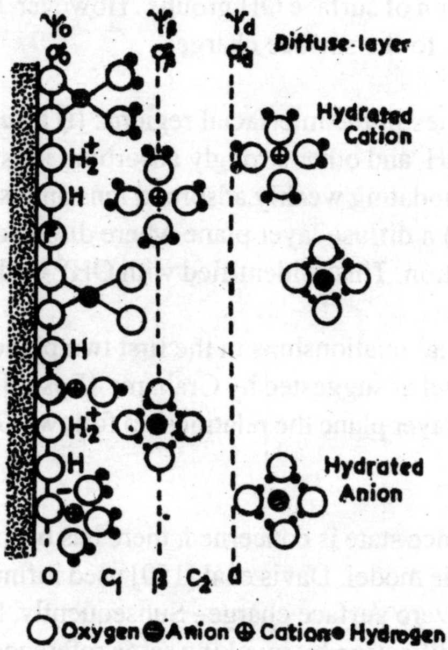

Fig. 8.4 An illustration of a triple layer model showing the specific adsorption of hydrated cation which undergoes partial dehydration as it approaches the negative surface sites to form inner sphere complex, or may retain its hydra tion sheath to form outer sphere complex on β-plane. Hydrated anion may adsorb as counter ions in the β-plane at a distance of at least one water mol ecule away from the surface. Both the planes are immobile. Diffuse layer contains both anions and cations distributed more randomly with increase in

The EDL correction factor in the mass law constant for the formation of inner sphere complex includes potential at the surface only, whereas for the formation of outer sphere complex the EDL correction factor includes potential difference between surface and β - plane. Hayes and Leckie [160] have observed that while adsorption of both cation and anion at the surface or o - plane is generally insensitive to the variation in ionic strength of the medium, adsorption of ions at the β - plane decreases with increase in ionic strength. This shows that weakly adsorbed ions are subjected to the influence of competition with electrolyte ions in contrast to strongly adsorbed ions. Thus, TLM provides a good method to differentiate between week and strong adsorption of ions on inorganic solids. The molecular hypotheses of TLM may be summarized as follows.

1. Surface charge originates mainly from the amphoteric dissociation i.e., protona tion and deprotonation of surface OH groups. However, strongly bound metal ions may also contribute to the surface charge.

2. There are three planes in the interfacial regions: (i) a surface plane for the adsorption of H^+, OH^- and other strongly adsorbing ions; (ii) a plane close to the surface for accommodating weakly adsorbed ions and is identified with IHP (β - plane), and (iii) a diffuse layer plane where dissociated counter ions occur in the bulk of the solution. This is identified with OHP or d - plane.

3. The charge - potential relationships in the first two planes are governed by extended Stern model as suggested by Grahame {Eqs .(5.19) and (5.20)}, while in the third or diffuse layer plane the relationship follows Gouy - Chapman model {Eq. (5.10)}.

4. As far as the reference state is concerned, there has been some differences among the proponents of the model. Davis et al [150] used infinite dilution relative to aqueous phase and zero surface charge. Subsequently, Hayes and Leckie [160] have suggested modification by invoking same reference state for both dissolved and surface species as 1.0 mole / litre at zero surface charge condition and no ionic interaction. This is based on one of the important tenets of surface complexation theory which treats surface functional groups as constrained ligands in solution. Hayes and Leckie's treatment of the thermodynamics of adsorption reaction has been discussed in section 5.7.

The basic differences between DDLM and TLM are as follows:

1. While the DDLM requires two or maximum three adjustable parameters, TLM needs five to six adjustable parameters to fit the adsorption data. For TLM, these are as follows: K_{a1}^{int}, K_{a2}^{int}, $^{*}K_{C+}^{int}$, $^{*}K_{A-}^{int}$, C_1 and binding constants of cations or anions with the surface. This renders the implementation of TLM more complex than DDLM.

2. Although DDLM also predicts increase in surface charge (σ_0) with increase in ionic strength, it does so through the increase in σ_d which is directly related to ionic strength by Eq. (5.10).In the TLM σ_d is evaluated as dissociated counter ion charge only, making it possible to estimate electrical potential ψ_d at the OHP. In the model simulation these estimates compared favourably with the experi - mental values of zeta (ζ) potential, though further improvement is possible by adjusting capacitance value between IHP and OHP i.e., C_2.

3. The TLM allows specificity in describing the development of proton surface charge in different electrolyte solutions. For example, surface charge of rutile in KNO_3 solution is different from that found in $LiNO_3$ solutions [104]. While TLM can explain this phenomenon, DDLM can not .

4. The TLM can provide a better physical picture than DDLM of the effect of ionic strength on the adsorption of cation or anion on oxides by invoking inner and outer sphere surface complex formation . These two terms are, in fact, related to strong and weak adsorption sites proposed in the DDLM. However, unlike DDLM the electrostatic terms in the mass law equations for the two types of complex formation are different in TLM .

8.2.2.4. Four Layer Model

It has been noted by Sposito [140] that the capacitance (C_l) between surface and IHP varies with pH of solution. At pH below PZC (positive surface charge), the capacitance value is less than that observed at pH above PZC (negatively charged surface). To explain this observation Bousse and Meindle [188] proposed a model with two separate β_- planes corresponding to the adsorption of cation (β_+) and anion (β_-) as outer sphere complexes. Since β_+ plane is more closer to the surface than β_- plane, this model may explain the observed difference in the interfacial capacitance values as observed by different authors [140,155,309,310]. However, such extension of models is not the outcome of any proposed radical changes leading to improvement in the model. The observed variation in the capacitance value for anion and cation adsorption is attributed to larger radii of anions coupled with associated hydration. The anticipated variation in capacitance can be incorporated while anion or cation adsorption instead of suggesting a separate plane, thereby complicating the model calculation.

Triple layer model has been extensively used for the modeling of adsorption of both cations and anions and a large body of literature has already been accumulated since 1978 when it was first proposed by Davis et al [150,151] . After some early attempts by James and Parks [60] and Sposito [140] no attempt has so far been made to review the massive literature that has been published so far. However, Cornell and Schwertmann [59] have made an extensive review of the literature on the adsorption of inorganic and organic ions on iron oxides up to 2000. Recently Bradle [221] has made a review of the adsorption of metal ions on soil and its constituents.

8.2.2.5 The CD - MUSIC Model

The CD - MUSIC model is an extension of the MUSIC (an acronym for multi site complexation) model developed by a group of surface and colloid chemists at the Agricultural university of Wageningen, The Netherlands, to describe the multi site complexation which allows differentiation of surface groups in terms of their formal charges [119,311]. Such differentiation led to the development of different concept on pH dependent surface charge and potential at mineral - water interface, where surface oxygen may bind zero, one or two protons depending upon their proton affinity. The proton affinity is believed to be related to formal charge calculated from bond valence of cation around a central anion i.e., oxygen. The background of the entire concept of MUSIC model and its ramification on the use of single or one pK model to describe surface protonation and deprotonation reaction has been discussed in section 3.2.

The proton affinities of singly, doubly and triply coordinated surface oxygen in the goethite structure have been calculated using a refined form of the original MUSIC model [311]. Based on this approach the pK values of doubly coordinated surface oxygen lie outside the normal pH range (pH 3 - 11) and therefore Fe_2OH^o groups are considered to be inert. The pK values of protonation / deprotonation of singly coordinated surface oxygen though lie also outside the normal environmental pH range, but lead to the formation of $FeOH^{1/2-}$ and $FeOH^{1/2+}$ as the surface oxygen species which contribute to the surface acidity of goethite [312]. For triply coordinated surface oxygen only a single protonation is possible, because only one unbounded oxygen orbital is available. Therefore, the surface acidity of goethite is controlled by only two reactions; one for the protonation of surface OH group and the other for the protonation of triply coordinated oxygen [312 , 313]. The pK value of both the surface protonation reactions was set equal to the experimentally determined pH_{PZC} of goethite as has been the common practice in the application of CD - MUSIC model [109, 314-316]. The theoretical basis of this practice has been discussed briefly in section 3.2 and in more elaborately by Koopal [117]. Typical log K values of protonation, electrolyte - surface ion pair formation and inner sphere complexation reactions of goethite with Cu^{2+} (typical metal ion) and PO_4^{3-} (typical oxyanion) according to CD - MUSIC model [312] are presented in Tables 8.3 and 8.4.

8.2.2.6 Charge Distribution of Adsorbed ion

The structure of EDL as envisaged by the MUSIC model is essentially similar to the TLM. The interface develops pH dependent charge because of the excess or deficiency of protons attached to the surface oxygens exposed to the liquid phase. Consequently, adsorption of any charged species contribute to the surface charge which in turn is balanced by a surrounding layer of equal but oppositely charged electrolyte ions constituting the diffuse layer. Protons and specifically adsorbed ions form inner sphere complex by ligand exchange, bound covalently to metal ion in the bulk mineral

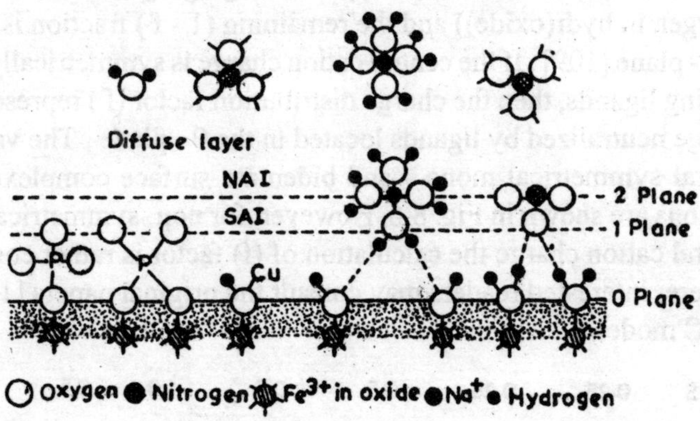

Fig. 8.5 Placement of hydrated cations (e.g., Cu^{2+}) and anions (e.g., PO_4^{3-}) in different planes of the electrical double layer according to the CD - MUSIC model. Protonated surface oxygen atoms lie in the 0-plane. Ions that are adsorbed specifically as inner sphere complexes are distributed between 0- and 1-planes and ion-pairs are placed in the 2-plane. Adapted after ref. 312.

or solid phase through common oxygen ligand. Non specifically adsorbed ions including electrolyte ions are displaced from the surface by one or more intervening water molecules forming outer sphere complexes and therefore occur farther away from the mineral surface than specifically adsorbed ions. The is illustrated in Fig. 8.5.

In the CD - MUSIC model the charge associated with all the adsorbing species including those forming surface-electrolyte ion pairs is located in one of the three discrete electrostatic planes designated as 0, 1 and 2 planes. The placement of charge associated with the adsorbing species in the three planes is based on the manner in which the ion is either strongly adsorbed or weakly associated with the surface and the physical arrangement of ligand(s) which may surround the central anion, for example , oxygen in PO_4^{3-} or OH^- in $CuOH^+$. The charge associated with protons is always placed in the 0 - plane because they are covalently bonded with the surface oxygen. The charges of non specifically adsorbed ions and hydrated electrolyte ions are placed in the 2 - plane. Depending upon the physical orientation of the associated ligands in specifically adsorbed species like oxyanions and hydroxo complexes of cations, the charge is distributed between 0- and 1- electrostatic planes.

The apportionment of certain type of ionic charges in more than one electrostatic planes is the unique feature of the CD - MUSIC model. By extension of the Pauling bond valence principle to charge neutralization in specifically adsorbed oxyanions and hydroxo complex of metal ions, a certain fraction (f) of the central cation charge of the adsorbed species is attributed to common bridging ligands which lie in the 0 - plane (e.g., oxygen in hydr(oxide)) and the remaining (1 - f) fraction is attributed to ligands in the 1 - plane [109]. If the central cation charge is symmetrically neutralized by all surrounding ligands, then the charge distribution factor (f) represents the fraction of the charge neutralized by ligands located in the 0 - plane. The values of the f factor for several symmetrical mono - and bidentate surface complexes involving PO_4^{3-} and Cu^{2+} ions are shown in Fig. 8.6. However, for non -symmetrical neutralization of the central cation charge the calculation of (f) factor is rather complex and is not discussed here; interested readers may consult the original papers [109,313-315] on CD - MUSIC model.

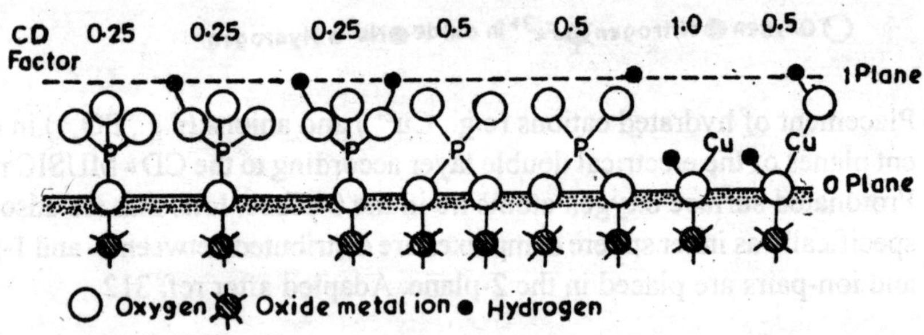

Fig. 8.6 Fractional(f) charge distribution(CD) of the adsorbed species of cations (e.g., Cu^{2+}) and anions (e.g., PO_4^{3-}) on the surface of goethite assuming sym metrical distribution of central cation charge among the surrounding ligands according to Pauling's valence rules. The CD-f values represent the portion of central cation charge which is neutralized by surface ligands. For sym metrical neutralization cation charge is calculated as the fraction of the total ligands situated in the 0-plane. Adapted after ref. 312

Because of its inherent complexity the CD - MUSIC model is difficult to fit to various equilibrium programs commonly used for numerical solution of adsorption problems, despite being based on more sound theoretical foundation than any other model. However, this problem is now being tackled satisfactorily by modifying the procedure for coding appropriate mass and charge balance components [312]. Such modifications necessitates assigning non- zero charge species as reference states (e.g., $FeOH^{1/2-}$ and $Fe_3O^{1/2-}$ for goethite). The electrostatic component i.e., Boltzmann factor represents ion charge rather than species stoichiometry, an accurate charge balance for 0 - plane requires that the coefficients of the electrostatic component reflects the actual charge of the reference adsorption site or its multiples. For FITEQL implementation of 3 plane CD - Music model, all charges are associated with protons, specifically adsorbed ions (SAI) and non-specifically adsorbed ions (NAI) and DDL ions must be accounted for ions of the three planes. Although the charge in the DDL due to electrolyte ions is not physically located in the 2 - plane, for the purpose of the computation of charge balance it is included in this plane ($\sigma_2 = \sigma_d + \sigma_{NAI}$) to satisfy the overall neutrality of the electrical double layer as follows :

$$\sigma_0 + \sigma_1 + \sigma_d + \sigma_{NAI} = 0 \tag{8.97}$$

Thus, for fitting the CD - MUSIC model to the FITEQL program , three electrostatic factors , namely $\exp(F\psi_1 / RT)$, $\exp(F\psi_2 / RT)$ and $\exp(F\psi_3 / RT)$ are also designated as components along with Na^+, NO_{3-} and relevant cations or anions added for adsorption. Interested readers may consult ref. 312 for more details. However, it needs more time for the CD - MUSIC model to gain similar popularity as DDLM or TLM.

In the CD - MUSIC model the electrostatic factors are separately defined for 0 - , 1 - and 2 - planes which relates the electrostatic energy change associated with a reaction that forms a species from constituent components

$$\Delta G_{el} = \exp(-\sum \frac{\Delta z_i e\psi_i}{kT}) \tag{8.98}$$

where Δz_i is the net change in charge on the electrostatic plane i and ψ_i is the is the potential on plane i relative to bulk solution potential. In the case of specifically adsorbed ions, the distribution of charge between 0 - and 1 - planes is made by defining the changes in charge Δz_0 and Δz_1 on these planes as follows

$$\Delta z_0 = \Delta n_H z_H + f\, z_{me} \tag{8.99}$$

and

$$\Delta z_1 = (1 - f)\, z_{Me} + \Sigma m_j z_j \tag{8.100}$$

where, nH is the net change in protons in the 0 plane , z_H is the proton charge, f is the CD factor , z_{Me} is the charge in the central cation of the adsorbed species , m_j is the number and z_j is the charge of adsorbed species ligands placed in the 1 - plane .

Table 8.3 Equilibrium constants for protonation and surface complexation re actions according to the CD - MUSIC model for adsorption of Cu^{2+} onto goethite surface* .

Reactions	log K
Surface protonation rections[1]	
$FeOH^{1/2-} + H^+ \leftrightarrow FeOH_2^{1/2+}$	9.2
$Fe_2O^{1/2-} + H^+ \leftrightarrow Fe_2OH^{1/2+}$	9.2
Electrolyte - surface ion pair formation reactions[2]	
$FeOH^{1/2-} + Na^+ \leftrightarrow FeOH - Na^{1/2+}$	-1.0
$FeOH^{1/2-} + H^+ + NO_3^- \leftrightarrow FeOH_2 - NO_3^{1/2-}$	8.2
$Fe_2O^{1/2-} + Na^+ \leftrightarrow Fe_2O - Na^{1/2+}$	-1.0
$Fe_2O^{1/2-} + H^+ + NO_3^- \leftrightarrow Fe_2OH - NO_3^{1/2-}$	8.2
Inner - sphere surface complexation reactions [3]	
$FeOH^{1/2-} + Cu^{2+} \leftrightarrow FeOHCu^{1/2+}$	12.06
$FeOH^{1/2-} + Cu^{2+} + H_2O \leftrightarrow FeOHCuOH^{1/2+} + H^+$	0.285

*Compiled from Ref . 312

1. Obtained from potentiometric titration [312] , log K_H = 9.2
2. Ion - pair formation constants for goethite : log K_a = -1.0 for NO_3^- , log K_c=-1.0 for Na^+ , log $_{KNa}$ = log K_a + log $K_{H,1}$.
3. CD - MUSIC optimized values .

Table 8.4 Equilibrium constants for protonation and surface complexation reactions according to the CD - MUSIC model for adsorption of PO_4^{3-} onto goethite surface*.

Reactions	log K
Surface protonation rections	
$FeOH^{1/2-} + H^+ \leftrightarrow FeOH_2^{1/2+}$	9.2
$Fe_2O^{1/2-} + H^+ \leftrightarrow Fe2OH^{1/2+}$	9.2
Electrolyte - surface ion pair formation reactions (1)	
$FeOH^{1/2-} + Na^+ \leftrightarrow FeOH - Na^{1/2+}$	-1.0
$FeOH^{1/2-} + H^+ + NO_3 \leftrightarrow FeOH_2 - NO_3^{1/2-}$	8.2
$Fe_2O^{1/2-} + Na^+ \leftrightarrow Fe_2O - Na^{1/2+}$	-1.0
$Fe2O^{1/2-} + H^+ + NO_3^- \leftrightarrow Fe_2OH - NO_3^{1/2-}$	8.2
Inner - sphere surface complexation reactions (2)	
$FeOH^{1/2-} + H^+ + PO_4^{3-} \leftrightarrow FeOPO_4^{2.5-} + H_2O$	19.94
$FeOH^{1/2-} + 2H^+ + PO_4^{3-} \leftrightarrow FeOPO_4OH^{1.5-} + H_2O$	27.22
$FeOH^{1/2-} + 3H^+ + PO_4^{3-} \leftrightarrow FeOPO_4(OH)_2^{0.5-} + H_2O$	30 00
$2FeOH^{1/2-} + 2H^+ + PO_4^{3-} \leftrightarrow Fe_2O_2PO_4^{2-} + 2H_2O$	29.71
$2FeOH^{1/2-} + 3H^+ + PO_4^{3-} \leftrightarrow Fe_2O_2PO_4(OH)^- + 2H_2O$	33.46

*Compiled from Ref. 312

(1) Ion pair formation constants as explained in the preceding table.

(2) CD - MUSIC optimized values . Log K^{int} values for bidentate adsorption reaction were determined from FITEQL optimization according to the method suggested by Tadanier and Eick [312]

CHAPTER 9

ION EXCHANGE ADSORPTION ON SOILS AND CLAYS

9.1 General Properties of Ion Exchange Reactions of Soils and Clays

One of the most important chemical properties of soils is their ability to retain and exchange positively charged ions on their surfaces. Primarily, it is the clays present in the soils that impart this unique property. To understand the cation exchange properties of soil colloids it is necessary to understand the electrical double layer at soil / electrolyte interface which has been discussed briefly in section 7.5. Basically, there is no difference in the behavior of electrical double layer at soil / electrolyte interface from that of pure hydr(oxide) / electrolyte interface discussed in detail in chapters 5 and 6. However, as far as the nature of surface charge is concerned there is some difference. While the charge on hydr(oxide) surface is considered to be distributed uniformly and is dependent totally on pH of the electrolyte solution , there are two different types of surface charge on soil clays; one is permanent (or pH independent) negative charge residing on basal surface, the other is variable (or pH dependent) residing on the edges . Since the area occupied by edges is very small compared to the total or overall surface area , soils or clays bear predominantly negative charge.

A negatively charged soil colloid when dispersed in an aqueous electrolyte solution will form an electrical double layer with cations concentrating very close to the surface, beyond which a diffuse layer consisting of randomly distributed cations and anions will extend to a distance till the eletrostatic potential asymptotically approaches zero. The cations within the double layer neutralize not only the negative surface charge, but also the dissolved anions within the double layer. The cations neutralizing the surface charge is exchangeable with other cations under suitable conditions. The exchange reaction is rapid, stoichiometric and generally reversible with change in conditions.

Since the existence of positively charged clay is rare (if occurs at all) and therefore do not exhibit anion exchange. But the interaction of clays with some anions, especially phosphate has generated special interest in the field of anion exchange. The adsorption of anions on soils may occur either by electrostatic interaction, or by non-electrostatic (specific) reaction with he surface. Electrostatic adsorption of anions like Cl^-, SO_4^{2-} or NO_3^- takes place either on pH dependent charged surfaces of hydrous oxides present in the soils or on the edges of kaolinite, if present in soil in substantial amount. However, in both cases adsorption occurs only at pH < 6.0 on kaolinite rich soil and at pH < 3.5 on montmorillonite rich soil. Though positively charged hydrous oxides of iron and aluminum adsorb most of the phosphates in soil by electrostatic interaction, some adsorption does occur through ligand exchange or anion penetration mechanism. Phosphorus adsorbed by this mechanism can be exchanged with other anions. According to this mechanism, clay minerals possess OH groups which are exposed to external reacting species. They are attached to crystalline aluminosilicate structure of clay minerals and are liable to be dissociated either as:

$$-(Si, Al)-OH \longleftrightarrow -(Si, Al)^+ + OH^-$$

or accept a proton to attain to attain positively charged $-(Si, Al)-OH_2^+$. However, ligand exchange can occur on surface carrying either positive or negative or even neutral charge. Ligand exchange may explain why weak acid anion shows maximum adsorption at pH values almost equal to their pK_a values. At pH = pK_a the amounts of anion (dissociated acid) available for ligand exchange and amounts of proton donors (undissociated acid) capable of neutralizing the liberated OH are highest.

9.2 Effect of Anion on the Completion of Cation Exchange Reaction

Because of the reversible character, the cation exchange reactions can be made to shift either in the forward or in the backward direction depending upon the concentration of reactants and products. The techniques commonly used to drive the reaction forward include (i) use of high concentration of exchanging cations, (ii) maintaining low concentration of product or released cation and (iii) removing the product from the reaction surroundings as precipitate or volatile gases. However, it is the properties of the products containing anions associated with the replacing cations play more important role in driving the reaction further towards completion. These properties are: 1) more weak dissociation, 2) less solubility and 3) more volatility. Indeed, these properties are complementary to the three techniques mentioned above to drive the reaction forward. The following examples will clarify each of the three properties of the products containing anions. Let us take the first property by considering the following exchange reaction where H_2O is the weakly dissociated product.

$$HX + Na^+ + OH^- \longleftrightarrow NaX + H_2O \qquad (A)$$

where X is the exchanger i.e., soil or clay . On the other hand, the progress of the reaction towards completion will be hindered if HCl is the product of the exchange reaction.

$$HX + Na^+ + Cl^- \longleftrightarrow NaX + HCl \qquad (B)$$

This is because H^+ generated from strongly dissociated HCl will tend to drive the reaction (B) towards left. An example of property (2) is the formation of $CaCO_3$ as precipitate which will remove Ca^{2+} from the surrounding

$$CaX + 2Na^+ + CO_3^{2-} \longleftrightarrow Na_2X + CaCO_3 \qquad (C)$$

On the other hand, the completion of the above reaction will be poor if the product of the reaction is $CaCl_2$.

$$CaX + 2Na^+ + 2Cl^- \longleftrightarrow Na_2X + CaCl_2 \qquad (D)$$

The third property may be exemplified by the following exchange reaction

$$NH_4X + NaOH \longleftrightarrow NH_4OH + NaX \xrightarrow{\text{Bubbling air or } N_2} NaX + NH_3\uparrow + H_2O \qquad (E)$$

Ammonia can be removed continuously by bubbling air or N_2 in the solution. But both NH_4Cl and $(NH_4)_2SO_4$ do not have such volatility. Anions such as CO_3^{2-}, SO_4^{2-} and HCO_3^- can also hinder cation exchange reaction by forming complex ions and thereby lowering activity of free cation.

9.3 Cation Exchange Selectivity

The adsorption of cations on a negatively charged soil surface though governed by coulombic attraction, it cannot explain the difference in adsorption within cations of similar charge or valence state. Obviously, this can be explained by the relative size of the hydrated ions. Ions of smaller radii have greater charge density per unit volume. Hence they attract water more strongly. An ion of larger hydrated radius is bound less tightly by soil colloid than is an ion of smaller hydrated radius, because the latter can be more easily accommodated within a more closely packed surface structure. Partially dehydrated ions are able to approach the surfaces even more closely and generally are retained more tightly by the surface. As early as 1935 , Jenny and Reitemeier [317] carried out a set of interesting experiments. They saturated montmorillonite suspensions separately with cations of different valence state and then attempted to replace or exchange them with a symmetrical amount (i.e., corresponding amount equal to the cation exchange capacity of the clay) of NH_4^+, or K^+ ions. It was observed that within ions of similar charge or oxidation state the fraction of cation released by NH_4^+ or K^+ decreased not only with increase in ionic radius but also with increase in the oxidation or valence state. Thus, while 67% and 31% of Na^+ (r = 0.97A°) and Cs^+ (r = 1.67A°) could be released respectively,29%, 14% and 2% of Ca^{2+} (r = 0.99A°), La^{3+} (r = 1.02A°) and Th^{4+}(r = 1.02A°) were released respectively by exchanging with NH_4^+ or K^+ . The above experimental results reveal that valence state of cations plays the most important role in determining the adsorption of cations onto surfaces of clay minerals or soils. Divalent ions, in general, are retained more strongly than monovalent; trivalent ions are retained more strongly than divalent and quadrivalent ions are most strongly retained on the surfaces of clay minerals or soil colloids though exchangeable cations posses similar radii.

In view of the above observation on relative exchangeability of the cations or the ease of their removal from clay or soil surfaces may be arranged in a series known as "Lyotropic Series".

$$Li^+=Na^+>K^+=NH_4^+>Rb^+>Cs^+=Mg^{2+}>Ca^{2+}>Sr^{2+}= Ba^{2+}>La^{3+}=H^+(Al^{3+})>Th^{4+}$$

The reverse of the above series represents the relative affinities of the ions for the soil surface. In this series monovalent H^+ is clubbed La^{3+} or Al^{3+}. This is because H^+ clays are unstable and decompose to release Al^{3+} from lattice sites to produce Al^{3+} saturated clays (see section 7.5.6).

9.4 Cation - Exchange Behavior of Vermiculites

Certain clay colloids show unusually high affinity for specific cations. For example, under certain conditions vermiculites exhibit high affinity for Mg^{2+} over many other divalent ions. The hydrated Mg^{2+} apparently fits so well in the water network between the expanded layers of vermiculites that the ion is preferentially exchanged over a wide range of concentration. At low Mg^{2+} ion concentration vermiculite prefers Ca^{2+} over Mg^{2+} because of smaller radius of the former ion. However, as Mg^{2+} concentration in solution increases beyond a certain threshold limits (>40%), the clay mineral shows a dramatic shift in its affinity for Mg^{2+}, which plays a significant role in stabilizing the interlayer water structure of vermiculite [196].

Normally, soil solutions have relatively high Ca^{2+}/ Mg^{2+} ratio, but under certain conditions the ratio may be low due to the release of high Mg^{2+} content in the solution. Such conditions may arise either due to the removal of Ca^{2+} by precipitation as $CaCO_3$ or due to sharp increase in Mg^{2+} concentration in the environment resulting from the dissolution of marine sediment or to the weathering of Mg- rich micas to vermiculites. In such a situation, vermiculite acts as an efficient scavenger of Mg^{2+} ion even in the presence of Ca^{2+} and /or K^+. To replace Mg^{2+} from magnesium saturated vermiculites, repeated leaching with high concentration of replacing ion is necessary. Besides Mg^{2+}, vermiculites also exhibit preferential adsorption of K^+, NH^{4+}, Cs^+ and Rb^+. But, unlike Mg^{2+} these ions are adsorbed on the surface where they fit perfectly in the hexagonal holes of the silicate network. This is known as "Fixation of Ions" which unlike Ca^{2+} and Mg^{2+}, cannot be exchanged with other ions. This is one of the most widely studied subject in soil or clay chemistry. Such fixation decreases with soil acidification and increases with soil liming. This is attributed to the formation of $Al(OH)_3$ or $Fe(OH)_3$ in between mica and vermiculite.layer lattices under acid condition. The formation of interlayer hydroxides prevents the collapse of clay lattice structure. But fixation reaction tends to bring about lattice collapse which is necessary for the stability of fixed cations .The process is believed to be activated by the dehydration of large weakly hydrated ions as adjacent silicate sheets approach one another during drying [196] .

The above explanation leads to the correlation between hydration energy and affinity or relative strength of the retention of ions on vermiculite surface. Ions of low hydration energies such as K^+, Cs^+, Rb^+ etc get easily dehydrated and therefore exhibit strong affinity for fixation. The hydration energy theory more readily explains the fact that Ba^{2+} ion having essentially same radius as NH_4^+, is not fixed by vermiculite or weathered trioctahedral micas. Barium ion with larger hydration energy cannot be readily dehydrated and entrapped by adjacent mineral lattices. Moreover, it readily rehydrates, forcing the lattices apart when mineral surfaces are rewetted after drying [196].

9.5 Valence Dilution and Ratio Law

In the case of exchange between cations of unequal valence states, dilution of the equilibrium solution favours the retention of cation of higher valence state. Let us consider exchange reaction of the following type and represents soil matrix.

$$CaX + 2NH_4^+ \longleftrightarrow (NH_4)_2X + Ca^{2+} \qquad (9.1)$$

The equilibrium constant K_R is given by

$$K_R = \frac{\left[(NH_4)_2X\right]\left[Ca^{2+}\right]}{\left[CaX\right]\left[NH_4^{\ +}\right]^2} \qquad (9.2)$$

where, the terms within the brackets represent concentration in moles / l or moles/g, if the activity terms are neglected for the moment. Equation (9.2) may be rearranged as follows

$$\frac{[(NH_4)_2X]}{[CaX]} = K_R \frac{[NH_4^+]^2}{[Ca^{2+}]} \tag{9.3}$$

This is commonly known as Kerr type exchange equation, wherein the activity terms are replaced by concentration terms by tacitly assuming that concentration is proportional to activity. However, we shall discuss later on the problem of estimating activity coefficients of species in the above equation.

Because of the occurrence of squared term in the RHS of Eq .(9.3) the ratio of NH_4^+ to Ca^{2+} in the double layer of soil colloid changes with total as well as relative salt concentration in bulk solution . For example, consider a solution containing 1.0 mmole/ l of both NH_4^+ and Ca^{2+}. The ratio is 1.0. Upon tenfold dilution the ratio becomes $(0.1)^2$ / 0.1 or 0.1. Hence the ratio of NH_4^+ to Ca^{2+} will decrease .The dependence of cation exchange on cation valence is called valence dilution effect. This may be further clarified from the following experimental data as quoted from ref. 196.

Table 9.1 : Replacement of exchangeable Ca^{2+} from 1 meq of exchange sites in montmorillonite by similar amount of Ba^{2+} and NH_4^+ at varying concentration of replacing ion*.

Solution added			% Ca^{2+} replaced by	
ml	N	Total	Ba^{2+}	NH_4^+
25	0.04	1meq	49.7	29.8
100	0.01	1meq	50.2	20.8
200	0.005	1meq	50.8	16.6

* Quoted from Ref . 196

It can be noted that while the replacement of Ca^{2+} by NH_4^+ decreases with dilution there is no change in the corresponding replacement by ion of same valence such as Ba^{2+}.

On the other hand the Ratio Law is concerned with the constancy of concentration ratio of mono and divalent cations present in the solution even after dilution [318].

Soil may contain both mono and divalent cations in exchangeable form with respect to corresponding cations in solutions. It has been observed that the pH of soil suspension decreases in monovalent electrolyte (1:1) solution such as 0.1 N KCl solution, because K^+ exchanges H^+ which is released in solution. If $CaCl_2$ solution is used instead of KCl, pH value remains unchanged over a wide range of concentration ratio of the two ions, i ,e., H^+ and Ca^{2+}. For example, suppose $a_{H+} = 0.004$ and $a_{Ca^{2+}} = 0.01$, and assuming the activities of solid species as unity, then

$$K_R = (\text{activity ratio of solid phases})\ a_{H+} / (a_{Ca^{2+}})^{1/2} = 0.04$$

If the activity of H^+ is reduced by half, the activity of Ca^{2+} would have to be reduced by the same factor squared to maintain the ratio 0.04. Thus,

$$K_R = \frac{0.5 x 0.004}{\{0.01 x (0.5)^2\}^{1/2}} = 0.04$$

The law, therefore, states that if a soil holds both monovalent and divalent cations in readily exchangeable form, the ratio of the amounts held is dependent on the ratio of the activity of mono valent cation to the square root of the activity of divalent cation or cube root of trivalent cation in solution.

9.6 Cation Exchange Equations

Several equations have been used to describe cation exchange process. Each has its own characteristics, merits and limitations. Therefore, the choice of a particular equation often appears to be subjective. Lack of familiarity with the usage, units and numerical value of exchange coefficients for other equations are the possible constraints in the use of an uniform approach to cation exchange equation. The limitations, which are inherent in most of the exchange equations, are as follows:

1. In ion exchange equation for cations, the involvement of simultaneous exhange of anion is seldom considered , even for highly acidic system.

2. The cation and / or anion exchanger is assumed to possess constant exchange capacity. However, the capacity varies with concentration of the salt and pH of the solution.

3. Simple stoichiometry (1:1) as assumed in most exchange equations, is not followed when simultaneous adsorption of molecules or formation of complex ion takes place.

4. All the equations assume complete reversibility. But, sometimes this may not happen due to complexity in the composition of soil clay.

The simplest and possibly the earliest exchange equation is the Kerr type exchange equation as given by Eq.(9.3). However, this equation is used only for homovalent exchange which means replacement of an ion of particular valence state by another ion of similar valence state. As already stated, the main problem with Kerr type equation is the use of concentration term instead of activity term. As a result, the equation follows only for a small range of concentration for many exchange reaction.

The activities of ions in solution can be estimated from Debye - Hückel theory or by using ion selective electrode system. However, the problem regarding activities of the adsorbed state of the ions is to be solved satisfactorily. The purpose of the different equations that have been proposed so far is mainly to simplify the problem of estimating activities of different ionic species so as to obtain uniform exchange constant or exchange coefficient. One such equation is the Gapon equation. Like the Kerr type equation, concentration terms are used instead of activities of soluble ions. The special feature of the equation is that mass law is expressed with chemically equivalent quantities both for exchange sites and exchanging cations. For example,

$$(Ca)_{1/2}X + Na^+ \longleftrightarrow NaX + \tfrac{1}{2}Ca^{2+} \tag{9.4}$$

which yields the following mass law equation in the form of (9.3)

$$\frac{[NaX]}{[Ca_{1/2}X]} = K_G \frac{[Na^+]}{[Ca^{2+}]^{1/2}} \tag{9.5}$$

It differs from Eq .(9.3) by including the term $[Ca_{1/2}X]$ rather than $[CaX]^{1/2}$. But, like Eq.(9.3) it suffers from applicability over a wide range of ion concentration. An important example is sodium - calcium exchange which is of great importance in irrigated soils. The Gapon equation is unsatisfactory if applied over the active range of Na - Ca composition, but works reasonably well over the range from 0 to 40% exchange which falls within most of the irrigated soils.

In another attempt to deal with the activities of surface species in mass law equation, Vanselow assumed that the activities of exchangeable cations were proportional to their mole fractions , which is equivalent to the assumption that ion adsorbed on soil surface behaves as if in an ideal solution . Thus, mole fraction of species 'a' in association with species 'b'(binary system) is $n_a/(n_a + n_b)$ where n is the number of moles per unit volume or per unit mass. Substituting this assumption in Eq.(9.3) results in the Vanselow's equation involving monovalent - divalent cation exchange.

$$\frac{[NaX]}{[CaX][NaX+CaX]} = K_V \frac{(Na^+)^2}{(Ca^{2+})} \tag{9.6}$$

where [ ] represents exchangeable concentration in mmole/g and ( ) represents activities of soluble ions in mmole/l. Since the LHS of Eq. (9.6) is a dimensionless quantity, K_V has the unit of l/mmole. This reaction has been extensively used to calculate the so-called thermodynamic exchange constants which are actually the average values of constants. But, these values rarely describe ion distribution at any specific composition with reasonable precision, especially beyond the range of uniform K_V values.

Krishnamurty and Overstreet and also Davis [quoted in ref.199] independently derived an equation for 1 - 2 valence exchange reactions on the basis of electrostatic attraction between charged surface and adsorbed cation. The calculation was performed on the assumption that the charged sites on the colloid surface are arranged in three different configurations, namely linear, open packed and closed packed as illustrated in Fig. 9.1. Therefore, the model resembles closely to that of Helmholtz double layer model, where the number of cations sorbed being equal to the number of charged sites. The resultant equation is a slightly modified form of the Vanselow's equation and is given by

$$\frac{[NaX]}{[CaX][NaX+q_i CaX]} = K_D \frac{(Na^+)^2}{(Ca^{2+})} \tag{9.7}$$

where,

$$q_i = Z_i - (2Z_i/Y) + 2/Y$$

The specific form of the above expression for Ca is however given by

$$q_{Ca} = Z_{Ca} - (2Z_{Ca}/Y) + 2/Y \tag{9.8}$$

For monovalent cation the value of $q_i = 1$

Y = the number of nearest neighbor (closest) charged sites

Z = exchangeable cation valence

The difference between the Davis equation and the Vanselow's equation is the specific ion factor q_i for the divalent cation. The values of q_i are: 1.0 for a linear array of charged sites (i.e., two nearest neighbor of exchanged sites), 1.5 for open packed

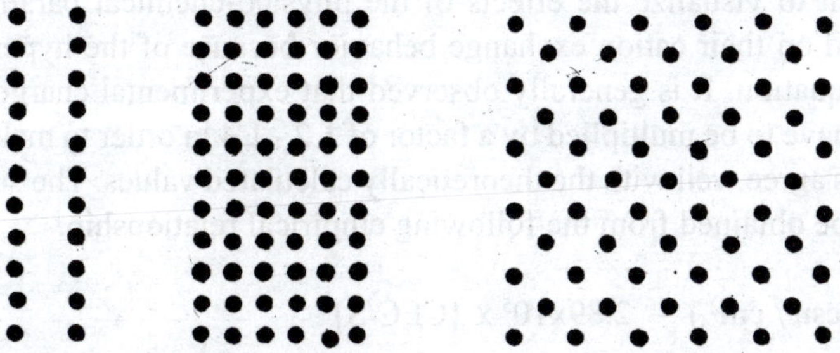

Fig. 9.1 Arrays of cation exchange sites on solid surfaces as configured from statistical thermodynamics in the Davis equation (Eq. 9.7). Adapted after ref. 199.

(i.e., four neighbor exchanged sites) and 1.67 for closed packed (i.e., six neighbor exchanged sites). When adsorption results in the neutralization of linear array of charges Eq. (9.7) reduces to Eq.(9.6).

Bolt [319] introduced some refinements of the exchange equation developed earlier by Eriksson [320] based on the interaction between electrical double layers of charged particles. The refined equation is given by

$$\frac{\Gamma_i}{\Gamma_s} = \frac{r}{\Gamma_s(\beta)^{1/2}} \sinh^{-1}\left(\frac{\Gamma_s(\beta)^{1/2}}{r + 4\vartheta_d (M_2)^{1/2}}\right) \tag{9.9}$$

where, Γs is the colloid charge density (me/cm2); Γi is the portion of the charge neutralized by the mono valent cation ; r is the reduced ratio (= $M_1/(M_2)^{1/2}$, M_1 and M_2 are the respective activities of mono and divalent cations in the bulk solution); β is the constant equal to 1.08×10^{15} cm/mmole for aqueous solution at 25°C and ν_d= cosh{exp(-Zeψ / kT)}at the mid plane between two adjacent colloid particles. In practice ν_d is normally set to unity, which is equivalent to no interaction between adjacent particles i.e., when two particles are separated by an infinite distance. In reality, this distance of separation is only a few hundred Angstroms. For this equation, instead of exchange coefficient, results are expressed in terms of per cent or fraction of the total sites exchanged by the mono-valent cation.

The advantage of Eriksson and Bolt equation is that it permits calculation of ion distribution entirely from measurable quantities such as exchange capacity, surface area and concentration of ions in the solution phase. The only disadvantage is that experimental determination of certain parameters is rather complex and that it is rather difficult to visualize the effects of the physico-chemical parameters of soil colloids based on their cation exchange behavior because of the hyperbolic cosine term in the equation. It is generally observed that experimental charge densities of soil colloids have to be multiplied by a factor of 1.2 - 1.4 in order to make the experimental results agree well with the theoretically calculated values. The surface charge density may be obtained from the following empirical relationship.

$$\Gamma s\ (esu / cm^2) = 2.89 \times 10^5 \times \{CEC/S\}$$

where, CEC = cation exchange capacity in meq / 100g
S = specific surface area in m^2 /g
and 1 esu = 0.3336×10^{-9} coulomb

This minor discrepancy arises from the errors in surface area measurement or non-uniform distribution of charge on the surface. Even then Eq.(9.9) shows excellent agreement between calculated and experimental values of exchange capacity over a wide range of concentration of exchanging ions. In contrast, for all other exchange equations it is necessary to estimate exchange constants under different sets of experimental conditions before comparing experimental and calculated values. Table 9.2 provides comparison of the values of the coefficients NH_4^+-Ca^{2+} exchange reaction obtained from different exchange equations.

Table 9.2 : A comparison of the exchange coefficients derived from several cation exchange equations for NH^{4+}-Ca^{2+} exchange reaction. (Reference 196)

Equations	Exchange oefficients at an ammonium concentration of 0.04N	0.01N	0.005N
Kerr (using concentration terms)	1.91×10^{-3}	1.81×10^{-3}	1.58×10^{-3}
Kerr (using activity terms)	1.44×10^{-3}	1.51×10^{-3}	1.37×10^{-3}
Gapon (using concentration terms)	1.17	1.07	0.97
Vanselow	2.22	2.50	2.35
Davis (q = 1.5)	1.75	1.88	1.73
Davis (q = 1.67)	1.63	1.74	1.59
	Exchangeable NH_4^+ in percentage		
Erikkson (calculated using S = 800m² / g)	26.1	20.2	17.9
,, (Experimental)	29.8	20.8	16.6

The Gapon equation (Eq. (9.5)) has been modified by the U S salinity Laboratory by substituting exchangeable Ca with a sum of Ca plus Mg. The corresponding Ca^{2+} in the bulk solution has also been replaced by $Ca^{2+} + Mg^{2+}$. This modification becomes necessary as most clay minerals contain both Ca and Mg as exchangeable cations The modified equation therefore takes the following form:

$$\frac{[NaX]}{[CaX + MgX]} = K_S \frac{[NaX]}{[Ca^{2+} + Mg^{2+}]^{1/2}} \tag{9.10}$$

The LHS of the above equations is called exchangeable sodium ratio (ESR), while the RHS is termed as sodium adsorption ratio (SAR). However, the true SAR is written as $[Na^+] / [(Ca^{2+} + Mg^{2+})/2]^{1/2}$ where the unit of concentration is meq /l. The reduced ratio in Eq.(9.9) is equal to $SAR/(1000)^{1/2}$. The empirical relationship between ESR and SAR is generally as follows for most of the irrigated soils.

$$ESR = -0.01 + 0.015\ SAR \tag{9.11}$$

This is equivalent to Gapon equation with an exchange constant of 0.015 .

9.7 Determination of Cation Exchange Capacity (CEC)

Since soil contains more than one exchangeable cations, the exchanging ion or the indexing ion must be used in such an amount that it can replace all the exchangeable cations in soil quantitatively. For example , if an exchanging ion B^+ is used to replace exchangeable ions A1 , A2 , A3 etc then either the sum of meq of A1, A2, A3 in leached solution or meq of B^+ which is exchanged for all the above cations should give the CEC of soil colloid or of any clay mineral . If the analysis of a cation (say A1) in the presence of other cations (say A2 and /or A3) poses problem due to mutual interference, then B^+ is to be analyzed by exchanging with another cation, say C^+. The following example will clarify the problem stated above.

Let us consider a soil sample containing Ca^{2+}, Mg^{2+}, K^+ as exchangeable ions. These ions may be exchanged with either NH_4^+ or Ba^{2+} ion by leaching the soil sample with their chlorides.

(1) Soil (Ca, Mg, K) + $NH_4Cl \leftrightarrow$ Soil - NH_4^+ + (Ca^{2+}, Mg^{2+}, K^+) Cl^-

(2) Soil (Ca , Mg , K) + $BaCl_2 \leftrightarrow$ Soil - Ba^{2+} + (Ca2$^+$, Mg^{2+}, K+) $2Cl^-$

In either case, the solution obtained after leaching contains Ca^{2+}, Mg^{2+} and K^+ in the presence of excess of either NH_4Cl or $BaCl_2$. In the case of exchange system (1),

Ca^{2+}, Mg^{2+} and K^+ can be analyzed by volatilizing NH_4Cl which is a tedious job. However, it is possible to free the residual solid from all soluble matter by repeated washing with aqueous solution of ethyl alcohol (to prevent peptization of soil particles) and distil NH_3 from the solid by heating with lime or magnesia and absorbing the evolved NH_3 in an excess of standard (0.1N) sulphuric acid back titrating the excess acid with an standard alkali solution . The difference gives amount of H_2SO_4 consumed by distilled NH_3.

In the exchange system (2) the Ba exchanged soil sample is also washed thoroughly with aqueous alcoholic solution until free from all soluble ions and the washed sample is treated with dilute HCl solution to exchange Ba^{2+} with H^+. Barium ion released in the solution can be easily determined either gravimetrically as $BaSO_4$ or more precisely by Atomic absorption Spectrophtometric (AAS) method. If 10g of oven dried soil sample is taken for each of the above two exchange systems , CEC may be calculated from two typical examples as follows :

1. Suppose back titration of excess 0.1N H_2SO_4 indicates that acid consumed by distilled NH_3 is 16 ml . Therefore, the amount of NH_4^+ exchanged for all cations present in the soil sample is (16 x 0.1 x 10) /10g = 16 meq / 100g . This is equivalent to 16mmole / 100g or 16 centimole / kg.

2. In the case of exchange system (2) the amount of BaSO4 is 0.1867g . Since 1 mole of $BaSO_4 \equiv 1$ g ion of Ba which is equivalent to 2 g ion of H , therefore H^+ exchanged = (0.1867/233.3) x 2 x104 = 15.76 meq /100g. This shows that both the methods give almost identical results .

The salt used to provide index or replacing cation should be fairly soluble in aqueous alcohol used for sample washing. Very low solubility of sodium chloride in ethanol is frequently overlooked as a source of error. If the solubility of the salt is low or poor, the salt precipitates and remains in the sample during the washing step. The salt, however, dissolves during the subsequent extraction step to give anomalous CEC values. Some soils may contain salt solution entrapped within the micropores of finely divided hydrous oxides. This may also lead to erroneous results. To overcome the problem of washing excess salt, some authors [321] have suggested simultaneous estimation of anions associated with index cations to verify the loss of cation during extraction and decrease in anion concentration before and after extraction.

Besides the conventional $BaSO_4$ method, copper complex of ethylene diamine has been successfully used for determining the CEC values of clay minerals [470]. In this method , 50 ml of 1M $CuCl_2$ solution is mixed with 102 ml of 1M aqueous ethylene diamine solution to allow the formation of $[Cu(en)]2^+$ complex ion. The mixture is diluted to 1 litre to give 0.05M solution of copper complex. About 0.5g of accurately weighed dry clay sample is placed in a stoppered 100ml reagent bottle and 5ml of 0.05M complex solution along with 25 ml of double distilled water are transferred to the bottle which is shaken in a constant temperature bath for about 30- 40 min. The solid is then separated from aqueous mixture either by membrane filtration or by centrifugation. Copper is determined in the supernatant liquid by iodimetric method. For this purpose, 5 ml of supernatant liquid is transferred to an iodine flask, acidified with 5 ml of 0.1M HCl to destroy the complex ion and the solution is diluted with water. Requisite quantity (about 1.5g) of solid KI is added and the flask is left in a dark place for 5-7 min and the liberated iodine is titrated with 0.02M $Na_2S_2O_3$ solution using starch as indicator. The CEC is then calculated from the following formula:

$$CEC(meq/100g) = MSV(x-y)/1000m \qquad (9.12)$$

where, M = molar mass of the complex
S = strength of $Na_2S_2O_3$ solution (moles/l)
V = volume of complex solution taken for iodimetric titration (ml)
m = mass of clay mineral (g)
y = volume of $Na_2S_2O_3$ solution required for titration of the solution (ml)
x = volume of $Na_2S_2O_3$ solution required for blank titration i.e., without solid (ml)

CHAPTER 10

ADSORPTION OF CATIONS ON INORGANIC ADSORBENTS

10.1 Characteristic Features of Cation Adsorption on Inorganic Adsorbents

The literature on the adsorption of cations from aqueous medium is considerably vast. As the importance of the role of adsorption phenomenon in understanding the problems associated with treatment of water for drinking and other purposes, treatment of waste water, protection of natural water from pollution, formation of minerals in aqueous environment, soil fertility, plant nutrition etc. is more appreciated the research activities in these fields are gaining more momentum. As a result, a large volume of publications on basic and applied aspects of adsorption is appearing every year. In the preceding chapters as well as in the following sections under the present chapter the terms like "adsorption", "precipitation" and "coprecipitation" have been used without providing their proper connotation or explicit meanings. At the outset of this chapter it is worthwhile to define the significances of these terms and their distinction from one another. This is particularly necessary as we shall come across the terms even more frequently to describe the sorption behavior of cations in various forms onto solid surface depending upon the conditions under which the process is carried out.

As broadly defined, sorption is a process by which chemical species from solution (here aqueous) accumulates on the surface of a contiguous solid phase. If the process of sorption leads to two dimensional arrays of species on the solid surface, it is known as "adsorption". On the other hand, if the accumulation leads to nucleation and growth of a new phase with three dimensional arrays of species, the process is known as "precipitation". The process may involve participation of one or more identical species derived from the solid phase. The essential characteristic of precipitation is the development of a solid phase which is intrinsically three dimensional in contrast to adsorption which is characterized by two dimensional arrangements of molecules even if the process involves mixed adsorbates. In his excellent and authoritative review Corey [322] has listed the following possible consequences when chemical species gradually accumulates on a solid surface.

1. Crystal growth : This occurs if the adsorbate is a component of the solid adsorbent.
2. Crystal growth and / or diffusion into the solid phase : This will take place if the adsorbate is not a component of the adsorbent but can substitute isomorphously for a component of the adsorbent and form a stable , three dimensional solid solution .
3. Formation of a stable surface compound (two dimensional solid solution) : This takes place if the adsorbate is not capable of forming a three dimensional solid solution with the adsorbent .
4. Stabilization of metastable , polynuclear ions : This occurs by adsorption onto oppositely charged surfaces of the adsorbent .
5. Heterogeneous nucleation of a new solid phase : This involves the formation a new solid phase composed of the adsorbate and a component from the solution (hydroxide , carbonate etc.) .
6. Heterogeneous nucleation of a new solid phase : This refers to a new phase composed of the adsorbate and a component of the adsorbent resulting in the dissolution of the adsorbent .

Generally, adsorption reactions belong to consequence no 3. However, this does not mean that the other consequences are of less importance and can be ignored. Depending upon the conditions of adsorption, e.g., higher pH or higher sorbate / sorbent ratio precipitation of stable new compound with growth in all directions also takes place. As a precursor to these consequences, cation must interact strongly with surface molecules. This type of adsorption is known as chemisorption or specific adsorption. As discussed earlier, such stronger interaction is possible when fully hydrated metal ion undergoes partial dehydration as it approaches the surface and coordinate directly with surface functional groups. This is also known as inner-sphere complex. Alternatively, if the metal ion is adsorbed by indirect coordination, such as hydrogen bonding with surface oxygen, the complex formed is called outer-sphere complex. Consequentially, such adsorption takes place little away from the surface because of longer bond distance than that of inner-sphere complex. Desorption of this type is generally reversible with change in solution conditions and mainly governed by electrostatic interaction, such as ion exchange adsorption on soils / clays etc. Schematic representations of the typical bonding patterns of inner and outer-sphere complexes are illustrated in Fig. 8.4 in chapter 8 showing the placement of adsorbed species in different planes of the electrical double layer.

Adsorption of heavy metal ions is generally irreversible and takes place through inner-sphere complex formation with surface functional groups , such as inorganic OH , silanol (-SiOH) or organic functional groups (in the case of soils) which are negatively charged at higher pH . The hydrated state of the adsorbed metal ion may undergo hydrolysis which is facilitated with increase in pH and is retained on the surface more strongly than large hydrated ion . As the sorbate / sorbent ratio is increased , multilayer adsorption takes place as the precipitate itself acts as an adsorbent and a new phase grows in all the three directions [323, 324]. The phenomenon of the precipitation of metal ion hydroxide at higher pH is shown in Fig. 10.1a. The newly adsorbed phase may even interact with one or more ions of the adsorbent phase through its dissolution and reprecipitation and finally the product grows as a mixed three dimensional solid solution involving both sorbate and sorbent ions. The phenomenon is called "coprecipitation" and is illustrated in Fig.10.1b. The simplified reaction steps involved in surface precipitation model are as follows:

a) Adsorption through surface complex formation with hydrolyzed metal ion species

$$SOH^{o} + M^{2+} + H_2O \longleftrightarrow S\text{-}O\text{-}MOH_2^{+} + H^{+} \qquad (10.1)$$

b) Precipitation of metal ion in the adsorbed phase

$$S\text{-}O\text{-}MOH_2^{+} + M^{2+} + 2H_2O \longleftrightarrow S\text{-}O\text{-}MOH_2^{+} + M(OH)_2\,(s) + 2H^{+} \qquad (10.2)$$

The water molecules in the above reactions are derived from the coordination sphere of the hydrated metal ion. The occurrence of surface precipitation is generally identified by deviation from ideal Langmuirian isotherm as discussed in context with increase in sorbate / sorbent ratio, later in this chapter.

Having defined and clarified the significances of some important terms relating to the adsorption phenomenon , we shall consider certain parameters such as pH, concentration of metal ions, ionic strength of electrolyte solution , presence of complexing ligands or anions , sorbate / sorbent ratio etc. which have profound influence on the adsorption of cations on inorganic solids or substrates.

10.1.1 Effect of pH

Keeping all other parameters constant , pH value of solution has immense influence on the adsorption of cation on inorganic solids . Adsorption generally increases sharply from 0 to 100 % of the total uptake within 1 - 2 pH units. The sigmoidal shape of the curve representing adsorption vs. pH is termed as "adsorption edge". The influence of each of the above mentioned parameters or factors on cation adsorption is gauzed by the shift of the adsorption edge . The shift towards acidic or alkaline side represents increase or decrease in the adsorption, respectively. The pH at which half of the adsorption occurs (pH_{50}) decreases with decrease in the first hydrolysis constant (K_{MOH}) of the adsorptive metal ion [214, 221]. With increase in the pH value of suspension the concentration of hydrolytic complexes of the adsorptive cation increase rapidly, but the actual hydroxy species formed depend upon the type of cation and its concentration and sometimes on the precise conditions of their formation [79]. Adsorption, however, increases sharply even when the concentration of dissolved hydroxyl species like MOH^+ is only very minor, especially when the concentration of free M^{2+} ion is very low. This suggests that hydroxyl species is preferred to free or unhydrolyzed ion [325]. Since adsorption of cation is accompanied by the net release of H^+, an analogy is frequently drawn between adsorption and hydrolysis of divalent metal ions, especially those belonging to transition metal group. However, sharp rise in the adsorption of metal ion begins at pH value much below the actual pH observed for the onset of hydrolytic precipitation from solution [325-327]. It has been argued that microscopic electrical field at the charged surface of the adsorbent is sufficiently strong to lower the activity and also the dielectric constant of viscinal water [328]. This results in lowering of the solubility product of metal hydroxide at the interface.

However, the observed increase in the adsorption of metal ion with increase in pH value is confronted with other factors also. As the adsorption edge approaches saturation, $M(OH)_2^{\circ}$ (aq) tends to be an important adsorptive species. Many metal ions

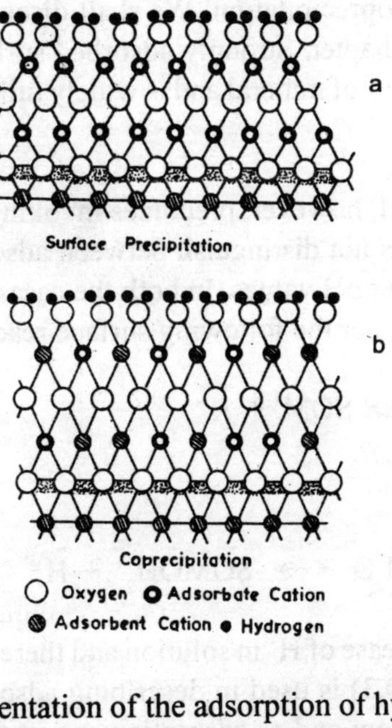

Fig. 10.1 Schematic representation of the adsorption of hydrated metal ion on oxide surface at higher pH value and at higher sobate/sorbent ratio, wherein increased adsorption results in the surface precipitation as hydroxide of the metal ion(a). At very high sorbate / sorbent ratio metal ion tends to form a new solid phase with sorbent through redissolution and coprecititation processes(b).

form hydrolytic complexes at lower OH^- / M^{2+} ratios by precipitation as a basic salt or as a solid product of varying composition wherein different anions or ligands may take part to balance the net positive surface charge [214]. These precipitates are generally metastable and redissolve and reprecipitate in a more cryatalline form. Thus, at high cation concentration it is difficult to determine whether cation is adsorbed or precipitated on the surface. The close relationship between adsorption and precipitation is reflected in the behavior of adsorption isotherm which usually does not show any inflection at the onset of precipitation , at least not before the bulk precipitation is appreciable. The close relationship is also manifested in the dependency of cation selectivity for adsorption on the solubility of the corresponding metal hydroxide. The higher the solubility (or solubility product constant, K_{SP}) the higher is the adsorption [214].

Surface precipitation is facilitated if the structure of the adsorbent phase is closely similar to the structure of the metal hydroxide phase, at least if the ionic radii of metal ions in both the phases are closely similar. In other words, hydrous metal oxide surface as adsorbent might act as template for the formation and growth of the precipitate. If metal ion is released by the dissolution of mineral or solid phase also participate in the growth of the surface precipitate, the process leads to what is defined in the preceding section as coprecipitation. We shall discuss more about surface precipitation later on in this chapter. Strongly adsorbed surface coating can drastically change the surface chemistry of mineral and is widely utilized in the mineral processing industry.

Surface precipitation model, however, precludes invoking hydrolysis for adsorption of cation as the model does not distinguish between adsorption of free cation or its hydrolyzed species at higher pH values. In both the cases equivalent amount of proton is released. Let us consider the following surface reactions for cation adsorption

$$SOH^{o} + M^{2+} \leftrightarrow SOM^{+} + H^{+} \qquad (10.3)$$

or equivalently

$$SOH^{o} + M^{2+} + H_2O \leftrightarrow SOMOH_2^{+} + H^{+} \qquad (10.1)$$

Both the reactions show release of H^+ in solution and therefore strongly dependent on pH. Generally, reaction (10.3) is used in describing adsorption at pH values below the state of surface saturation; and as adsorption approaches saturation region with increase in pH reaction (10.2) is used. The later reaction is also used to describe cation adsorption at high sorbate / sorbent ratio when surface saturation is readily attained.

On the basis of the above argument an attempt has been made to relate the affinity of the hydroxylated surface of adsorbent towards cation (i.e., equilibrium constant for the binding of cation with surface OH groups) to the position of the cation in the periodic table, vis - a - vis charge of cation [329]. It has been observed that cations on the left side of the periodic table are most resistant to hydrolysis compared to easily hydrolysable cations on the right side [79]. Thus, alkaline earth and trivalent metal ions on the left side of the periodic table are classified as "hard" Lewis acids, because they have high effective nuclear charge i.e., higher charge/radius ratio. These ions exhibit low polarizability and therefore tend to form ionic bonds. On the other hand, transition and post transition metal ions on the right side have low charge / radius ratio because of the relatively larger ionic radii. They are classified as either "soft" or "borderline soft" acids and are highly polarizable and tend to form covalent bonds. Though according to hard and soft acid base theory (HSAB) theory hard acids combine preferentially with hard bases (e.g., OH^-) and soft acids to soft bases [91,330], there is no dearth of examples violating the HSAB theory. This is because interactions of Lewis acids and bases are affected by strength as well as hardness. The strength of acid or base may be more important than the hard - soft characteristics; both the factors must be considered at the same time [331]. Strength of Lewis acid or base behaves in an opposite manner with that of corresponding hardness .Thus, soft cations are much stronger acids than hard cations of similar size [331, 332]. It is therefore possible that strong acid can replace weaker one, even though such a reaction appears to violate the HSAB theory. For example, the stronger but softer acid Cd^{2+} can replace H^+ which is a weak but hard acid from hard base OH^-.

$$\underset{\text{soft}}{Cd^{2+}} + \underset{\text{hard hard}}{H\,OH} \rightarrow \underset{\text{soft hard}}{Cd\,OH^+} + \underset{\text{hard}}{H^+} \tag{10.4}$$

Since factors other than hardness influence reactions of many cations, the first hydrolysis constants (K_{MOH}) for many cations do not correlate well with hardness indicator. It is observed [329] that while for alkaline earth metals (hard acids) there exists good correlation between K_{MOH} and HSAB indicator, softer cations , especially those belonging to transition metals, do not show good correlation . Even HSAB parameters are not consistent for many softer cations. It is therefore difficult to apply universally HSAB theory to the hydrolysis of metal ions in solution and even more difficult to apply to the reactions between cations and surface OH group. Hardness or softness of Lewis acid or base is dependent on its environment. It is difficult to know the nature of hardness of surface OH groups as these are affected by the structural environment of solid as well as the nature of solution which accounts for the development of surface charge . The net effect of all the complicating factors for HSAB reaction is included implicitly in the stability constants of similar reactions. This forms the basis of "Linear Free Energy Relationship" or LFER between reactions of cations with OH^- in solution and OH groups on hydrated oxide surface {329]. The derivation of the present LFER in its simplest form consists of the following steps: (1) Expressing log K values as differences in the standard molar free energies (μ_o) between products and reactants

$$RT \ln K^{int}_{SOM} = -(\mu^o_{SOM} + \mu^o_H - \mu^o_{SOH} - \mu^o_M) \quad (10.5)$$

$$RT \ln K_{MOH} = -(\mu^o_{MOH} - \mu^o_{OH} - \mu^o_M) \quad (10.6)$$

(2) Taking the difference between the two equations and rearranging

$$2.3RT(\log K^{int}_{SOM} - \log K_{MOH}) = (\mu^o_{SOH} + \mu^o_M - \mu^o_{OH} - \mu^o_H) + (\mu^o_{MOH} - \mu^o_{SOM}) \quad (10.7)$$

(3) Assuming that the value of the non-constant term in Eq (10.7) is negligibly small or covaries with K_{MOH} as they are "related", which means that the nature of the binding of metal ion (M^{2+}) with OH^- in solution is similar to that of its binding with surface OH groups, we get direct relationship between log K^{int}_{SOM+} and log K_{MOH}. Figure 10.2 illustrates log-log plot of binding constants of some selected cations with surface OH groups of HFO, γ-Al_2O_3, δ-MnO_2 and silica against their corresponding first hydrolysis constants i.e., reaction with OH group in aqueous solution. Satisfactory linear relationships are observed for all the four oxide substrates with slope values either close to unity (γ-Al_2O_3) or slightly greater (HFO) or less than unity (δ-MnO_2 and SiO_2). Slope value higher than unity suggests stronger coordination of surface OH groups with cations than with OH^- ions in solution. On the other hand, lower slope values indicate opposite trend i.e., OH^- ions in solution exhibit stronger coordination than surface OH groups with adsorbing cations. This is probably the behavior with oxides having low ΔpK_a like δ-MnO_2 and SiO_2, compared to those having high ΔpK_a like HFO and γ-Al_2O_3. The linear relationships in Fig.10.2 can be used to predict surface binding constants for similar divalent cations for which limited or no sorption data are available in literature.

10.1.2 Effect of Ionic Strength of Medium

Ionic strength of the background electrolyte solution plays a very important role in controlling the adsorption of cations on different inorganic solids. Natural water systems contain a variety of inorganic ions in widely varying concentrations, for example, less than 10^{-3} M for fresh water lake to greater than 0.5 M for sea water and

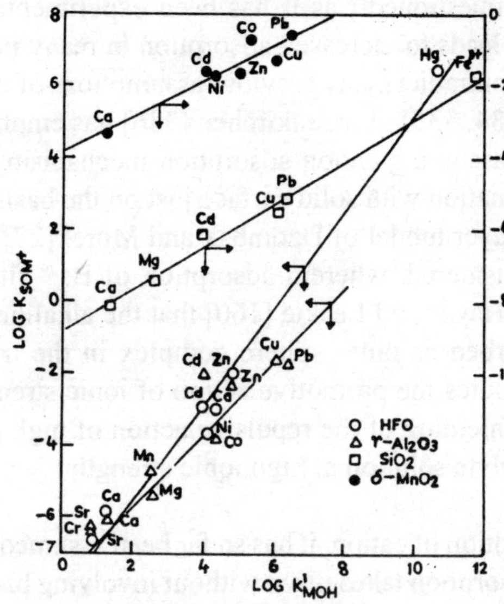

Fig.10.2 Linear free energy relationship(LEFR) between the first hydrolysis constants (K_{MOH}) of different metal ions and their corresponding binding constants K_{SOM^+} with oxides having high and low pH_{PZC} values. The two categories of sorbents can be distinguished depending upon the slope values of the linear plots. Data derived from refs. 283, 284, 329, 345, 377 435 and 450

hydrothermal brines . However, laboratory experiments on adsorption have been carried out mostly in 1:1 electrolytes such as $NaNO_3$, KNO_3 and $NaClO_4$ solutions, mainly because NO_3^- and ClO_4^- ions bind very weakly with transition or post transition heavy metal ions. On the other hand, NaCl, though apparently inert to many adsorption systems, may form chloro complexes with certain heavy metal ions like Pb^{2+}, Cd^{2+}, Hg^{2+}, Zn^{2+} etc. thereby affecting their adsorption behavior [333] .

Although the effect of the nature and concentration of electrolyte solution on the adsorption of cation was known for quite some time, Hayes and Leckie [160] were the first to propose explanation for it . The authors have observed that while the adsorption of transition metal ions on goethite from $NaNO_3$ solution is independent of ionic strength, adsorption of alkaline earth metal ions like Mg^{2+} and Ba^{2+} is highly sensitive to the variation of ionic strength. According to the authors, heavy metal ions lose their hydration sheath while approaching the solid surface and are adsorbed as inner sphere complex . On the other hand , metal ion like Ba^{2+} retain its hydration sheath and is therefore adsorbed little away from the surface as outer sphere complex [160] . Theoretically, increase in ionic strength leads to increase in surface charge and therefore should not promote adsorption of cations. This hypothesized mechanism has been seriously questioned , as it has been experimentally observed that increase in ionic strength leads to increased adsorption in many instances [33] , and spectroscopic evidences contradict many previous assumptions of inner / outer sphere complex formation [31, 334, 335] . Lützenkirchen [336] has emphasized the lack of proper discrimination in assigning cation adsorption mechanism to either inner or outer sphere complex formation with solid surface just on the basis of model fitness. For example, in the two layer model of Dzombak and Morel [27] only one layer of specific adsorption is considered, wherein adsorption of Ba^{2+} fits well. This goes against the hypothesis of Hayes and Leckie [160] that the alkaline earth metal ions, especially Ba^{2+}, are adsorbed as outer sphere complex in the triple layer model. Lützenkirchen [336] attributes the promotive action of ionic strength on cation adsorption to the effective shielding of the repulsive action of high surface charge by counter ions of the electrolyte solution at high ionic strength .

While modeling the adsorption of cation, it has so far been assumed that regardless of the nature of substrate, adsorption takes place without involving background electrolyte ions in the process. However, comparison of the adsorption behavior of metal ions from different types of single or mixed electrolyte solutions of varying concentrations provides evidences that electrolyte ions play important role in the adsorption process. The participation of electrolyte in the adsorption of many cations has been established not only from model fitting over a wide range of electrolyte concentration, but also from spectroscopic studies [335]. Therefore, the effect of ionic strength variation is dependent on various direct and indirect factors, such as type of metal ion, type of solid and the net effective charge carried by it, the type of electrolyte ion etc. Criscenti and Sverjensky [333] have made an exhaustive and critical review on this subject. There is no point in reproducing the results / data compiled by the authors here. The important findings by the authors will only be summarized below.

(i) Type of metal ion: Generally, the transition and post transition metal ions do not show any effect in their adsorption with increase in ionic strength. However, Cd^{2+}, Pb^{2+}, Zn^{2+} and Cu^{2+} exhibit both increasing and decreasing tendencies besides no effect with variation of strength, depending upon the type of solid as well as the nature of electrolyte solution . On the ther hand, alkaline earth metal ions do consistently show a decrease in adsorption with increase in ionic strength.

(ii) Type of solid: Except Cd^{2+}, Co^{2+} and Pb^{2+} which exhibit decrease in adsorption on solids like hydrous iron oxides , hematite , γ- Al_2O_3 and gibbsite, other transition metal ions either show no change or tendency to decrease with increase in ionic strength irrespective of the type of solid. Indeed, the properties of bulk solid are not the dominant factors in controlling adsorption with respect to the concentration of electrolyte solution .The most important property which governs cation or anion adsorption is the surface charge at the solid - liquid interface. While SiO_2, quartz (α- SiO_2) and δ- MnO_2 carry negative charge at pH above 3.0 - 3.5, hydrous oxides of Fe, Al and Ti carry positive charge in the entire pH range of adsorption. Adsorption on such solids shows either no change or shows decreasing trend. For alkaline earth metal ions adsorption occurs near the pH_{PZC} or on negatively charged surface and always exhibit decrease in adsorption with increase in ionic strength.

(iii) Type of electrolyte: Extensive literature survey carried out by Criscenti and Sverjensky [333] reveals the following features regarding the type of electrolyte on cation adsorption. (a) In $NaNO_3$ solution, adsorption of transition and heavy metal ions shows little or no effect with increase in ionic strength. In $NaClO_4$ solution, while adsorption in most cases tends to increase there are instances where adsorption shows either no change or decrease with in crease in ionic strength. In NaCl solution, adsorption invariably tends to decrease with increase in concentra-tion of electrolyte solution . (b) Adsorption of alkaline earth metal ion decreases with increase in concentrations of both $NaNO_3$ and NaCl solutions , but the decrease is sharper in NaCl solution. This clearly suggests possible interaction of anions with alkaline earth metalions resulting in the formation of complexes of lower charge which consequently show decrease in adsorption .

The critical review of Criscenti and Sverjensky [333] reveals that the effect of ionic strength on the shifting of adsorption edge cannot be used as an indicator for inner and outer sphere surface complex formation. The authors have used the "Extended Triple Layer Model"(ETLM) developed by Sahai and Sverjensky [337] to demonstrate that many of the earlier adsorption data, which show lack of ionic strength effect, do show some decrease with increase in ionic strength. In the conventional TLM the ratio of the activity coefficients of both dissolved and surface complex species is considered to be equivalent to electrostatic factor (see, ref.160 and also section 5.7 of the present monograph). In the ETLM, both activity coefficients and electrostatic factor are included in the mass law equation. Thus, for the surface reaction (10.3) the equilibrium constant according to ETLM is given by:

$$^{*}K_M^{\text{int}} = \frac{[SOM^+]\ [H^+]\gamma_{SOM^+}\gamma_{H^+}}{[SOH][M^{2+}]\gamma_{SOH}\gamma_{M^{2+}}} 10^{-F\psi_o/2.3RT} \tag{10.8}$$

The ratio of the activity coefficients of the surface complex species is, however, assumed to be unity.

To fit ETLM to metal ion adsorption data, Criscenti and Sverjensky have assumed the formation of various surface complexation species involving metal and electrolyte ions. While the neutral (SOH°) and deprotonated (SO^-) surface species are considered to reside in the bulk solid, dissolved ionic species such as M^{2+}, MOH^+, ML^+ or L^- adsorb onto the surface to form surface complexes. If the surface complexes are represented without any space between the surface site (S) and the dissolved ionic species, then the latter are supposed to occur in the o - plane closed to the surface. For example, when the complexes are expressed as SOM^+, $SOHM^{2+}$, SOML, $SOHML^+$ and SOMOH the corresponding dissolved ionic parts i.e., M^{2+} , ML^+ etc are supposed to occur in the o- plane having potential ψ_o. When the surface complexes are expressed with a hyphen or double dots between surface site and dissolved ionic species, the latter are are placed in the β - plane having potential ψ_β. For example, in surface complexes like SO-- M^{2+}, SOH - M^{2+}, SO-. MOH^+, SOH..ML^+ the species after hyphen or double dots are placed in the β- plane which is little away from the surface. For surface complexes of the mixed type such as SOM^+- L^-, $SOHM^{2+}$ - L^-, $SOHM^{2+}$- OH^- etc. M^{2+} is placed is placed in the o- plane and L^- and OH^- are placed in the β- plane. Accordingly, electrostatic factor in the mass law expression shall undergo modification like the conventional TLM. A few examples given below will illustrate the assignment of potential in the electrostatic factor. It is assumed here that dissolved or aqueous ionic species do not contribute to the electrostatic factor in mass law equation as they are considered at "infinite" distance from the surface.

(1) $$SOH + M_{aq}^{2+} + L_{aq}^{-} \leftrightarrow SOML + H_{aq}^{+} \qquad (10.9)$$

$$K_{M1}^{\text{int}} = \frac{a_{SOML}\, a_{H^+}}{a_{SOH}\, a_{M^{2+}}} \qquad (10.10)$$

As the surface complex does not carry any charge ψ_o = 0; the value of the electrostatic factor is therefore unity.

(2) $$SOH + M_{aq}^{2+} + L_{aq}^{-} \leftrightarrow SOM^{+} - L^{-} + H_{aq}^{+} \qquad (10.11)$$

$$K_{M2}^{\text{int}} = \frac{a_{SOM^+-L^-}\, a_{H^+}}{a_{SOH}\, a_{M^{2+}}\, a_{L^-}} 10^{F(\psi_o - \psi_\beta)/2.3RT} \qquad (10.12)$$

In this example the surface complex has two unit charges each on o - and β - planes.

This is further clarified in example 3.

(3) $$SOH + M^{2+}_{aq} + L^{-}_{aq} \leftrightarrow SOHM^{2+} - L^{-} \quad (10.13)$$

$$K^{int}_{M3} = \frac{a_{SOHM^{2+}-L^{-}}}{a_{SOH}\ a_{M^{2+}}\ a_{L^{-}}} 10^{F(2\psi_o - \psi_\beta)/2.3RT} \quad (10.14)$$

10.1.3 Effect of the Presence of Complexing Agent

The effect of the presence of complexing ligands in solution on the adsorption of metal ion has been studied quite extensively. Broadly, there are three types of effect depending upon the nature and extent of coordination of the concerned ligand .Several authors [338-340] have postulated that a non - symmetrical metal-ligand system may interact with the adsorbent surface forming three types of bond configurations. The type I configuration would result in a "metal - like" adsorption behavior i.e., adsorption increases with increase in pH value; type II configuration would imply "ligand - like" behavior i.e., adsorption decreases as pH value increases; and type III configuration postulates a combination of both metal - like and ligand - like adsorptive behavior. Type III is a limiting case in which the complex adsorbs much less strongly than either free metal or free ligand. These are schematically illustrated in Fig.10.3. The resulting surface complex may sometimes provide additional bonding possibilities for metal- ligand complex in solution due to certain specialized metal surface interaction like hydrogen and hydrophobic bonds as in the case of organic ligands [341] .This is exemplified by type II(2) in Fig.10.3. The value of x in ML_x may vary from 0 to 5, depending upon the coordination number of the metal ion and

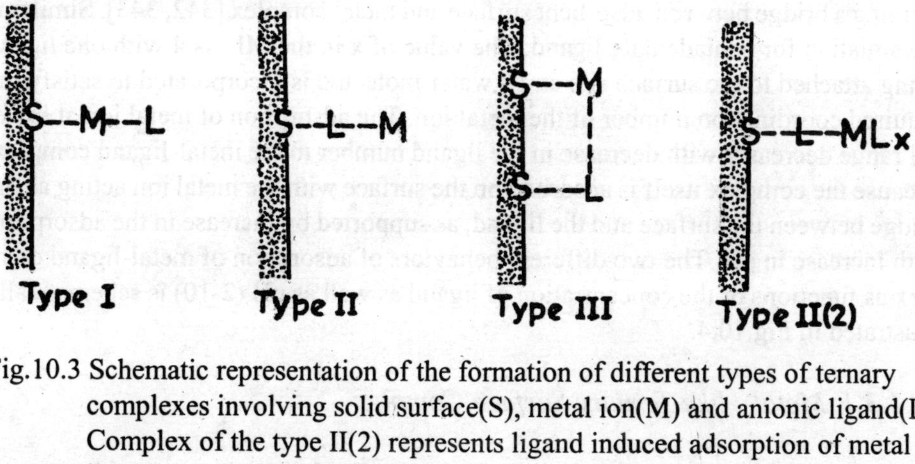

Fig.10.3 Schematic representation of the formation of different types of ternary complexes involving solid surface(S), metal ion(M) and anionic ligand(L). Complex of the type II(2) represents ligand induced adsorption of metal - ligand complex in solution.

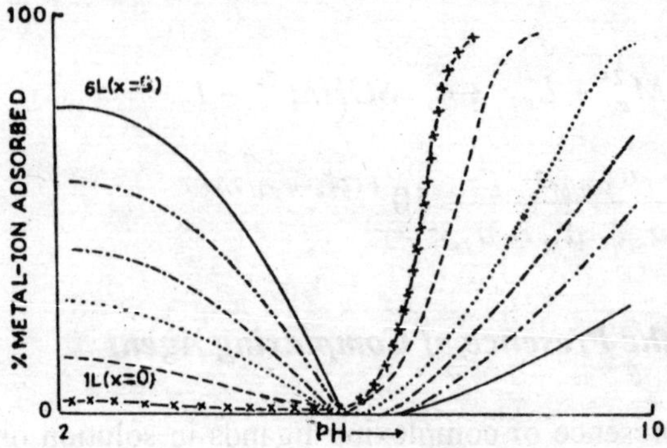

Fig.10.4 Conceptualized behavior of the adsorption of metal ion as function of pH in the presence of different concentrations of ligand. At low ligand concentration adsorption is predominantly metal like. As concentration increases adsorption gradually changes to ligand like i.e., higher adsorption at lower pH value. An opposite trend can be observed for the adsorption of free metal ion. Adapted after ref. 344.

the complexation equilibrium.

This is because, if the coordination shell of the metal ion is completely filled by atoms of the ligand L then ML_6 (assuming hexa-coordinated) cannot react with SO^-, except in the case where such deprotonated surface group can successfully compete with the ligand atoms of L. Since this is generally unlikely to occur, a ligand atom must first adsorb onto the surface and then the resultant surface complex can react with ML_x. Therefore, to increase and promote metal adsorption ligand must be adsorbable on solid surface, form complex with metal ion in solution and be capable of forming a bridge between adsorbent surface and metal complex [342, 343]. Similar is the situation for pentadentate ligand. The value of x in the ML_x is 4 with one ligand being attached to the surface site and a water molecule is incorporated to satisfy the required coordination number of the metal ion. The adsorption of metal ion at lower pH range decreases with decrease in the ligand number in the metal-ligand complex, because the complex itself is adsorbed on the surface with the metal ion acting as the bridge between the surface and the ligand, as supported by increase in the adsorption with increase in pH. The two different behaviors of adsorption of metal-ligand complex as functions of the concentration of ligand as well as pH (2-10) is schematically illustrated in Fig.10.4.

10.1.3.1 Metal - like Ternary Surface Complex

The formation of surface - metal - ligand ternary complexes, where metal ion acts as bridge between surface and ligand, can occur either by sequential bonding of the ions or by adsorption of the metal - ligand complex present in solution as such. Thermodynamically, these two mechanisms are equivalent and equilibrium adsorption data can no way distinguish between them. If metal like complexes are formed in a system, the adsorption edge shifts to alkaline side parallel to the ligand free system. This means that metal - like complex (type I) leads to decrease in adsorption. The formation of type I complex may be represented by the following reaction

$$SOH + M^{n+} + H_mL \leftrightarrow S-O-ML^{(n-m-1)} + (m+1)\,H^+ \qquad (10.15)$$

where H_mL is the fully protonated ligand. Complex formation of type I will lead to increase in adsorption of L at higher pH values; concomitantly, the adsorption of metal ion will decrease if metal ligand complexes are formed in solution. Consequently, slope value of the adsorption edge will decrease i.e., will be smeared (see Fig.10.4). A large volume of work has been carried out using organic complexing ligands, but only a few studies have been carried out on the adsorption of metal complexes with inorganic ligands. The presence of SO_4^{2-} and Cl^- has been reported to decrease Cd^{2+} adsorption onto goethite surface relative to the presence of inert NO_3^- ion in the pH range 5.5 - 8.0, but shows increase in the pH range 3-5 [344]. This is attributed to the formation of metal like ternary surface complex as shown below:

$$SOH + Cd^{2+} + Cl^- \leftrightarrow SO\text{-}Cd^{2+}\text{-}Cl^- + H^+$$

$$SOH + Cd^{2+} + SO_4^{2-} \leftrightarrow SO\text{-}Cd^{2+}\text{-}SO_4^{2-} + H^+$$

The effect of Cl^- and SO_4^{2-} on the adsorption of Cd^{2+} on goethite is shown in Fig. 10.5 which indicates that Cl^- is a stronger complexing agent than SO_4^{2-}. Analogously, adsorption of Ag^+ on ferrihydrite (amorphous HFO) is lowered by the formation of complexes of $S_2O_3^{2-}$, Cl^- and CN^- ions [338].

10.1.3.2 Ligand - like Ternary Surface Complex

In ligand -like adsorption of metal, complex ligand species adsorb more strongly than free metal ion at lower pH value to occupy substantial portion of the total available surface sites. With increase in pH value, free metal ion tends to adsorb more strongly, but the availability of adsorption sites has already been made limited due to the adsorption of complexed ligand species at lower pH and therefore adsorption decreases. In other words, the complex is the dominant surface species at lower pH and free metal ion at higher pH values. Whatever adsorption of metal that may occur with increasing pH is only metal - ligand complex with ligand acting as bridge between the surface and the metal. Therefore, the adsorption edge of metal - ligand complex looks similar to that of ligand alone. The ligand can act as bridge when it is multidentate so that it can coordinate with both surface and metal.

Let us assume that L is a tertradentate ligand coordinated to a hydrated metal ion of coordination number six i.e., $[M(H_2O)_2L]^{z+}$. If all the four ligand atoms are coordinated to the metal ion, the complex has two H_2O molecules which can be exchanged for surface ligands. Again, if the geometrical arrangements of the ligand atoms permit the simultaneous coordination of just two of the four ligand atoms along with the two H_2O molecules to the metal ion, then an alternative situation arises where the other two ligand atoms can replace surface OH groups [343]. The reaction which is facilitated at lower pH (< 5) is presented below.

$$2SOH + M(H_2O)_2L^{z+} \longleftrightarrow S_2 - L - M(H_2O)_2L^{(z+2)+} + 2OH^- \qquad (10.16)$$

Alternatively, if L is a bidentate ligand, then in the complex ML, two sites of the coordination shell of the metal ion are occupied by the ligand L and four sites by H_2O molecules. One of these four sites when coordinated with the surface, it results in the formation of ternary metal - like surface complex at higher pH values [343].

$$SOH + M(H_2O)_4L^{z+} \longleftrightarrow SO - M(H_2O)_3L^{(z-1)+} + H_3O^+ \qquad (10.17)$$

One of the classical and earliest examples of ligand - like adsorption of cation was demonstrated by Davis and Leckie [338] for the adsorption of Ag^+ on hydrous ferric oxide(HFO) in the presence of $S_2O_3^{2-}$ ion and is shown in Fig.10.6. The decrease was attributed to the formation of Ag $S_2O_3^-$ complex which shows poor adsorption at pH < 7. Subsequently, Balistrieri and Murray [345] observed that the presence of SO_4^{2-} ion increased the adsorption of trace metal like Cu, Pb and Cd on goethite. Similarly, Hoins et al [346] and Lamy et al [347] observed that Cd adsorption on goethite increased in the presence of sulphate and oxalate respectively. Adsorption of Cu on goethite was also found to increase in the presence of either chelidamic acid or sulphate ion, but in the presence of both the ligands adsorption of Cu was found to decrease [348]. It was proposed that SO_4^{2-} competed with Cu - ligand complex for surface sites and in addition, adsorbed SO_4^{2-} reduced the positive charge which also resulted in the suppression of Cu - ligand adsorption.

Mercury(II) ion does not exhibit quantitative adsorption on oxide adsorbents even in the absence of complex forming ligands, unlike divalent transition metal ions . The weak affinity of Hg^{2+} is attributed to its strong tendency for hydrolysis yielding $Hg(OH)_2$ which is complete even at pH little above 4.0. It is the pH value at which adsorption also attains saturation or plateau region [48]. In the presence of Cl^-, adsorption edge shifts considerably towards alkaline side indicating metal - like adsorption. The authors have shown that solution complexation constants of Hg^{2+} with OH^- and Cl^- are much higher than its binding constant with goethite surface. The values of constants of some relevant surface and solution phase reactions as determined by Gunneriesson and Sjöberg [48] are given below .

Table 10.1 Equilibrium constants of the formation of solution and adsorbed species of Hg^{2+} on goethite.

Reactions	log K
$FeOH^{o} + Hg^{2+} \leftrightarrow FeOHg^{+} + H^{+}$	-7.22
$Hg^{2+} + H_2O \leftrightarrow HgOH^{+} + H^{+}$	-2.59
$FeOHg^{+} + OH^{-} \leftrightarrow FeOHgOH$	6.56
$HgOH^{+} + OH^{-} \leftrightarrow Hg(OH)_2$	11.19
$FeOHg^{+} + Cl^{-} \leftrightarrow FeOHgCl$	2.98
$HgOH^{+} + Cl^{-} \leftrightarrow HgOHCl$	7.43

The stability constants of solution equilibria are greater by factors of $10^{4.6}$-$10^{4.4}$ than the surface binding constants. This large difference can possibly be the outcome of the influence of neighboring surface sites. The linear coordination between Hg^{2+} and surface OH ($FeOH^{o}$) is then disturbed by the attraction between Hg^{2+} ion and adjacent surface groups which resulted in the weaker attraction to that in aqueous medium [48].

In natural aquatic systems, the concentrations of adsorbent surface sites far exceed the concentration of Hg^{2+}. Under this condition and in the presence of low chloride concentration, adsorption increases considerably.

10.1.3.3 Ternary Surface Complexes with Organic Ligands

The influence of organic complexing ligands has been studied much more extensively than that of inorganic ligands . One of the most common and widely studied ligand is ethylene diamine tetra acetic acid (EDTA) and related chelating agents. This compound is used as a complexing agent in some electroless metal plating baths. EDTA or similar compounds are used in a large number of pharmaceutical and cos-

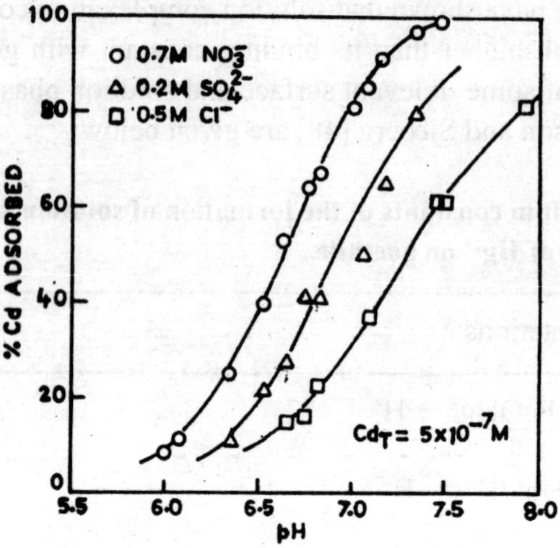

Fig.10.5 Effect of inorganic anions like NO_3^-, SO_4^{2-} and Cl^- on the adsorption of Cd^{2+} on HFO. While NO_3 is considered to be truly indifferent, Cl^- seems to have stronger complexation effect than SO_4^{2-} as indicated by the relative shift of the adsorption edge towards alkaline side. Redrawn using the data from ref. 344.

metic products, such as shampoos and toothpastes, as well as in a variety of food products. Various metal salts of chelating agents have been suggested as agricultural

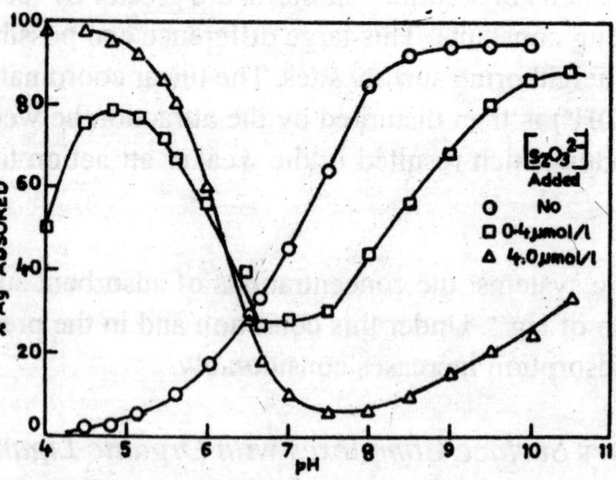

Fig.10.6 Effect of the addition of thiosulphate ion ($S_2O_3^{2-}$) on the adsorption of Ag+ as function of pH. While adsorption of pH < 7 indicates ligand like behavior, at pH > 7 adsorption is distinctly metal-like depending upon the concentra tion of ligand in the system. Redrawn after data from ref. 338.

additives to mobilize micronutrients in soils and render them more available for plant uptake [349]. As a result, its content in sewage and in some natural water systems has increased considerably in recent time. The structural formula of the compound is as follows:

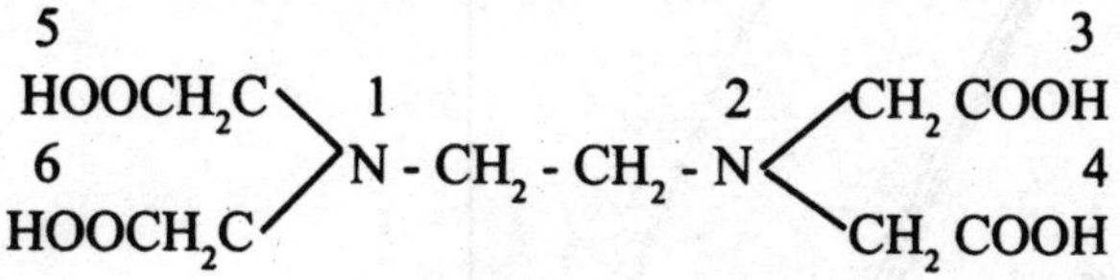

The divalent metals generally form pentadentate complexes, while trivalent metals like Co(III), Cr(III), Fe(III) etc tend to form hexadentate complexes. The two soluble complex species with M^{2+} ions are $MHEDTA^-$ and $MEDTA^{2-}$. While the concentration of the former decreases to almost zero with increase in pH value from 3 to 6, the concentration of the later species gradually reaches maximum at the pH value 6.5 - 7.0 [343]. The structural changes that take place during the deprotonation of the former is shown below:

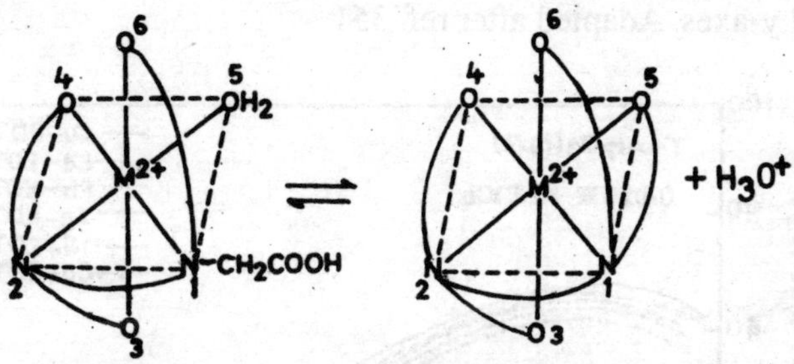

Fig.10.7 Structural change that takes place with change in the coordination (pentadentate to hexadentate) of M^{2+}- EDTA complex with increase in pH from 4 to 7.

The adsorption of metal - EDTA complexes on hydrous oxides of iron, aluminum, titanium and certain aluminosilicate compounds has generally been found to be ligand like with significant adsorption taking place at low pH and decreasing to almost zero adsorption at pH near PZC of the adsorbent. However, at very low pH values (< 3) the complexes are unstable and metal ions are released into the solution. The most significant and noteworthy characteristic of the adsorption on hydrous oxides of iron and aluminum is that most of the complexes with divalent metal ions belonging to transition and post transition groups that form stable complexes with EDTA (log K_{st} >16-17) show similar values of adsorption with respect to pH i.e., adsorption edges form cluster due to superimposition on each other [342, 350-352] Adsorption onto goethite surface decreases gradually from the maximum value at pH around 5.5 to almost zero adsorption at pH about 8.3. The adsorption of Pd - EDTA complex was

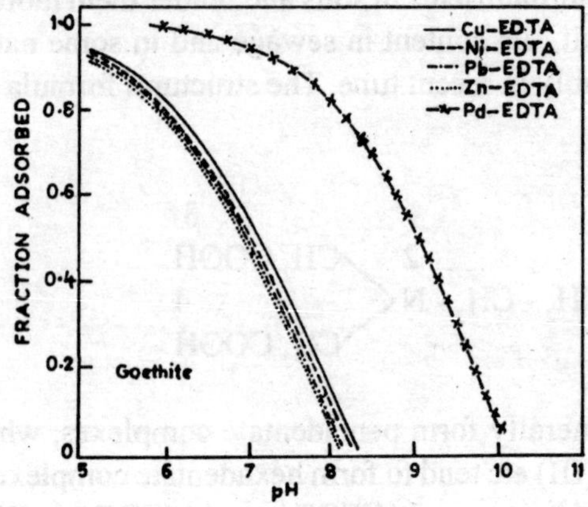

Fig.10.8 Characteristics of the adsorption of different metal ions on goethite in the presence of EDTA as function of pH. The closeness of the ligand-like adsorption edges of some transition metal ions can be noted. The curves are schematic in nature and do not necessarily correspond to the scales for x- and y-axes. Adapted after ref. 351.

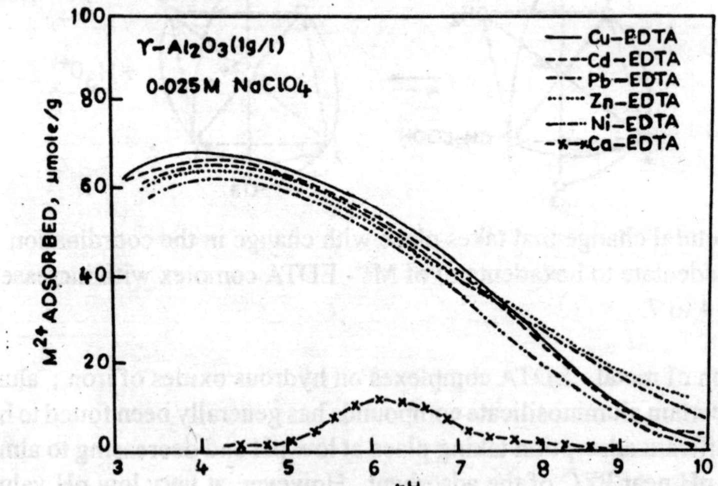

Fig.10.9 Characteristics of the ligand-like adsorption of some metal ions on γ-Al_2O_3 as function of pH in the presence of EDTA. The deviations of alkaline earth metal ions like Ca^{2+} may be noted. The curves are schematic in nature and do not necessarily correspond to the scales for x- and y- axes. Adapted after ref. 342.

found to be an exception with high adsorption [351] at relatively higher pH range. Figures10.8 and 10.9 illustrate schematic representations of such superimposed adsorption edges on goethite [351] and $\gamma - Al_2O_3$ [342] surfaces, respectively. It may be

Table 10.2 Equilibrium constant of the formation of some metal - EDTA complexes (Ionic strength = 0.1) *

Metal ion	log K$(MHEDTA)^-$	log K$(MEDTA)^{2-}$	Metal ion	log K$(MHEDTA)^-$	log K$(MEDTA)^{2-}$
Ag^+	3.07	7.32	Cr^{3+}		24.0
Ba^{2+}	2.07	7.78	Cu^{2+}	11.54	18.80
Ca^{2+}	3.51	10.57	Fe^{2+}	6.86	14.33
Mg^{2+}	2.28	8.69	Fe^{3+}	16.2	25.1
Cd^{2+}	9.10	16.59	Mn^{2+}	6.9	14.04
Co^{2+}	9.15	16.21	Hg^{2+}	14.6	21.80

*Source: Lurie, Ju., Handbook of Analytical Chemistry (English Transln.), Mir Publishers, Moscow, 1975, pp 297 - 298.

noted from Fig.10.9 that the adsorption edges on $\gamma - Al_2O_3$ are rather smeared and extend from pH 3 to 10, but does not reach zero value unlike adsorption on goethite.It may also be noted from Fig. 10.9 that the adsorption of Ca - EDTA complex is almost zero at pH about 5.0, but reaches maximum at pH around 6.5, before decreasing again to almost zero at pH 9 - 10. The results indicate that stability of Ca - EDTA complex is low (log $K_{st} \leq 11.4$) at lower pH values and therefore, adsorption of metal or metal like complex is poor. As the stability of the complex increases with increase in pH value , the surface sites compete with EDTA for free metal ion , rendering the Ca-EDTA complex the predominant aqueous phase species, the adsorption of which is normally poor, especially at low metal ion concentration.

Two important questions that may be raised from the study of the pH effect on adsorption of metal - EDTA complexes on hydrous oxides of iron and aluminum are: i) Which one of the two metal complex species, viz, $MHEDTA^-$ and $MEDTA^{2-}$ interact with the surface, or do they interact at the same time? ii) Which surface species interact preferentially with the metal - EDTA complex and under what conditions? Bowers and Huang [342] from their studies on speciation of Ni -EDTA complex as function of pH and temperature, have shown that at pH >5 $NiEDTA^{2-}$ predominates, while at pH <5 $NiHEDTA^-$ is the predominant species especially at elevated temperature (> 40°C). Consequently, adsorption species also follow similar behavior. Determination of the heat of adsorption at pH 3.0 shows that while free EDTA adsorbs exothermally, temperature has little or no effect on the adsorption of $NiHEDTA^-$ complex, suggesting that the latter is the preferred species at this pH.
However, analysis of both solution and adsorbed species shows that metal /EDTA

ratio in the adsorbed phase is generally 1:1 over a wide range adsorption condition, suggesting the possibility of the adsorption of both $MEDTA^{2-}$ and $MHEDTA^{-}$. At very low concentration of metal ion i.e., when $[M^{2+}] / [EDTA] < 1$, metal - EDTA complex tends to break down . Under this condition the surface competes for either "free ligand " or "free metal" species , depending upon the stability of metal - ligand complex relative to the corresponding metal - surface binding constant [342] . At low metal ion concentration surface-metal binding constant is stronger than the corresponding stability constant of metal - ligand complex. In this situation, metal - like adsorption will predominate, as excess free surface sites would out-compete EDTA to metal ion. At higher metal ion concentration, the stability constant of metal - EDTA complex becomes stronger than the surface binding constant and therefore ligand like adsorption will predominate by out competing metal ion to ligand. For the adsorption of Ni -EDTA complex on γ - Al_2O_3, low concentration refers to less than 10^{-5} M and high concentration signifies $>10^{-4}$M [342]. Besides, Fig. 10.7 suggests that penta-coordinated $M(H_2O)HEDTA^{-}$ is structurally more flexible for coordinating with surface groups than hexacoordinated $MEDTA^{2-}$ where all the four acetate groups are coordinated to the octahedral M^{2+} ion .

As far as the adsorptive surface group is concerned, the positively charged group predominates at pH below PZC. A good correlation between decrease in the concentration of positive surface charge and decrease in the adsorption of metal complex, both with respect to pH, can be noticed [342,351]. Further, when adsorption values of both free EDTA as well as its complex with metal ion are normalized by dividing with the fraction of positively charged surface species (in moles) an excellent correlation is obtained between the two quantities [342]. This quantitative evidence suggests that metal complexes exhibit an overwhelming preference for positive surface, with very little participation of either neutral or negatively charged groups. Metal - EDTA complexes are attached to SOH_2^+ through hydrogen bonding [353,354] . Spectroscopic (FTIR and EXAFS) studies do not provide any indication of the formation of inner sphere complex between ligand and surface metal ion for the adsorption of Pb - EDTA complex on goethite [355]. Therefore, hydrogen bonding is the only plausible mechanism of the adsorption of metal-chelate complexes. Consequently, all the divalent metals show similar adsorption behavior.

A hydrogen bond may form between two functional groups, one serving as proton donor or Broensted acid group (class A) and the other group serving as electron donor or Lewis basic group (class B). In addition, there is another group which may act as both proton and electron donors (class AB). These groups with requisite functional properties are present either on the surface of an adsorbent or in an approaching sorbate ion, which may be a cation or an anion. For an idealized structure of a partially coordinated oxide surface these are illustrated in Figs. 2.1(a - c) in chapter 2.

However, these are also schematically illustrated below as proposed by Boehm [80] for a hydroxylated oxide surface. According to him, surface species of types 1 and 3 should be the strongest generators of hydrogen bonding with nitrogen of amine or oxygen of carboxylic acid in EDTA . Though type 2 is relatively week generator of H^- bonding, but can form or participate actively in H - bonding if necessity arises. Therefore, modeling of the adsorption of metal - EDTA complex solely as $MEDTA^{2-}$ without stating pH range is not appropriate. The results, however, may not differ much [356] even though the electrostatic factor will differ because of the difference in the charge of the two metal - complex species.

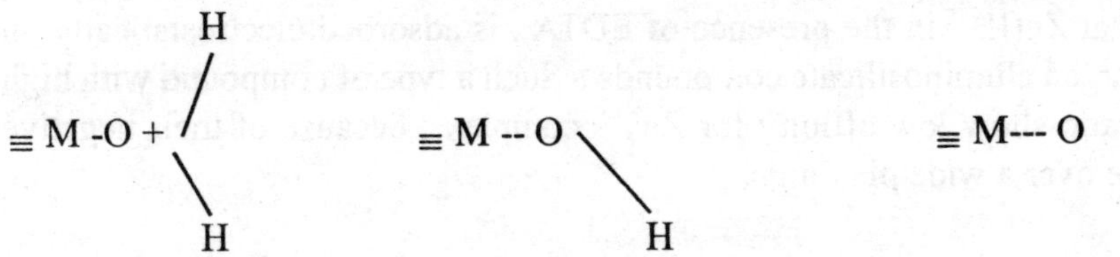

Class A (Type 1) Class AB (Type 2) Class B (Type 3)

The high adsorption of divalent metal ions in the presence of EDTA on to the surfaces of hydrous oxides of iron and aluminum in the pH range where positively charged surface groups are predominant, suggest ligand -like adsorption with EDTA acting as bridge between surface and metal ion. If, indeed, SOH_2^+ groups were solely responsible for the adsorption of metal - EDTA complex , and if this mechanism were independent of the solid type, then adsorption of complex would not be expected on those solids whose pH_{PZC} values are low, such as silica, zeolites, mordenites etc.This has actually been observed by Bowers and Huang [342]. Adsorption of metal - EDTA complexes on such oxides generally behaves metal - like with metal ion acting as bridge between surface and ligand.

Zachara et. al. [357] have observed that Co-EDTA complex (CoY^{2-}) is stable above pH 6 , but dissociates to some extent below pH 6 , promoted by the surface dissociation of adsorbents like goethite , gibbsite etc. A non-electrostatic surface complexation model was found to describe adequately the sorption using surface complexes with CoY^{2-}, FeY^- and the dissociated product Co^{2+} . For adsorption on subsurface sands consisting of hydrous oxides of Al , Fe , and Si , the dissociation of CoY^{2-} yielded multi component mixtures of Co^{2+}, CoY^{2-}, AlY^- and FeY^- sorbing onto sediment surface to varying extents.

The study of the adsorption of Co-EDTA complex assumes special importance from environmental aspect of the mobilization of ^{60}Co, a highly hazardous nuclear waste, in ground water. Though the EDTA complexes of both Co(II) and Co(III) show ligand like adsorption at lower pH , Co(III) complex (CoY^-) show much less adsorption than the corresponding Co(II) complex (CoY^{2-}) under similar conditions on subsurface aluminosilicate compounds [358] . Therefore, it was observed that CoY^- rather than the CoY^{2-} was responsible for enhanced migration of ^{60}Co in ground water. This is because, at low pH and high Eh (potential) values Co(II) is oxidized to Co(III) which in complexated state with EDTA exhibit enhanced mobilization [359].

Nagy and Kónya [360,361] have observed ligand like adsorption of Pb^{2+} and Co^{2+} on Ca - montmorillonite in the presence of EDTA . It was observed that the stability of PbY^{2-} increases with increase in pH , but that of $PbHY^-$ decreases . Huang et al . [362] have found that Zn(II) , in the presence of EDTA , is adsorbed electrostatically on positively charged aluminosilicate compounds . Such a type of compound with high Si /Al mole ratio show low affinity for ZnY^{2-} complex , because of their negative surface charge over a wide pH range.

Güclu and Apak [363] have studied adsorption of Cu^{2+}, Cd^{2+} and Pb^{2+} on acid treated red mud in the presence of EDTA . The treated red mud (a waste product of aluminum industry) is a typical composite adsorbent and its main composition as reported by the author is as follows : Fe_2O_3 41.2% , Al_2O_3 19.4% , SiO_2 23.8% , TiO_2 7.2% . In view of the complex nature of the adsorbent it is not possible to distinguish the adsorptive power of the individual oxides towards the total adsorption capacity, though most of the adsorption sites are confined to Al_2O_3 and Fe_2O_3. Therefore, not only a single representative surface site (SOH) , but also single values of surface acidity constants (K_{a1} and K_{a2}) were assumed for model calculation. Consequently, only one pH_{PZC} value was determined by the conventional acid - base titration method; but the estimation of other EDL parameters for such a composite material is confronted with considerable uncertainty. Therefore, the authors used non - electrostatic surface complexation model by excluding electrostatic factor in the equilibrium constant of the surface reaction. Since the three metal ions adsorb very strongly on treated red mud, the coulombic contribution to the overall free energy change of adsorption is assumed to be very small [364]. Accordingly, Güclu and Apak [363] proposed three different modes of adsorption of metal ions . These are as follows:

1. Adsorption of free or uncomplexed metal ions (M^{2+}) on SOH^o as covalently bonded surface complex (SOM^+) which is also known as inner sphere complex

$$M^{2+} + SOH^o \leftrightarrow SOM^+ + H^+ \quad (10.3)$$

$$K_{is.1}^{M} = \frac{[SOM^+][H^+]}{[M^{2+}][SOH^o]} \quad (10.18)$$

2 . Adsorption of negatively charged metal - EDTA complex (MY^{2-}) onto positively charged surface via electrostatic interaction as outer sphere complex .

$$SOH_2^+ + MY^{2-} \leftrightarrow SOH_2^+..MY^{2-} \tag{10.19}$$

The above reaction may be represented more realistically as follows:

$$SOH^o + H^+ + MY^{2-} \leftrightarrow SOH_2^+..MY^{2-} \tag{10.20}$$

$$K_{OS.2}^M = \frac{[SOH_2^+ \cdot \cdot MY^{2-}]}{[MY^{2-}][SOH^o][H^+]} K_{a1} \tag{10.21}$$

where, K_{a1} is the protonation constant according to Eq . (3.1)

3. Adsorption of MY^{2-} complex on SOH^o through partly covalent bonding to form outer sphere complex either by interaction between partially uncoordinated metal complex with surface oxygen , or through hydrogen bonding between oxygen of the carboxylic acid groups of the metal - EDTA complex and surface hydrogen .

$$MY^{2-} + SOH^o \leftrightarrow SOHYM^{2-} \tag{10.22}$$

$$K_{0S.3}^M = \frac{[SOHYM^{2-}]}{\left[MY^{2-}\right]\left[SOH^0\right]} \tag{10.23}$$

The process explains the observed low uptake of MY^{2-} at pH greater than pK_{a1} , and thus represents the covalent contribution to outer-sphere complex formation.

The equilibrium constants for the above three adsorption mechanisms were determined analytically by rearranging Eq. (10.18), (10.21) and (10.23) to appropriate linear forms based on the conservation of mass. No dissociative adsorption of MY^{2-} complex to yield SOM^+ or partial dissolution of the constituent oxides of red mud by MY^{2-}, within the pH and concentration range, to yield FeY^- and AlY^- aqueous complexes were assumed to takes place. The concentration of free metal ion, surface complex and other relevant species involved in the mass law equations can be obtained from the quadratic equations derived also on the basis of the conservation of mass. For example, Eq. (10.21) can be rearranged as follows if $SOH^{2+}.. MY^{2-}$ is considered as equivalent to metal ion adsorbed (M_{ads}) in moles/g.

$$\frac{\{M_{ads}\}}{\{N_s - M_{ads}\}[H^+]} = K^M_{OS.2}\ K^{-1}_{a1}\ [MY^{2-}] \tag{10.24}$$

Taking logarithm on both sides of the equation one gets

$$\log\frac{\{M_{ads}\}}{\{N_s - M_{ads}\}} = \log(K^M_{OS.2}\ K^{-1}_{a1}) + \log[MY^{2-}] - pH \tag{10.25}$$

According to Güclu and Apak [363] the linear plot of LHS of Eq. (10.25) against log $[MY^{2-}]$- pH gives log as intercept, from which one can calculate $K^M_{OS.2}$ provided K^1_{al} value is already known from potentiometric titration of treated red mud sample.

Since $\{M_{ads}\} = 1/r\,(M_o - [MY^{2-}])$, where M_o is the initial concentration (analytical) of MY^{2-} in moles/l and r is the solid concentration (g/l), the following set of relationships can be easily derived:

$$\{M_{ads}\} = \{SOH_2^+\}[MY^{2-}] \tag{10.26}$$

$$\{SOH_2^+\} = \{[H^+]/(K_{a1} + [H^+])\}\{N_s - M_{ads}\} \tag{10.27}$$

$$[MY^{2-}] = [M_o - r\{M_{ads}\}] \tag{10.28}$$

Combining Eqs. (10.24) , (10.26) , (10.27) and (10.28) and on rearranging the following relationship can be obtained.

$$\{M_{ads}\} = \frac{K^M_{OS.2}\ \{N_s - M_{ads}\}[H^+]}{(K_{a1} + [H^+])[M_o - r\{M_{ads}\}]} \tag{10.29}$$

Transposition and further rearrangement of the above equation lead to the following quadratic equation for M_{ads}.

$$r\{M_{ads})^2 - \{M_{ads}\}\left\{\frac{1}{K^M_{OS.2}}\left(\frac{K_{a1}}{[H^+]}+1\right)+[M_o]+rN_s\right\}+M_oN_s = 0 \tag{10.30}$$

Solution of the above quadratic equation gives the amount of metal ion adsorbed, once the values of $[M_o]$ and $[MY^{2-}]$ are determined experimentally. Using the non - electrostatic surface complexation model, the authors were able to show satisfactory fitting of experimental adsorption data with those calculated for adsorption , either as free M^{2+}or metal-EDTA complex or combination of both, depending upon the experimental conditions [363].

Nitrilo triacetic acid (NTA) is also a well known complexing agent for transition and other heavy metal ions, but is not as strong as EDTA. Studies of the adsorption of Co(II) - NTA on gibbsite [364] and Pb -NTA on TiO_2 show that complexation has little effect on the adsorption of metal ions. Obviously, the oxide surfaces compete for metal ions by forming weak mono and bi-nuclear metal-like complexes. Schindler and others [343] have also studied adsorption of metal complexes with various ligands, such as ethylene diamine, oxalate, glycine, alanine, 2-2' bipyridine etc onto silica surface.

10.1.4 Effect of Cation Concentration (Sorbate / Sorbent Ratio)

The effect of increasing cation (sorbate) concentration, or in other words, increase in sorbate / sorbent ratio, is manifested in the nature of Freundlich type isotherm. On a log-log plot of adsorption density vs. concentration, the ideal Langmuirian (or pseudo-Langmuirian) behavior is characterized by unit slope value at low concentration which on increasing beyond a certain range results in a sharp decrease to almost zero slope value (see Fig .8.1). Deviations from this ideal behavior such as lower slope value (<1.0) at low concentration , gradual decrease instead of sharp decrease in slope value at higher concentration followed by a rise in adsorption density etc, as illustrated in Fig .10.10, indicate either multiplicity of site type or surface precipitation due to high sorbate / sorbent ratio. In the case of cation adsorption, its hydroxide may precipitate at the surface prior to its actual formation in the bulk solution and thus contribute to total adsorption. This has been clearly demonstrated by spectroscopic studies [365,334] and also by model calculation [323]. The contribution of the surface precipitation increases with increase in the sorbate / sorbent ratio. The situation also manifests itself in the shifting of adsorption edge to higher pH value i.e., more alkaline side parallel to the adsorption edge at low concentration of cation. Plateauing of pH adsorption edge is seldom observed, because of the difficulty in saturating the binding sites before the onset of precipitation. When precipitation in the bulk solution occurs, the adsorption edge becomes almost vertical.

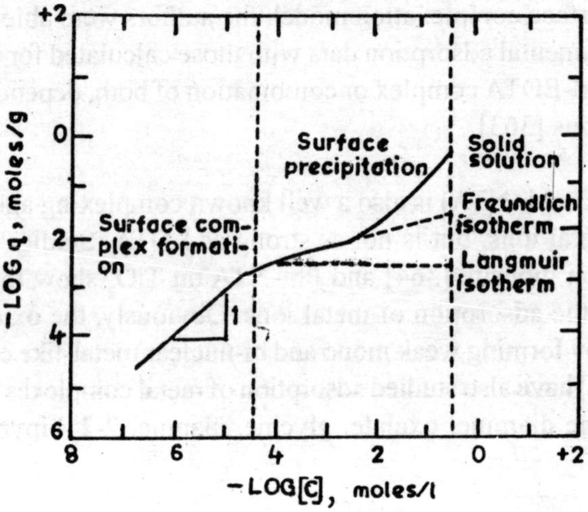

Fig.10.10 Different types of non-Langmuirian behavior of adsorption isotherm at high sorbate / sorbent ratio. The scales of x- and y-axes are meant to

However, some authors have argued that the above phenomena regarding the changes in the nature of adsorption isotherm or shift in the adsorption edge with increase in the sorbate / sorbent ratio do not provide conclusive evidence that precipitation has taken place. For example, according to Sposito [366] the adsorption isotherms should be regarded simply as curve fitting devices, *albeit* with predictive capability; and provide no evidence as to the mechanism of the adsorption process. The dichotomy between multisite adsorption and surface precipitation, as manifested by similar changes in Freundlich type isotherm, has not yet been cleared. Similarly , the shift in the adsorption edge with increase in the metal ion concentration does not provide direct evidence of the adsorption of hydrolyzed or even precipitated species , unless the ion activity product (IAP) for a proposed surface precipitate exceeds the corresponding solubility product constant (SPC) . This is briefly explained as follows:

By knowing the composition of the aqueous solution phase, the activity of an electrolyte having the same chemical formula as the assumed precipitate can be calculated [367]. This calculation may utilize either mean ionic activity coefficients and total concentration of ions in the electrolyte solution, or single ion activity coefficient and free concentrations of the different ionic species in the solution. The latter approach to calculating the activity of an electrolyte is called ion activity product or IAP. The calculated activity is compared with the SPC of the assumed precipitate (K_{SP}) as a test for the existence of a solid phase. If the calculated IAP is smaller than the K_{SP} of the precipitate, the corresponding solid phase is considered to be absent, at least during the period of solubility measurement. However, this procedure is also confronted with the problem of precision with which the activity product can be determined. This is because IAP includes not only the Gibbs free energy change in a solid dissolution process, but also the activity of the solid itself. Let us consider for example, the dissolution of $CdCO_3$ (s) for which IAP is given by

$$IAP = a_{Cd^{2+}} \times a_{CO_3^{2-}} = K_{SP}\, a_{CdCO_3\,(s)} \tag{10.31}$$

where, $K_{SP} = 10^{-11}$ at 298K . If Cd(II) is coprecipitated with another metal e.g., Ca(II) to form a solid solution then $a_{CdCO_3(s)} < 1$, and therefore IAP < K_{SP}. Thus, a homogeneous or mixed surface precipitate can be expected to yield typically low IAP values and if chemical species of interest occurs in trace quantities, the discrepancy between IAP and K_{SP} can be of the order of one. A low IAP value of this order suggests the absence of surface precipitation as under- saturation exists in the solution phase. However, the method is not full-proof as the error in the calculation of IAP or K_{SP} is apparent [366].

The other methods used to distinguish adsorption from surface precipitation involve determination of zeta (ζ) potential and reaction kinetics. The application of ζ - potential in this regard is due to the experimental work of James and Healy [328] who have shown that in the presence of hydrolysable cations, ζ - potential of an oxide colloid (mainly silica) varies with pH in this way: At first, the potential decreases gradually to a minimum value in the negative side, followed by gradual rise to a maximum value (above zero or on positive side) and decreases again even to a negative value (see Fig.10.11). According to the authors, the initial decrease to a minimum value with increase in pH is attributed to gradual neutralization of negatively charged colloid

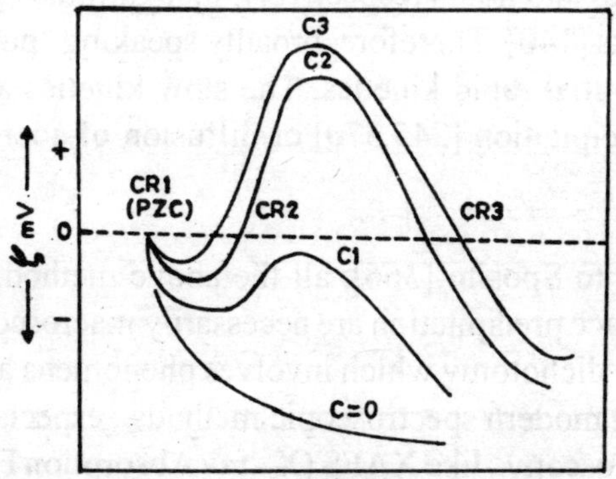

Fig.10.11 Schematic representation of typical nature of the variation of zeta(ζ) potential with pH for some oxides (e.g., SiO_2 and δ-MnO_2) and clay minerals (e.g., kaolinite) whose PZC lie at relatively lower pH values.

loid surface due to adsorption of cation, which may be in partially hydrolyzed form, until the surface is saturated with the adsorbed species. Zeta potential also increases accordingly until maximum value is obtained through a second charge reversal (CR2) Subsequent decrease in ζ-potential with increase in pH is attributed to the reversal of charge (CR3) caused by coating of hydroxy polymers of cation and not to the original adsorbent surface. This interpretation is applied to any bivalent, trivalent or even tetravalent metal ions that tend to hydrolyze above pH 6 in aqueous solution. The concentration of hydrolytic species is not relevant even if it is quite small. Its adsorption is determined by the affinity of the adsorbent for it and total metal ion concentration (C1, C2, C3, etc.) in aqueous solution [328]. If a precipitate of metal hydroxide polymer is formed on an adsorbent, the above pattern of the variation of ζ- potential with pH should resemble closely with colloids consisting of pure hydroxide or hydroxyl polymer of the relevant sorbate metal ion [328,368].

However, question has been raised on two counts. First, surface precipitation is postulated to occur at pH below the experimentally found onset of metal ion hydrolysis. Second, the observed ζ potential - pH relationship can be explained even without invoking surface precipitation. Adsorption of fully hydrated cation can be enhanced by decreasing the surface charge with increase in pH and concomitant increase in ζ-potential by bringing the positive charge at the interface. Subsequent decrease in ζ-potential is attributed to the formation of hydrolytic species at the expense of the adsorbed cation which begins to desorb [328,369].

The other macroscopic method that attempted to distinguish between adsorption and precipitation is the study of the kinetics of adsorption. It has generally been observed that the rates of adsorption of cations and anions are initially very fast, operating in time scales of minutes and hours respectively, then diminish gradually to time scales of days or even weeks [140]. Therefore, broadly speaking "pure" adsorption of cation is reflected by the initial rapid kinetics. The slow kinetics are interpreted either in terms of surface precipitation [245,370] or diffusion of adorbate into the adsorbent [371].

However, according to Sposito [366], all the above methods aiming to distinguish adsorption from surface precipitation are necessarily macroscopic in nature and therefore cannot solve the dichotomy which involves phenomena at the molecular level. It is recommended that modern spectroscopic methods , especially those relating to X-ray absorption spectroscopy , like XAFS (X - ray Absorption Fine Structure) , EXAFS (Extended X-ray Absorption Fine Structure) , XANES (X-ray absorption Near Edge Spectroscopy) etc. should be used as far as possible to explain the mechanism of either cation or anion adsorption on inorganic minerals.

Farley et al [323] were the first to incorporate surface precipitation in the modeling of cation and anion adsorption on amorphous hydrous ferric oxide (HFO) at high sorbate / sorbent ratio. The essential aspects of their postulation are presented below, However, for more details the readers may consult the original article by the authors [323] and the monograph by Dzombak and Morel [27] for numerical method of calculation to fit adsorption data at high sorbate / sorbent ratio.

According to this model, at high cation concentration adsorption takes place through the precipitation of its hydroxide. The model is an extension of the usual surface complexation model for the adsorption of cations at low or moderate concentration by considering the formation of a new surface phase which is a solid solution of the hydroxides of the metals of both sorbent and sorbate(or solute). The activities of the constituent hydroxides are assumed to be equal to their mole fraction in the solid solution. At low cation concentration the mole fraction of the surface precipitate in the solid solution is very low. As the concentration increases, the proportion of surface precipitate increases while that of surface complex decreases until the surface precipitation becomes the dominant mechanism. At very high concentration of cation, the mole fraction of the surface precipitate becomes high due to the onset of bulk precipitation in the solution phase. Farley et al [323] postulated that besides reactions (10.1) and (10.2) metal ion is also adsorbed as $M(OH)_2$. Consequently, mass balance requires mass law equations i.e., solubility product constant for calculation of the activities of all the surface species. The reactions proposed by the authors using HFO as adsorbent are as follows:

(a) Adsorption of M^{2+} on $Fe(OH)_3$ (s)

$$\equiv FeOH^{o} + M^{2+} + 3H_2O \overset{K_M^{App}}{\longleftrightarrow} Fe(OH)_3\,(s) + {>}MOH_2^{+} + 2H^{+} \qquad (10.32)$$

(b) Precipitation of M^{2+} as $M(OH)_2(s)$

$$> MOH_2^{+} + M^{2+} + 2H_2O \overset{1/K_{SPM}}{\longleftrightarrow} M(OH)_2(s) + {>}MOH_2^{+} + 2H^{+} \qquad (10.33)$$

(c) Precipitation of Fe^{3+} as $Fe(OH)_3$ (s)

$$\equiv FeOH^{o} + Fe^{3+} + 3H_2O \overset{1/K_{SP\,Fe}}{\longleftrightarrow} Fe(OH)_3\,(s) + \equiv FeOH^{o} + 3H^{+} \qquad (10.34)$$

The reaction for the adsorption of Fe^{3+} on $M(OH)_2(s)$ is not shown as it can be derived from reactions (10.33) and (10.34) and hence is not an independent reaction. When the surface species $>MOH_2^{+}$ and $\equiv FeOH^{o}$ are divided out from mass law equations of the reactions (10.33) and (10.34) the constants become the solubility products of $M(OH)_2(s)$ and $Fe(OH)_3$ (s) respectively with respect to bulk solution. Therefore, if the activity of any solid species in the above two equations is less than unity then it can precipitate at pH value lower than under "normal" conditions i.e., when the activity of any of the solid species is equal to unity.

10.2. Proton Stoichiometry or Ratio of Proton Released / Metal ion Adsorbed

This is one of the most widely discussed topics in the field of cation adsorption on inorganic solids , especially on hydrous oxides , but without arriving at any definite conclusion on the value of H^+ released per mole of metal ion adsorbed which is also known as proton stoichiometry. This is because the ratio is such a quantity that depends highly on the nature of metal ion and its concentration, pH, adsorption density, ionic strength of the solution, nature of any complexing agent present in the solution, nature of solid adsorbent and its composition and so on. In fact, the release of proton is directly related to change in the mechanism of adsorption which manifests itself in the change in the value of proton stoichiometry.

Theoretically, the ratio of H^+ released / metal ion adsorbed may be either one or two depending upon the formation of either mono- or bidentate inner sphere complex with the surface OH groups. In such a case the approaching hydrated cation undergoes partial dehydration so that the metal ion can form coordination bond with surface oxygen by displacing proton attached to oxygen [249]. But in the case of outer-sphere complex water molecules of the hydrated ion participate in the formation of hydrogen bonding with oxygen of the adsorbent surface. This process involves the release of lesser proton from the surface than that required for the formation of inner-sphere complex [285].

Further, there are myriad of possible surface reactions, especially in the presence of complexing agents [333]. These reactions do not follow the usual stoichiometry with respect to the release of H^+. Therefore, there is no single value of H^+/M^{2+} ratio which is characteristic of a particular cation or of a particular adsorbent. This is evident from the compilation of ratio obtained by different authors for a number of adsorption systems as shown in Table 10.3. It may however be noted from the table that alkaline earth metal ions generally show lower values than transition and post-transition metal ions. Similar trend is also observed regarding the selectivity of metal ions towards adsorption [214]. This suggests that affinity of a cation for adsorption is related to its ability to remove H^+ from the surface i.e., to form stronger bond with the surface (inner-sphere complex) even at pH below PZC. In many cases the H^+/M^{2+} ratio tends to increase with increase in pH value. This indicates that the dominant surface species changes from free metal ion to its hydrolyzed species which are known to adsorb more strongly than free ion . Since hydrolysis of metal ion releases more protons in solution, the proton stoichiometry therefore tends to increase with increase in pH.

The value of H^+/M^{2+} ratio is seldom found to be an integer or close to an integer value. In fact, non-integral or fractional values are more common as can be noted from Table 10.3. Proton release as represented by reaction between surface functional groups and M^{2+} is related to the stoichiometry at molecular level [27,40,372]. However, proton release as observed in experimental systems is the net macroscopic result of adsorption reaction accompanied by surface ionization and ion pair formation to a new equilibrium condition. This is evident from the shift in the acid base titration curve and consequently pH_{PZC} of the surface. For example, experimental measurement shows the release of one proton per mole of Pb^{2+} ion adsorbed on goethite as inner sphere complex which is also supported by spectroscopic studies [40]. However, model calculation by TLM indicated that best fit was obtained for the formation of outer-sphere complex with the release of two protons per mole of Pb^{2+} adsorbed [160]. Similar discrepancy between calculated and observed values was noted when DDLM was used for the theoretical estimation of proton stoichiometry.

According to Dzombak and Morel [373], the stoichiometry of proton release for the adsorption of metal ion (e.g., Zn^{2+} on HFO) is not as simple as it may appear from adsorption reaction. They have shown that it is the combination of surface composition and electrostatic factor that primarily controls the magnitude of the ratio. However, in this regard the authors have made certain assumptions for their calculations which may appear to be not easily comprehensible. Using DDLM they have plotted

Table 10.3 : Proton Stoichiometry (H^+/M^{2+} adsorbed) for Some Common Metal Ion and Adsorbent Systems

Adsorbent (solid)	Metal ion	pH range	(H^+ released / M^{2+} adsorbed) ratio	Reference
Goethite	Cu^{2+}	-	2.4 ! 0.5	46
(α- FeOOH)		3.8 - 5.9	1.1	276
		5.0 - 6 0	1.3 - 1.7	285
	Cd^{2+}	-	2.2 $\pm$ 0.6	46
		5.8 - 9.1	0.4	276
	Co^{2+}	-	2.3 $\pm$ 0.7	46
	Pb^{2+}	-	2.0 $\pm$ 0.5	46
		6.0 - 8.0	1.0	40
	Zn^{2+}	-	2.2 $\pm$ 0.3	46
Amor. HFO	Ag^+	6.5 - 9.0	0.9	286
(Fe_2O_3, nH_2O)	Ca^{2+}		0.95	214

	Cd^{2+}	6.6 - 7.2	1.67	286
		-	1.80	385
	Cu^{2+}	5.1 - 5.5	1.90	286
		-	1.88	385
	Pb^{2+}	5.0 - 6.0	1.2 - 1.6	380
		4.5 - 5.0	1.98	286
		-	1.65	385
	Zn^{2+}	6.0 - 7.5	1.04 - 1.07	283
		6.0 - 6.5	2.30	286
		-	3.20	385
	Ni^{2+}	6.2 - 8.2	1.1 - 1.3	283
	Co^{2+}	6.5 - 8.0	0.96 - 1.38	283
Akaganeite (β- FeOOH)	Zn^{2+}	6.0 - 8.2	1.03 - 1.36	283
	Ni^{2+}	5.5 - 8.5	0.43 -1.42	283
	Co^{2+}	6.0 - 8.0	0.86 - 1.78	283
β- MnO_2	Cd^{2+}		1.3	374
	Co^{2+}	-	2.10 - 0.26	296
	Pb^{2+}	6.0	1.4	374
	Mn^{2+}	-	1.1 - 1.7	375
	Zn^{2+}	6.0	1.1	374
		-	2.0 - 2.1	296
Birnessite	Ba^{2+} , Sr^{2+}	3.0 - 7.0	0.45 - 0.5	382
	Mg^{2+} , Ca^{2+}	,,	<0.25	382
	Mn^{2+} , Co^{2+}	,,	1.03	382
	Ni^{2+}	,,	1.18	382
	Zn^{2+}	,,	~ 0.14	382
γ - Al_2O_3	Ca^{2+}	-	1.0 - 1.3	377
	Cd^{2+}	6.0 - 8.0	1.2 - 2.0	285
	Pb^{2+}	4.0 - 7.0	1.5	39
SiO_2 (amor.)	Ca^{2+}	-	1.0	214
	Zn^{2+}	6.0 - 7.5	1.2	381
	Co^{2+}	-	1.1 - 1.7	379
	-	6.3 - 7.6	1.18 - 1.34	381
	Ni^{2+}	6.3 - 7.8	1.36	381
α - TiO_2	Cd^{2+}	5.0 - 7.0	0.9 - 1.1	285

logarithm of concentrations of various surface species including species like $FeOZn^+$ against pH. From the slope value of the linear region the authors obtained the stoichiometric ratio which was about 2.0 for the above adsorption system . According to the authors, the effect of pH on the adsorption of metal ion is not similar to the release of H^+ during the adsorption. In other words, both the phenomena are not mirror image of one another. The competition for surface sites as well as the coulombic effect, contribute to the proton exchange, so that the ratio is about 2 under all conditions. At low pH and at low ionic strength the dominant surface species is $FeOH_2^+$. Adsorption therefore takes place by displacing two H^+ ions from $FeOH_2^+$ sites according to the following reaction

$$Zn^{2+} + FeOH_2^+ \longleftrightarrow FeOZn^+ + 2H^+$$

With increase in pH, the removal of surface protons tends to reduce the competition between adsorbed cation and H^+ and adsorption takes place mainly on monoprotonated surface sites (FeOH) which also releases two protons for the adsorption of one mole of Zn^{2+}. But, these ideal arguments are seldom found to be valid in actual experiments due to many other factors cited in the beginning of this section.

There are mainly three methods of determining macroscopic proton stoichiometric ratio for the adsorption of metal ion on hydrous oxides. The first method is the one proposed by Kurbatov et al [256]. It is based on the linearization of mass law equation for the interaction of adsorbing ion on undissociated surface sites i.e., SOH^o. The relevant relation ship has been shown in Eq(8.26) . The experimental results, however, show considerable variation with increase in adsorption density and pH. For the adsorption of Ni, Co, Mn and Zn on HFO and akaganeite, Kanungo [283] has obtained fractional values varying from 0.8 to 1.5 depending upon the nature of electrolyte solution and pH range of adsorption. For the adsorption of Ni , Co , Cu and Zn on δ-MnO_2, Kanungo et al [284] have found much lower values (0.4 - 0.6) of proton stoichiometry , mainly because of high adsorption density in the concerned pH range i.e., 2.75 - 5.0 . Therefore, despite its usefulness as curve fitting parameter, the proton stoichiometry (designated as χ_p) values derived from Kurbatov type plots have deficiencies which restrict their uses beyond certain range of pH value and at high adsorption density where the linear relationship does not hold good. Therefore, slope values of Kurbatov plots are always expressed along with the explicit mention of pH range of adsorption.

The second method which was originally proposed by Hohl and Stumm [39], consists in titrating the solid adsorbent batch-wise with acid or base of known concentration in an electrolyte solution of known ionic strength. The procedure is essentially similar to batch titration method described in section 4.2, except that titration is carried out also in the presence of metal ion in addition to that carried out in absence of metal ion (blank) under the conditions exactly similar to that used in the presence of metal ion. The blank titration is held constant for a known amount of solid and ionic strength of the medium and need not be carried out repeatedly. These two titration data are plotted against corresponding equilibrium pH. The difference between the two sets of titrations carried out in the presence and in the absence of metal ion represents H^+ or OH^- ion released during adsorption. Since such titrations involve readjustment of surface charge balance due to adsorption, simple correlation between $\Delta(C_A - C_B)$ values and the moles of metal ion adsorbed will not provide true picture of proton stoichiometry . By using the TLM Hayes and Leckie [40] proposed a modification in the calculation of proton stoichiometry by taking into account readjustment of surface charge during metal ion adsorption. However, the authors have made certain assumptions like (a) electrostatic factor is not considered in the equilibrium or mass law constants of various surface reactions, but this error can be minimized using activity values of the dissolved species instead of their molar concentrations. (b) Metal ion is considered to be adsorbed on the surface plane which is identified with the IHP in the TLM, as inner sphere complex. The derivation of required relationship for readjustment of surface charge due to adsorption is given below.

10.2.1 Relationship Between Net Proton Release and Proton Stoichiometry (χ_p)

The following derivation of the relationship between net proton release and its ratio with moles of metal ion adsorbed slightly differs from the one presented earlier by Hayes and Leckie [40], mainly on two aspects: First, the sign convention followed in this book for the formation of surface species during batch titration of solid is opposite to that followed by the above authors. Secondly, the equilibrium constants for anion pair formation used in this book is expressed in the inverse form ($1/{}^*K_{a^-}^{int}$) of the one used by the above authors in their derivation.

Let us consider the TLM for the adsorption of M^{2+} on a hydrous oxide sample from NaCl solution . For batch titration in the absence of metal ion we have

$$C_A - C_B = [H^+] - [OH^-] + [SOH_2^+] + [SOH_2^+\text{-}Cl^-] - [SO^-] - [SO^-\text{-}Na^+] \qquad (10.35)$$

And in the presence of metal ion

$$C_A^* - C_B^* = [H^+] - [OH^-] + [SOH_2^+]^* + [SOH_2^+\text{-}Cl^-]^* - [SO^-]^* - [SO^-\text{-}Na^+]^* + [M^{2+}]_{Ads} \qquad (10.36)$$

In Eq(10.36) the nature of the adsorbed species has not been explicitly identified because of the possible formation of more than one type of species. Subtracting Eq(10.35) from Eq(10.36) gives the net release of H^+ during the adsorption of metal ion . However, the concentration of each surface species may also be expressed as the fraction of total concentration of all the surface species as follows:

$$\alpha_o = \frac{[SOH_2^+]}{S_T} = \frac{[SOH_2^+]^*}{S_T^*}$$

$$\alpha_1 = \frac{[SOH]}{S_T} = \frac{[SOH]^*}{S_T^*}$$

$$\alpha_2 = \frac{[SO^-]}{S_T} = \frac{[SO^-]^*}{S_T^*}$$

$$\alpha_3 = \frac{[SO^- - Na^+]}{S_T} = \frac{[SO^- - Na^+]^*}{S_T^*}$$

$$\alpha_4 = \frac{[SOH_2^+ - Cl^-]}{S_T} = \frac{[SOH_2^+ - Cl^-]^*}{S_T^*}$$

where,

$$S_T = [SOH_2^+] + [SOH_2^+ - Cl^-] + [SO^-] + [SO^- - Na^+] + [SOH] \quad (10.37)$$

And

$$S_T^* = [SOH_2^+]^* + [SOH_2^+ - Cl^-]^* + [SO-]^* + [SO^- - Na^+]^* + [SOH]^* + [M^{2+}]_{Ads} \quad (10.38)$$

Subtracting Eq(10.35) from Eq(10.36) we get

$$(C_A^* - C_B^*) - (C_A - C_B) = \{[SOH_2^+]^* - [SOH_2^+]\} - \{[SO-]^* - [SO^-]\} + \{[SOH_2^+ - Cl^-]^* -$$

$$[SOH_2^+ - Cl^-]\} - \{[SO^- - Na^+]^* - [SO^- - Na^+]\} + [M^{2+}]_{Ads}$$

Expressing concentration in terms of mole fraction

$$\Delta C_A = \{\alpha_o (S_T^* - S_T)\} - \{\alpha_2 (S_T^* - S_T)\} + \{\alpha_4 (S_T^* - S_T)\} - \{\alpha_3 (S_T^* - S_T)\} + [M^{2+}]_{Ads}$$

$$\Delta C_A = (S_T^* - S_T)\{\alpha_o - \alpha_2 + \alpha_4 - \alpha_3\} + [M^{2+}]_{Ads} \quad (10.39)$$

Again, by subtracting Eq(10.37) from Eq(10.38) we get

$$S_T^* - S_T = [M^{2+}]_{Ads} \quad (10.40)$$

Substituting in Eq(10.39)

$$\Delta C_A = [M^{2+}]_{Ads}\{\alpha_o - \alpha_2 + \alpha_4 - \alpha_3\} + [M^{2+}]_{Ads}$$

or, $\Delta C_A = [M^{2+}]_{Ads}\{1 + \alpha_o - \alpha_2 + \alpha_4 - \alpha_3\}$

or, $$\bar{n} = \chi_p = \Delta C_A / [M^{2+}]_{Ads} = \{1 + \alpha_o + \alpha_4 - (\alpha_2 + \alpha_3)\} \qquad (10.41)$$

where $\bar{n}$, is designated as ligand number i.e., number of surface ligands associated with each mole of metal ion and is equivalent to proton stoichiometry, since ΔC_A represents the net release of H^+. Eq.(10.41) suggests that the simplification of the fractional concentration terms on the RHS signify the value of proton stoichiometry. But, in practice, it is considered as the factor which readjusts the surface charge resulting from metal ion adsorption. Since this value varies with pH, it becomes necessary to normalize it by plotting the ratio $\Delta C_A/\{1 + \alpha_o + \alpha_4 - (\alpha_2 + \alpha_3)\}$ against metal ion adsorbed (all in moles/l). The slope value of the resulting linear relationship gives the value of χ_p.

The values of fractional concentrations of different surface species can be estimated using the equilibrium constant values of their formation as described in chapter 5. As an example, let us take the derivation of αo value. We have

$$\alpha_o = \frac{[SOH_2^+]}{S_T} = \frac{[SOH_2^+]}{[SOH_2^+]+[SO^-]+[SOH_2^+ - Cl^-]+[SO^- - Na^+]+[SOH]}$$

Expressing concentrations of surface species in terms of [SOH]

$$\alpha_o = \frac{[SOH]}{\frac{K_{al}}{[H^+]}\left\{\frac{[SOH][H^+]}{K_{al}} + \frac{K_{a2}[SOH]}{[H^+]} + \frac{[SOH][H^+][Cl^-]}{K_{Cl^-}} + \frac{K_{Na^+}[SOH]Na^+]}{[H^+]} + [SOH]\right\}}$$

Dividing numerator and denominator by [SOH] and simplifying

$$\alpha_o = \left(1 + \frac{K_{a1}K_{a2}}{[H^+]^2} + \frac{K_{a1}[Cl^-]}{K_{Cl^-}} + \frac{K_{Na^+}[Na^+]K_{a1}}{[H^+]^2} + \frac{K_{a1}}{[H^+]}\right)^{-1} \qquad (10.42)$$

Similarly, we may derive the required relationships for α_2, α_3 and α_4 for other surface species

$$\alpha_2 = \left(1 + \frac{[H^+]^2}{K_{a1}K_{a2}} + \frac{[H^+]^2[Cl^-]}{K_{Cl^-}K_{a2}} + \frac{K_{Na^+}[Na^+]}{K_{a2}} + \frac{[H^+]^2}{K_{a2}}\right)^{-1} \quad (10.43)$$

$$\alpha_3 = \left(1 + \frac{[H^+]^2}{K_{a1}K_{Na^+}[Na^+]} + \frac{K_{a2}}{K_{Na^+}[Na^+]} + \frac{[H^+][Cl^-]}{K_{Cl^-}K_{Na^+}[Na^+]} + \frac{[H^+]}{K_{Na^+}[Na^+]}\right)^{-1} \quad (10.44)$$

$$\alpha_4 = \left(1 + \frac{K_{Cl^-}}{[Cl^-]K_{a1}} + \frac{K_{a2}K_{Cl^-}}{[H^+]^2[Cl^-]} + \frac{K_{Na^+}K_{Cl^-}[Na^+]}{[H^+]^2[Cl^-]} + \frac{K_{Cl^-}}{[H^+][Cl^-]}\right)^{-1} \quad (10.44)$$

Assuming that the values of activity coefficients of Na^+ and Cl^- ions in NaCl solution are closely similar, their concentrations may be cancelled out between numerator and denominator in some terms of the above equations.

The present author has used some the unpublished results of his work on the adsorption of trace metals on HFO and akaganeite to determine proton stoichiometry or ligand number from Eq(10.41) . Figure 10.12 illustrates a typical example of the of the plots of batch titration of akaganeite in absence (blank) and in the presence of metal ion. If x ml 0.1N HCl or 0.1N NaOH represents the difference between the two plots at any pH value, then ΔC_A is 0.x mmole /l of proton released at that pH. The calculated values of $1 + \alpha_o + \alpha_4 - (\alpha_2 + \alpha_3)$ for the adsorption of metal ions on the present HFO sample vary from 1.5 to 1.0 in the pH range 6.0-7.5. The corresponding value for the adsorption on β -FeOOH in the same pH range is consistently about unity.

Fig.10.13 depicts the linear plots of $\Delta C_A / \{1 + \alpha_o + \alpha_4 - (\alpha_2 + \alpha_3)\}$ vs. $[M^{2+}]_{Ads}$ (in mmole/l) for the adsorption of Ni^{2+} and Co^{2+} on both HFO and akaganeite. It may be noted that at pH< 7.0 the slope values for the adsorption of Ni^{2+} and Co^{2+} on the HFO sample are less than 0.5 , whereas at pH >7.0 the slope value is about 0.9. In the case of akaganeite, the slope value is about 1.0 for the adsorption of Ni^{2+}, whereas for Co^{2+} the slope value is about 0.6. The adsorption of Zn on both the samples does not follow linear relationship as envisaged by Eq (10.41), but non-linear regression analysis suggests that the ligand number is about 0.5 - 0.6. Low values of χ_p for the adsorption of cations on these two hydrous iron oxide samples suggest that considerable frac-

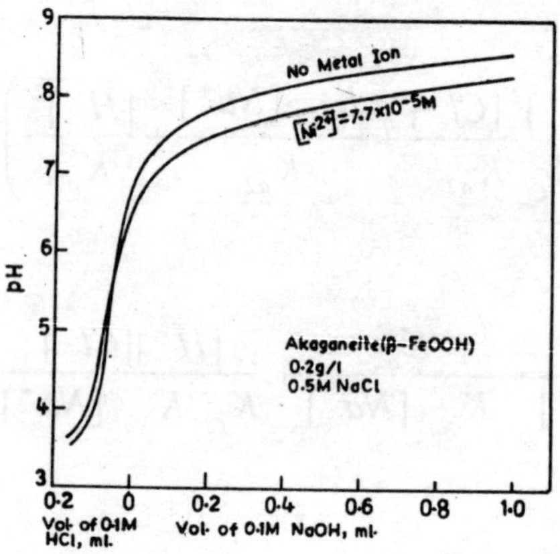

Fig. 10.12 An example of batch titration in NaCl solution of a hydrated iron oxide (β-FeOOH) in absence as well as in the presence of metal ion like Ni^{2+}.

tions of surface sites are possibly in complexated state, especially in 0.5M NaCl solution. Hayes and Leckie [40] have shown good linear relationship for adsorption of Pb on goethite with a slope value of unity. However, the authors have used high solid concentration (30g/l) in $NaNO_3$ medium for nearly similar metal ion concentration suggesting low adsorption density of the system, in contrast to the high adsorption density of the present system.

The third method of determining the χ_p value of an adsorption reaction is based on the principle of "isotherm subtraction" suggested by Perona and Leckie [286].The procedure involves determining adsorption isotherm for a metal ion at two or three constant pH values which should not differ from each other by more than 0.5 unit. Kurbatov coefficients (see Eq 8.27) may be used to interpolating to pH values for which no adsorption data are available .The isotherms are plotted in the same log - log scale for metal ion adsorbed vs. concentration , both in moles /l, so as to yield parallel straight lines for each corresponding pH value . If isotherms of two consecutive pH values are not parallel over a wide equilibrium concentration of metal ion, proton stoichiometry will vary with adsorption density. Isotheram subtraction illustrated schematically in Fig 10.14 where dotted vertical lines indicate two log (concentration) values, say C1 and C2 for two corresponding pH values say pH1 and pH2, at a constant adsorption density i.e., log Γ. Similarly, two other sets of log (C) values are noted at different places of isotherms to determine proton stoichiometry using the method of Perona and Leckie [286]. Theoretical derivation of the final form of relationship with detailed thermodynamic background has not much relevance in the present context and is therefore not reproduced here. Interested readers may consult the original paper for

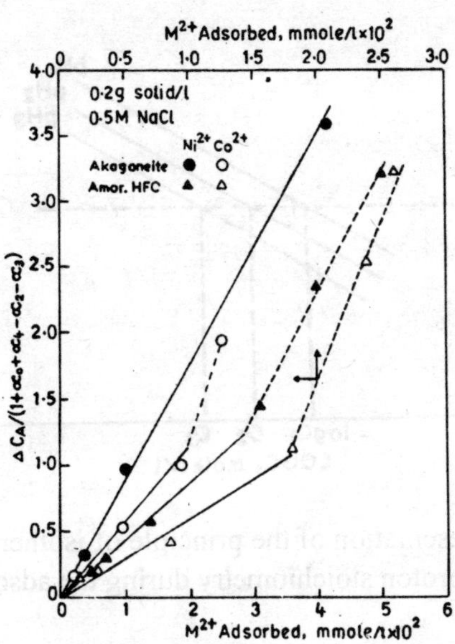

Fig. 10.13 Plot of $\Delta C_A / [1+\alpha_o+\alpha_4 - (\alpha_2+\alpha_4)]$ vs. $[M^{2+}]$ads for the adsorption of Ni^{2+} and Co^{2+} on hydrous ferric oxide(HFO) and β-FeOOH also in 0.5M

more details. The final form of the relationship is as follows:

$$\chi_p = \left(\frac{\partial H^+}{\partial \Gamma_{M^{2+}}}\right)_{pH} = \left(\frac{\partial \log[M^{2+}]_{sol\,n}}{\partial\, pH}\right)_{\Gamma_{M^{2+}}} = \left(\frac{\partial(\log C_2 - \log C_1)}{\partial(pH_2 - \log pH_1)}\right)_{\Gamma_{M^{2+}}} \quad (10.45)$$

The main disadvantage of the method is that it involves carrying out tedious and time consuming adsorption isotherm experiments at two or three constant pH values. While the maintenance of constant temperature does not pose any problem, maintenance of constant pH value is rather cumbersome as the value tends to fall as adsorption increases with time. Therefore, it becomes necessary to adjust pH continuously with the addition of either acid or alkali. As a result, the total volume of solution changes even if it is by a small amount. To avoid this, buffer solutions of different pH values may be used [284] at the risk of incorporating complex forming anions in the solution. Even after all these efforts, there is no guarantee that isotherm plots at different pH values will be parallel to each other. Therefore, whichever method is used proton coefficient (χ_p) will always tend to be a function of adsorption density and pH.

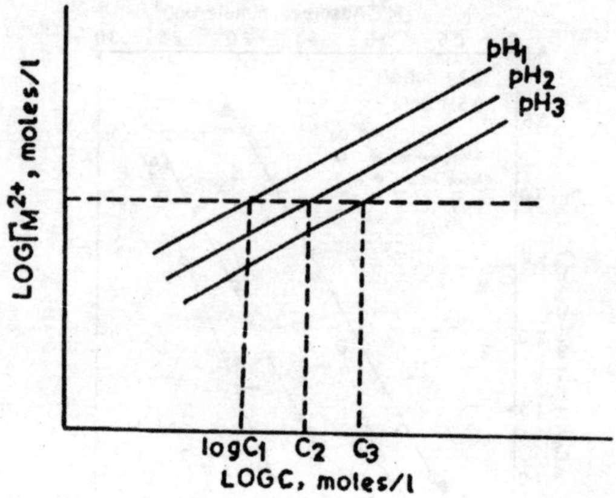

Fig.10.14 Schematic representation of the principle of isotherm subtraction for the determination of proton stoichiometry during the adsorption of metal ion on hydrous oxides.

10.3 Selectivity and Competitive Adsorption of Cations

The characteristic features of cation adsorption on inorganic solids, especially oxides and hydrous oxides of certain heavy metals, have been discussed in the preceding sections of the present chapter as well as in different foregoing chapters. The main objective of this discussion is to assess the binding strength of cations with the adsorbent surface. Despite our assumptions (which has however been established both theoretically and experimentally) that many cations form complexes with surface sites acting as ligands, the bondage may not be purely chemical in nature and electrostatic factor also plays important role in such bonding. The reversibility of adsorption is an important criterion in assessing the strength of bonding. Long range electrostatic or van der Wall's force leads to reversible cation exchange reaction, whereas structural substitution / replacement or co-precipitation results in strong chemical bonding with the surface which is irreversible in nature. X-ray absorption spectroscopy (XAS) has been extremely useful in studying the nature of bonding of the adsorbed species with the adsorbent surface [334,335]. In either case, the intrinsic binding constant determined at zero surface charge (or electrostatic) condition is a valid thermodynamic quantity. The free energy of adsorption calculated from binding constant values suggests spontaneity of adsorption reaction.

Table 10.4 provides binding constant values for the adsorption of selected cations on some commonly used oxides / hydrous oxides as compiled from available literature which is also abound with many examples wherein the extents of adsorption have been calculated using different models and compared with experimentally found values without explicitly mentioning the binding constant values. However, binding constant values provide explanation for the so-called selectivity of adsorbents towards cation adsorption. Selectivity or affinity of an adsorbent towards cations and /or anions depends upon the nature of both sorbate and sorbent and their relative quantities i.e., sorbate/sorbent ratio. For example, Pretorius and Linder [435] have observed that at pH 4 and sorbate /sorbent ratio1:30 the adsorption sequence on δ-MnO_2 follows Cd<Zn≅Ni<Cu≅Pb. At sorbate/ sorbent ratio1:3 the predicted sequence is as follows: Cd<Zn<Ni<Pb<Cu. It also depends upon the nature and concentration of electrolyte solution, pH etc. Consequently, different authors have observed different sequence of cation affinity for an oxide surface. In fact, there is no unifying principle governing the selectivity of an oxide for divalent cations. Benjamin and Leckie [385] have proposed that an oxide surface (in fact, most adsorbents) is composed of a variety of adsorption sites with different binding energies. Each site may be considered as a potential "ligand" for complexation with a cation. All these ligands have many properties in common, but small differences among them lead to differences in energetics of adsorption and therefore observed selectivity. It was suggested that factors such as nature and proximity to various types of crystal defects, micro crystallinity, local field strength, pore size etc., possibly lead to differences in the energetics of the binding sites or ligands. The sequence of selectivity, as it appears from the differences in the experimental values of the binding constants of different cations with different types of adsorbents, is also included in the table. However, these sequences are tentative and may change with change in the adsorption system and experimental conditions. The selectivity sequence observed by the above authors for the adsorption on amor. HFO was as follows: Cd<Cu<Hg<Zn<Co. Some selectivity data pertaining to earlier literature have been presented by Kinniburgh and Jackson [214].

However, the greatest significance of the binding constant values lies in examining the mechanism of competitive adsorption of metal ions. It may be noted from Table 10.4 that the binding constants of transition and post transition metal ions with oxide surfaces are greater than those of alkaline earth metal (e.g., Ca, Mg) ions, which may, however, occur in higher concentrations in natural water systems. One of the two possible explanations for the apparent difference in the binding strength of the two groups of metal ions is that the heavy metal ions may get adsorbed more strongly on all the possible sites including those occupied by Ca and Mg. The other possibility is that there are some sites meant specifically for strongly adsorbing transition metal ions, but not available for weakly adsorbing alkaline earth metal ions. In the case of former situation the two groups of metal ions will compete with each other for the adsorption sites, and in the case of latter situation adsorption will occur independent of each other according to their respective binding strength [307]. Fortunately, the second situation or possibility generally prevails where the binding site of one metal

Table 10.4 : Binding Constants ($-\log K_{SOM+}^{int}$) of the Adsorption of Some Divalent Metal Ions on Selected Oxide Adsorbents at 25°C.

Adsorbent	Electrolyte	Metal Ion	$-\log K_{SOM+}^{int}$	Selectivity	Reference
γ-Al_2O_3	7.5x10^{-3} M	Cu	1.6	Cu>Pb>Zn>Co>Mn	450
	$NaNO_3$	Pb	1.8		,,
		Zn	2.3		,,
		Co	3.8		,,
		Mn	4.7		,,
	0.1N	Mg	5.4	Mg>Ca>Sr>Ba	377
	$NaNO_3$	Ca	6.07		,,
		Sr	6.27		,,
		Ba	6.59		,,
	0.1N	Pb	2.2		39
	$NaClO_4$	Cu	2.1		,,
Goethite	0.5M	Cd	1.3	Cd>Pb>Cu	345
(α-FeOOH)	NaCl	Pb	1.8		,,
		Cu	3.0		,,
		Pb	0.5		404
Amor HFO	0.1M	Mg	4.6	Mg>Ca>Sr>Ba	329
	$NaNO_3$	Ca	5.8		,,
		Sr	6.6		,,
		Ba	7.2		,,
		Ag+	5.3		,,
		Hg	-6.45	Hg>Cu>Pb>Zn>Ni	,,
		Cu	-0.6	>Cd≅Co>Mn	,,
		Pb	-0.3		,,
		Zn	2.0		,,
		Ni	2.5		,,
		Cd	2.9		,,
		Co	3.0		,,
		Mn	3.5		,,
	0.5–0.1M				
	NaCl	Zn	2.83	Zn>Ni ≅Co>Mn	283
		Ni	3.85		,,
Akaganeite	,,	Zn	3.25	Zn>Ni =Co>Mn	,,
(β-FeOOH)		Ni	3.62		,,
		Co	3.62		,,
		Mn	3.78		,,
δ-MnO_2	KNO_3	Pb	0.58	Pb>Cu>Ni ≅Zn>Cd	435

		Cu	1.31		”
		Ni	1.65		”
		Zn	1.62		”
		Cd	1.60		”
	0.5M	Zn	-0.43	Zn>Co>Ni>Cu	284
”	NaCl	Co	0.60		”
		Ni	0.76		”
		Cu	0.9-1.16		”
”	0.1M	Co	0.65	Co>Cu>Zn>Ni	”
	KNO_3	Cu	1.16		”
		Zn	1.38		”
		Ni	1.60		”
Amor. SiO_2	0.1M	Pb	5.1	Pb>Cu>Cd>Mg>Ca	132,343
	NaClO4	Cu	5.5		”
		Cd	6.1		”
		Mg	7.3		”
		Ca	8.1		”
TiO_2	0.1M	Pb	0.2-0.44	Pb>Cu>Cd	”
	$NaClO_4$	Cu	1.43		”
		Cd	3.32		”

ion is not the preferred site for the other. This is the direct outcome of the hypothesis of multiple adsorption sites proposed by Benjamin and Leckie [385] for amor. HFO. Though all the adsorption sites are susceptible to occupation by cations , it is the strongest binding sites that are occupied preferentially. At low adsorption density the binding constant value is independent of metal ion concentration . In other words, as long as the concentration of surface sites including those with stronger binding energy is very large compared to the metal ion concentration in solution, binding constant is independent of adsorption density. But, as the adsorption density increases beyond a certain limit, i.e., when the availability of stronger binding sites become limited, binding constant value tend to decrease. The phenomenon of competitive adsorption is based on the above principle of preferential occupation of increasingly higher energy sites.

Benjamin and Leckie [307] attempted to explain the phenomenon using the following example of multiple complexation in solution phase. Suppose, metal A forms monodentate complexes with ligands X, Y and Z of decreasing energy so that the speciation of the metal will follow the decreasing equilibrium or stability constants of the formation complexes AX, AY and AZ . This means that metal will preferentially bind first with ligand X, and then with ligand Y and finally with ligand Z.

Now let us consider the situation when a second metal B is added to the system. The concentrations of free ligands will change which may lead to a change in the relative concentrations of A complexes. This will also reflect how successfully B competes for the available ligands X , Y and Z . If metal B exhibits similar binding preferences as A, the resulting competition will have considerable impact on the speciation of A . On the other hand, if metal B prefers to combine with ligands which are not preferred by A, the impact will be minimal. If metals A and B have identical relative preferences for all the ligands X, Y and Z then it should satisfy the following equality of ratios $b_{AX} / b_{BX} = b_{AY} / b_{BY} = b_{AZ} / b_{BZ}$ where b_{AX} , b_{BX} etc represent equilibrium constants for the formation of corresponding complexes . Deviation from the equality of the ratios indicates lack of competition between the metals and the ligands . It has been observed [378] that the same principle is also applicable to the binding of metals to surface sites. If two metals compete for the same sites, then strongly bound metal forces the less strongly bound metal to weaker binding sites. This is reflected in shifting of the adsorption edge of less strongly metal to higher pH side relative to strongly bound metal. Otherwise, adsorption edge of either metal will remain unchanged in the presence and in the absence of the other. This has been demonstrated by Benjamin and Leckie [307] for the adsorption of weakly bound Cd on amor. HFO in the presence of equal amounts of strongly bound Pb, Zn and Cu . Earlier, Gadde and Laitinen [374] observed that the adsorption of Cd and Zn hydrous manganese dioxide (HMO) is decreased in the presence of Pb when they are present in almost equal concentration. Balistrieri and Murray [345] have also observed similar competition for the adsorption of weakly bound Cd on goethite in the presence of Cu, Pb and Zn in major ion sea water. Kanungo [283] has observed mutual competition among Mn, Ni, Co and Zn for adsorption on HFO in major ion sea water. The competition is high for Co and Ni, moderate for Mn and low for the adsorption of Zn. The above study reveals that the interfacial properties of goethite and amor. HFO in major ion sea water are distinctly different and therefore their relative affinities towards metal ions also differ. Similarly, other studies have shown that there exists competition between metal ions for adsorption on some oxides (e.g., γ-FeOOH and γ-Al_2O_3) even though the available surface sites are in great excess compared to metal ion concentration . This is in contrast to the observation made by the same authors for HFO [345]. Kanungo et al [284] have also demonstrated the existence of competitive adsorption of Ni, Co, Cu and Zn on δ-MnO_2 in major ion sea water.

Cowan et al [33] have observed that Cd adsorption is suppressed by the presence of equal concentrations of alkaline earth metal ions. On the other hand, the presence of Cr(III) has been found to increase the uptake of Zn and Ni on ferrihydrite [376] , otherwise the two metal ions adsorb independent of each other. It has been suggested that either mixed metal hydroxo surface species form or that the surface properties of the adsorbent have been altered by Cr(III) adsorption . It has been observed [402] that the adsorption of Cu on hematite is suppressed by P_b in the pH range 4.5 - 6.3, whereas the effect of Cu on the adsorption of pb is much less pronounced when both the ions are present in equal concentration.

10.4 Mn^{2+} Release in Solution During Adsorption of Certain Cations on MnO_2

Though this topic is presently confined to the adsorption of certain cations on hydrous manganese oxide (HMO), similar phenomenon concerning the release of metal ions constituting the other oxide adsorbents (e.g., HFO) may also occur. Such release, even in trace quantities, can have major influence on the surface properties of the adsorbents, especially in the region of low pH range. Therefore, this aspect has to be investigated systematically.

McKenzie [248], as early as in 1970, observed that adsorption of divalent metals like Cu, Ni and Co on HMO resulted in the release of substantial amounts of Mn^{2+} in solution, especially at lower pH values. It was also observed that adsorbed Co displaced more manganese in solution than Cu and Ni. McKenzie suggested that other things being equal there would be gain in crystal field stabilization energy (CFSE) by the substitution of Cu^{2+}, Ni^{2+}, Co^{2+} etc. for Mn^{2+} in the HMO structure, but it is the replacement of Mn^{3+} by low spin Co^{3+} that would result in more stable state. To explain the higher release of Mn^{2+} by Co^{2+} compared to that by other metal ions, he proposed that Co^{2+} substitutes structural Mn^{3+}, coupled with electron transfer to form Co^{3+} in the structure and corresponding release of Mn^{2+} in solution. Though the adsorption of Zn was not included in the investigation by McKenzie [248], Loganathan and Burau [296] argued that though Zn^{2+} might not substitute Mn^{3+} but could displace structural Mn^{2+} as both the ions do not possess any CFSE. On the basis of this observation, Loganathan and Burau [296] proposed that there are at least three adsorption sites in the HMO sample (possibly a birnessite though the authors designated it as δ-MnO_2) used by the authors depending upon the nature of metal ion adsorbed. While Ca^{2+} is adsorbed by displacing only H^+ (bound H site), Zn^{2+} gets adsorbed by both bound H and Mn^{2+} structural sites. On the other hand, Co is adsorbed at both Mn^{2+} and Mn^{3+} sites in addition to H^+. By estimating the maximum release of Mn^{2+} from Langmuir plot of Mn^{2+} release isotherm and subtracting this value from total metal ion adsorbed the authors re-estimated proton stoichiometry value which was found to be about 2.0 for the adsorption of Zn and Co on the HMO sample.

However, Burns [429] put forward little different argument to explain high release of Mn^{2+} during the adsorption of Co on HMO. Prior to putting his argument, Burns [429] discussed elaborately the literature information available at that time regarding the nature of fixation of Co in soil, weathered manganiferrous rocks, ocean-bed manganese nodules and also in various synthetic preparations of HMO . He clearly demonstrated that the occurrence of Co in natural resources is directly related to the extent of manganese oxide phases present in them. In other words, Mn and Co are directly interrelated wherever they occur together.

The crystal chemistry of Co^{2+} and Co^{3+} suggests that though both the ions occur in stable octahedral environment, it is only the low spin Co^{3+} shows exceptionally high CFSE relative to any other cations of the 3d series. Table 10.5 shows the ionic radii and the CFSE values for some 3d transition metal ions in high and low spin states. Both Mn^{2+} (high spin) and Fe^{3+} ions have no CFSE in octahedral field and their radii differ greatly from low spin Co^{3+} (0.53A°) so that the substitution of Co^{3+} for these two cations is highly unlikely. Indeed, the only cobalt oxyhydroxide (CoOOH) mineral "heterogeneite" has no structural similarity with any oxyhydroxide of either Fe(III) and Mn(III) . The close similarity in the ionic radii of low spin Co^{3+} (0.53A°) and Mn^{4+} (0.54A°) suggest that adsorbed Co^{2+} is oxidized on the surface of δ-MnO_2 and occupies Mn^{4+} site.

The structural features of poorly crystalline variety of HMO have been discussed in section 7.4.4. The occurrence of numerous and irregular vacancies in layers of edge shared $[MnO_6]$ octahedra in birnessite and δ-MnO_2 form the basis for explaining the adsorption of Co^{2+} on HMO and consequent release of Mn^{2+} in solution . Hydrated cations like Co^{2+}, Ni^{2+}, Cu^{2+} , Zn^{2+} etc. belonging to transition metal group are invariably adsorbed onto the surface of HMO in the vicinity of the octahedral vacancies . It is also possible that these transition metal ions may replace Na^+, K^+ and Mn^{2+} ions held between the octahedral sheets of HMO surfaces. The adsorbed Co^{2+} ion is however subjected to oxidation, especially at higher pH to give Mn^{2+} and $Co(OH)_3$ according to the following equation :

$$2Co(aq) + 2OH^-(aq) + 2H_2O(l) + \delta\text{-}MnO_2\,(s) \longleftrightarrow 2Co(OH)_3(s) + Mn^{2+}(aq)$$

The standard free energy change of the above equation is -17.1 kcal/mol which suggests that the process is spontaneous in nature. However, the applicability of the above reaction is restricted because of the two main reasons: (1) The reaction takes place in the alkaline medium, whereas, it has been observed [72] that the release of Mn^{2+}decreases sharply with increase in pH. (2) Spectroscopic studies of adsorbed cobalt do not provide any evidence for the formation of separate CoOOH phase. Therefore, oxidized Co^{3+} occupies the site vacated by Mn^{4+} ion involved in the redox reaction or fill up the vacant sites. Both the situations are favoured by ionic radius and high CFSE of Co^{3+}. As far as the other metal ions are concerned, adsorption takes place on Mn^{2+} sites or interlayer cations.

However, recent investigations [72] show that there occurs self release of Mn^{2+} from δ-MnO_2 when suspended in aqueous electrolyte solution, especially at lower pH value (< 5.0). This is because of the inherent instability of Mn(III) in acidic medium leading

Table 10.5 : Ionic Radii and Crystal Field Stabilization Energies (CFSE) of Some Transition Metal Ions in Different Spin States.

d^n configuration	Ion	Weak field or high spin state			Strong field or low spin state		
		$t_{2g}^p e_g^q$ configuration	Ionic radii[a] (A°)	CFSE[b] (kJ/mol)	$t_{2g}^p e_g^q$ configuration	Ionic radii[a] (A°)	CFSE[b] (kJ/mol)
d^3	Mn^{4+}	$t_{2g}^3 e_g^0$	0.54				
d^4	Mn^{3+}	$t_{2g}^3 e_g^1$	0.65	113.3	$t_2g^4 e_g^0$	0.58	302.2
d^5	Mn^{2+}	$t_{2g}^3 e_g^2$	0.82	0	$t_2g^5 e_g^0$	0.67	187.7
d^6	Co^{3+}	$t_{2g}^4 e_g^2$	0.61	66.9	$t_2g^6 e_g^0$	0.53	401.7
d^7	Co^{2+}	$t_{2g}^5 e_g^2$	0.74	80.3	$t_2g^6 e_g^1$	0.65	180.7
d^8	Ni^{2+}	$t_{2g}^6 e_g^2$	0.78	123.4	-	-	-
d^9	Cu^{2+}	$t_{2g}^6 e_g^3$	0.70	56.3	-	-	-
d^{10}	Zn^{2+}	$t_{2g}^6 e_g^4$	0.74	-	-	-	-

a) Taken from Lurie, Ju . *Handbook of Analytical Chemistry*, Engl . Transln ., Mir Publishers, Moscow , 1975, pp 22-26 ; and also from ref, 429.

b) Calculated from orbital splitting (Δo) values of hexa-aquo ions given in ref . 331 p. 363.

to its disproportionation to Mn(II) and Mn(IV) . The process is facilitated by the adsorption of divalent metal ions like Co^{2+}, Cu^{2+} and Zn^{2+}. The adsorption of Ni^{2+} does not help in this process because its large ionic radius and lower CFSE value (see Table 10.5). This has been clearly demonstrated [72] wherein the release of Mn^{2+} from δ-MnO_2 in presence of Ni^{2+} is found to be lower than in its absence.

10.5 Literature Survey on the Adsorption of Cations on Solid Inorganic Adsorbents

10.5.1 Adsorption on Pure Oxides and Hydrous Oxides

A large number of references have so far been cited in connection with discussions on various aspects of cation adsorption in the present as well as in the preceding chapters. Besides these, there are also a large volume of literature dealing with adsorption of cations on various inorganic adsorbents which include oxides and hydrous oxides of iron, manganese, aluminum, titanium, silicon etc. However, the number of published papers using hydrous oxides of iron and manganese as adsorbents exceeds far more than any other oxidic adsorbent as, mainly because of their ubiquitous presence in our environment. The other important inorganic adsorbents investigated for adsorption of cations are aluminosilicates compounds like clays , clay minerals , zeolites etc. The interest in this group of adsorbents is gaining momentum as they are relatively inexpensive and readily available in nature including soil. Their adsorption capacities can be enhanced after suitable treatment in acid or in certain salt solutions. The mode of cation adsorption on aluminosilicate compounds is found to be different from pure oxides or hydrous oxides. The same is applicable to soil which is a composite of oxides, hydrous oxides and alunimosilicate minerals. Therefore, it is proposed to discuss entire aspect of cation adsorption on soil and aluminosilicate minerals in a separate section.

Table 10.6 intends to provide more literature references in addition to those which have been cited during in the discussions in the present and also in other chapters. The references have been classified according to the adsorption of individual metal ion on different adsorbents. The table does not include mixed or composite oxides, nor any so-called "cheap or inexpensive" ill-defined adsorbents as they do not fall within the scope of the present monograph. Also, not included in the table are the references dealing mainly with spectroscopic studies of the adsorbed metal ions to examine their nature of bondage with the adsorbents.

10.5.2 Adsorption on Alumino-silicate Constituents of Soil

The presence of heavy metals and certain oxyanions in our environment, especially in soils has drawn particular attention because of two opposite reasons : (a) usefulness from agricultural point of view (for example, plant nutrition and soil fertility) and (b) non-usefulness from environmental point of view (for example, soil and water pollution). Some heavy metals are essential for plant nutrition when present in small quantities or micro level in soil, but when present in quantities much above their usefulness limit then it becomes a matter of concern for environmental hazard. Such hazardous situation is mostly man-made, except for a few instances of natural origin, such as dissolution of sulphide and carbonate minerals under oxidizing condition in

Table 10.6 : Literature References on the Adsorption of Metal Ions on Some Inorganic Oxides / Hydrous Oxides .

Oxides / Hydrous oxides	Metal ion*	References
Goethite (α-FeOOH)	Ag	412
	Au	401
	Cd	37, 38, 345, 352, 387, 392, 393, 398, 400, 411, 423- 425, 469
	Co	48, 390, 392, 395, 413, 428 410, 412, 426, 469
	Cr(III)	388
	Cu	345, 390, 392, 395, 410, 412, 426
	Hg	47, 48, 389,467
	Mn	390, 395
	Ni	387, 390 393, 412, 416
	Pb	40, 345, 384, 390, 392, 396, 400, 404, 412
	U(UO_2^+)	397, 468
	Zn	345, 384, 386, 387, 392, 393, 395, 412, 427
Lepidocrocite (γ-FeOOH)	Cd	38
	Mn	414
Akaganeite (β-FeOOH)	Cd	38
	Mn, Ni, Cu, Zn	283
Amor. HFO/ Ferrihydrite	Ag	419
	Ca, Sr	282, 326, 349, 399
	Cd	245, 383, 385, 391, 423
	Co	283, 383
	Cu	41, 281, 385
	Eu	407, 419
	Ga	405
	Hg	214, 422
	Lanthanides	403
	Mn	283
	Ni	283, 416
	Np	394
	Pb	41, 383, 385, 414
	UO2+	417
Hematite (Fe_2O_3)	Ag	412
	Ba, Ca	409, 421
	Cd	420
	Co	40, 52, 406, 412
	Cr(III), Mn, Sr	419
	Cu	52, 402, 412
	Eu	407
	Ga	408
	Hg	422

	Ni	52, 412
	Pb	402, 412, 420
	Zn	412, 420
Magnetite (Fe_3O_4)	Ag, Cd, Cr(III) , Mn	419
	Co	327, 415
	Ni	403
δ-MnO_2	Ag	441
	Ba, Ca, Mg	294, 382, 431, 433
	Cd	374, 431, 432, 435, 436, 448
	Co	72, 284, 327, 382, 429-431, 442
	Cu	72, 284, 382, 434-436
	Eu	443
	Mn	375, 431
	Ni	72, 284, 382, 431, 435, 444
	Hg	437
	Pb	374, 434, 435
	Sr	442
	Zn	72, 284, 374, 382, 431, 432, 434, 435, 446
Birnessite	Co	296, 429
	Ba, Cu	294
	Cr(III)	447
	Zn	296
Buserite	Ca	439
	Co, Ni	438, 440
	Cu, Zn	438-440
γ-MnO_2	Ba	294
	Bi, Pb	445
	Cd	449
	Co	295
	Cu	294, 295, 445
	Cr(III)	445, 447
	Ni	444
	Zn	446
∝-Al_2O_3	Ba, Ca, Mg	125, 132, 450, 451
	Cd	452, 455, 458
	Co	259, 453
	Cu	39, 125, 132, 259, 447
	Cr(III)	457
	Mn	451
	Ni	259
	Pb	39, 259, 455
	Zn	259, 459
AlOOH / Gibbsite	Alkaline earth and transition metals	326, 456, 460

∝-Al_2O_3	Cd / Citrate	454
Amor .SiO_2	Cd, Pb	302
	Cu	302, 466
	Hg	467
TiO_2	Cu/EDTA	461
	Cu/CN	462

* Unless otherwise stated the metal ions are of oxidation state +2

surface e.g., acid drain from pyrite mines as well as sub-surface water (e,g., arsenic, fluoride and Cr(VI) contamination). Even then, human activities are also partly responsible in the above instances of environmental problem. The man-made hazards originate mostly from unscientific and rackless disposal of industrial and domestic wastes on land and in water bodies. Adsorption is the major process of accumulation of hazardous elements / compounds in soils over the years. This particular aspect of adsorption process which is going to be discussed in the following sections, is of interest to all the concerned researchers including physical chemists, soil chemists, environmental chemists and even geochemists.

As major constituents of all types of soil, clays serve as important sinks of toxic heavy metals and oxyanions. The other constituents of soils, namely hydrous oxides of iron, aluminum and manganese also occur in intimate association with clay minerals, thereby providing additional sites for adsorption. Therefore, attempts have been made by many investigators to study adsorption behavior of heavy metals and certain oxyanions on clays / clay minerals to understand the retention of these inorganics in our environment, especially in soil.

The structure and other physico-chemical features of soil constituents have been discussed in chapter 7. These features clearly suggest the presence of two types of adsorption sites on soils. One is the negatively charged site on clay surface (basal plane) where the well known phenomenon of cation exchange occurs. They are also called "permanent sites" as they originate from inherent cation charge deficiency in clay structure and are independent of pH variation of the aqueous medium. The other sites, which occur mostly on edges of the clay particles, are generally pH dependent, being positive below a certain pH value (pH_{PZC}) and negative above that pH. The adsorption in this category of surface sites occurs through surface complexation reaction [221,234]. The bonding of the adsorbed ion with clay surface in this category is more stronger than the former wherein the adsorption takes place through electrostatic interaction. The overall adsorption capacity is the sum total of metal ion uptake by both the categories of adsorption sites at any pH value. The cation exchange behavior of clay minerals which is more relevant to agriculture and soil scientists, has been discussed in more details in chapter 9.

10.5.2.1 Modification of Clays

Various attempts have been made to improve adsorptive and catalytic properties of clay minerals using different techniques, of which the two most frequently used are: (1) Intercalation and pillaring and (2) treatment with acid.

Intercalation and Pillaring

Intercalation is the process of incorporating or inserting a guest species within the empty spaces of the host molecule in such a way that the structure of the latter is preserved. In the present case, certain large organic ligands or their complexes with metal ions are incorporated within the interlayer spacing of clay minerals without affecting their basic crystalline structures as revealed by the XRD measurement [471]. Pillaring of clay minerals by various inorganic (chloro- or hydroxo-) as well as organic compounds is well known in the field of catalyst design. In this technique certain refractory oxides in very finely divided state are inserted within the interlayer spacing of clay minerals, generally smectites, so that the interlayer thickness remain unchanged at a certain value to enable free access of reacting ions in solution phase or gases during catalytic reactions. Apart from their intrinsic catalytic activity, clays pillared with refractory metal oxides exhibit high surface area and high thermal stability [472,473].

Pillared clays are prepared by exchanging interlayer cations with large ionic species such as alkylammonium ions, polynuclear complex ions of inorganic ligands (e.g., Cl^- or OH^-), metal complexes with large inorganic ligands etc. The exchanged clays when heated the organic components are decomposed and the metal ions are converted to their corresponding oxides in clustered state. The most widely studied pillared clay is ZrO_2 bearing montmorillonite. The zirconium species is incorporated within the interlayer spacing of clay mineral as tetrameric species of zirconyl oxychloride from aqueous medium giving rise to a pillared clay of 0.7nm interlayer spacing. Pillared clays with higher interlayer spacings such as 1.0, 1.2 and 1.4nm can be prepared by incorporating hydrolytically polymerized $ZrOCl_2$ as three dimensional clusters [474-476]. Pillared clays containing clustered halides of other polycations such as Mo^{2+}, Al^{3+}, Ti^{4+}, Fe^{3+}, Cr^{3+}, Ga^{3+} etc have also been prepared [472,473,477,478].

A pillared clay is distinguished from an ordinary intercalate because of intra crystalline porosity of the former made possible by lateral separation of intercalated guest molecules. Intercalated compounds when heated at about 500°C, leads to the conversion of polyoxocations to corresponding metal oxide particles cross linked to the clay layers to form stable structure with high porosity [479].

Acid Treatment

Clay minerals when treated with high concentration of inorganic acid at elevated temperature for a predetermined period of time is known as acid treatment or acid activation. It is well-known that acid treatment of clay minerals leads to the partial dissolution of aluminum from octahedral layer and is replaced with H^+ along with exchangeable cations. However, tetrahedral Si^{4+} remains largely unaffected. The acid treatment generally increases surface area and acidity of clay minerals [480,481]. The surface acidity, especially Lewis acidity, shows sharp increase on heating the acid treated clay at about 500°C [482,483]. Adsorption capacity of kaolinite has been found to increase by acid treatment [484-486].

10.5.2.2 Metal ion Adsorption on Clay minerals and Zeolites

Clays are known for their ability to remove heavy metals by cation exchange, specific adsorption and hydroxide precipitation on their surfaces. In the case of soils, adsorption is enhanced due to the presence of organic matter which also contain surface functional groups for complex formation with metal ions [221]. The presence of hydrous oxides of iron and manganese further enhances the overall adsorption capacity of soil [221]. Literature survey reveals that only a few selected metal ions such as Cd, Cr(III), Co, Cu , Ni, Pb etc. have been used for adsorption studies on soils and clay minerals. A few studies have also been carried out on the adsorption of Mn and Zn. It is more appropriate to review the literature on the basis of important individual aluminosilicate compound as an adsorbent.

Kaolinite

As stated, the existence of two types of adsorption sites on different clay minerals including kaolinite was first emphasized by Schindler et al [233]. The first one involves adsorption onto permanent negative charged site resulting from isomorphous substitution of either Al^{3+} in octahedral sheet or Si^{4+} in tetrahedral sheet by cations of lower oxidation states. The second category of surface site originates from bonds at the edges and charge on this site is variable i.e., depends upon the variation in pH as well as ionic strength of the medium. This view has been subsequently supported by almost all researchers working on the adsorption behavior of metal ions on clay minerals including kaolinite [487-490].

However, isomorphous substitution in case of kaolinite is very low, about 0.01 per unit cell [140]. This results in the low value of CEC compared to other clay minerals (see chapter 7). The low permanent negative charge renders the ditrigonal cavities between two adjacent tetrahedral sheets unreactive [140]. The principal adsorption sites are therefore silanols (Si-OH), aluminols (Al-OH) and Lewis acid i.e., negatively charged oxygen or deprotonated oxygen attached to metal ion of hydrous oxide located at the edges of the layers. All the three sites are proton donor groups which can form complexes with metal ions in solution. Only the aluminol group can act as proton acceptor at lower pH values and therefore can form complexes with anions [140,221. The structural properties and their influence on metal ion adsorption as function of pH have been discussed by Miranda-Trevino and coles [491].

On the basis of two sites adsorption on clay minerals, the reactions between kaolinite and divalent metal ions may be classified as follows:

(A) At lower pH range (generally, 4 - 6) the value of which depends upon the nature of metal ion, the following exchange reaction takes place at the permanent negative site (X^-)

$$2(X\text{---}Na^+) + M^{2+} \longleftrightarrow (X^-)_2\text{--}M + 2Na^+ \tag{10.46}$$

Assuming that the background electrolyte is 1-1 sodium salt (e.g., $NaNO_3$) solution the exchange site occurs in the X--Na form. The dashed joint between surface site and metal ion signifies the formation of outer sphere complex held by electrostatic bond of longer distance. The adsorption edges of divalent metal ions exhibit distinct saturation plateau region at the above pH range indicating adsorption onto X- site prior to complexation with variable charge site at higher pH value (233,241,424,487-490). The plateau region is generally sharper for those metal ions whose hydrolysis occur at higher pH (e.g., Cd, Ni, Zn etc.) and suppressed for those which hydrolyze at relatively lower pH values (e.g., Cu, Pb, Mn, Cr(III), Fe(III), etc.). However, some investigators [483,486] have failed to recognize the occurrences of such breaks in their pH vs. adsorption curves for Cd, Co and Ni on kaolinite and montmorillonite and their acid treated forms.

(B) At higher pH values (>6.0- 6.5) adsorption occurs on surface hydroxyl groups which deprotonate (SO^-) to enable to form inner sphere complexes with metal ions. Although no distinction is made here between aluminol and silanol groups, these sites are mostly aluminol (Al-OH) groups which take part in this process [233]. However, the formation of different types of inner-sphere complexes has been proposed for different transition metal ions. Generally, those metal ions which tend to adsorb at lower pH values form monodentate complexes like SOM^+ or SOMOH; those which tend to adsorb at higher pH values form bidentate complex of the type $(SO)_2M$ [490]. But, it is difficult to categorize metal ions only on the basis of pH of adsorption as other factors like ionic strength and nature of suspending medium, other cations and anions present in the system affect the mechanism of adsorption and consequently the model parameters used for fitting the experimental data [490]. Therefore, x-ray spectroscopic technique like EXAFS is now-a-days used to ascertain the actual mechanism of adsorption of heavy metal ions.

Srivastava et al [490] have studied the adsorption of Cd, Cu, Pb and Zn on kaolinite individually (single element system) as well as in their mixture (multi element system) from aqueous 0.01M $NaNO_3$ solution. It has been observed that for single ion system adsorption on permanent negative charge site is predominant up to pH 5.7, 6.0, 7.3 and 8.0 for Cu, Pb, Zn and Cd respectively. For best model (constant capacitance) fit, the authors invoked the formation of $(SO)_2M$ type complex in case of Cd and Zn, whereas for Cu and Pb the formation of SOMOH type gave the best fit of the experimental data. The authors used pH_{50} (pH at which 50% adsorption takes place) as an index for selectivity sequence of metal ions on kaolinite surface. Thus, for single metal ion system the sequence observed was Cu>Pb>Zn>Cd , and for multielement system the observed sequence was Pb>Cu>Zn>Cd. Bhattacharyya and Sengupta [486] have used Langmuir capacity as an indicator for ascertaining selectivity order which was found to be Pb≃Co>Ni>Cd>Cu for adsorption on kaolinite. This is close to the order Mn>Co>Ni>Cu observed by Yavuz et al [492]. But, this type of order or sequence appears to differ with other investigators who have observed that Cu is adsorbed more strongly than Cd and Ni [233,424,493].

Literature survey reveals that Pb^{2+} and Cd^{2+} are the two most widely studied metal ions for adsorption on soil in general and on kaolinite in particular. This is because soil and its constituent clays act as important sink for these two environmentally hazardous toxic elements. The initial rates of their adsorption on kaolinite are very fast and at equilibrium adsorption follow Langmuir and Freundlich isotherms [485,494]. Both Ca and Zn compete with Cd for adsorption sites and may even desorb Cd from soil. On the other hand, on pure kaolinite Cd competes with Ca for adsorption sites [33].

Bhattacharyya and Sengupta [483.486,495-498] have routinely reported the kinetic aspects of adsorption of several metal ions including Cd and Pb on kaolinite and montmorillonite in their publications. The following five kinetic models were commonly used by the authors:

(a) Pseudo first order equation of Lagergren (Eq. 8.5)

$$\ln(q_e - q_t) = \ln q_o - k_1 t$$

(b) Pseudo second order equation (Eq. 8.8)

$$t / q_t = 1 / k_2 q_e^2 + t / q_e$$

(c) The Elovich equation in the following form

$$q_t = (1/\beta) \ln t + (1/\beta) \ln (1/\alpha\beta) \qquad (10.46)$$

It is , however, not clear how the above equation was obtained from its original form i.e.,

$$q_t = (1/\beta) \ln (t + 1/\alpha\beta) - (1/\beta) \ln (1/\alpha\beta)$$ (see Eq. 8.18).

Possibly, it was assumed that $1/\alpha\beta$ was very small compared to t. But, this appears to be doubtful considering the values of α and β obtained by the authors [486] for the adsorption of cations on kaolinite and montmorillonite.

(d) One dimensional diffusion equation through the pores i.e., internal mass transfer

$$q_t = k_i t^{0.5} \qquad (10.47)$$

(e) Diffusion of metal ion through the liquid film around clay particles i.e., external mass transfer

$$\ln \{(1 - q_t / q_e)\} = k_f t \qquad (10.48)$$

But Eq.10.48 is essentially similar to pseudo first order equation and the value of k_i is similar to that of k_f as the authors also found [486]. But, pseudo first order model was discarded as the calculated values of q_e differ widely from those of experimental values. Therefore, film diffusion or external mass transfer model also get discarded automatically.

Kinetics of a chemical reaction are generally studied to elucidate the reaction mechanism and therefore these two aspects of study are very much interrelated. But the kinetic studies carried out so far by the above authors do not reveal any additional information on the mechanism of uptake of metal ions by clay minerals. Another drawback of the above studies is that all the kinetic parameters were determined at a fixed temperature and therefore no attempt was made to determine the activation energy values which probably could have thrown some light on the mechanism of the adsorption process. However, kinetic studies have important bearings on the engineering aspect of adsorption, especially regarding the design of adsorption column or reactor.

Similarly, many authors [483,485,485,492,499,500] have observed good fit of the experimental data for the adsorption of heavy metals on kaolinite and montmorillonite to the linearized forms of Langmuir and Freundlich isotherms. From the slope values of the linear plots the authors have determined maximum adsorption (q_m) values which are also called "monolayer capacity" of the relevant metal ions. But such a conclusion is fraught with some problems. First, the applicability of the Langmuir equation is based on certain tenets as discussed in section 8.2.1. One of the important tenets is that adsorption must be proportional to the concentration of adsorption species. This suggests that all the adsorption sites should have same affinity for the adsorbate species, a condition which is valid only at low equilibrium concentration. The validity of true Langmuirian equation can be checked from the log - log plot of amount adsorbed (moles/g) vs. equilibrium concentration (moles/l), the slope of which should be close to unity. However, slope values obtained from such plots, in most cases, are much less than unity [483,486] indicating that the isotherms are not true Langmurian in the concentration range studied. This situation arises when the surface sites are heterogeneous in nature. The wide differences in the values of maximum adsorption observed by the authors for Langmuir and Freundlich type plots indicate complexity of the problem. Therefore, linearity of the Langmuir type plot does not necessarily suggest the validity of the Langmuir isotherm. The coefficients obtained from the linear plots can at best serve as model parameters which may be used to predict adsorption beyond the range of experimental data. But, to attribute the q_m values obtained from the linear plots to "mono layer capacity" without checking validity of the Langmuir model is a questionable proposition. Secondly, the existence of two types of surface sites on clay minerals has been established beyond doubt. Therefore, Langmuir parameters should also vary depending upon the pH of the medium [488,489].

The adsorption of Pb and Ni on kaolinite has been found to be exothermic, whereas Cu, Co and Cd adsorb on kaolinite endothermically [486]. Angove et. al. [488] have also observed endothermic nature of the interaction between Cd and kaolinite. The endothermic nature of the adsorption of Cu, Co, Ni, and Mn on kaolinite has been observed by Yavuz et. al. [492].

Montmorillonite

Because of high CEC value resulting from large permanent negative charge in the interlayer region, montmorillonite or its source mineral bentonite exhibit higher adsorption capacity for divalent cations than kaolinite. This has been demonstrated by Bhattacharyya and Sengupta in a comparative study for the adsorption of various metal ions on both the clay minerals [486,501]. The increase in adsorption on montmorillonite is sometimes more than three times the extent adsorbed on kaolinite un-

der similar conditions. Acid activated montmorillonite exhibits slightly more adsorption
(2-4 mg/g). than untreated clay [502]. However, ZrO_2 pillared clay shows much lower value, though TBA (tetrabutyl ammonium) pillared clay show only little lower (2 - 3 mg/g) than the acid treated clay [501]. In all, treated montmorillonite clay does not offer any extra advantage over untreated cly as far as adsorption of heavy metal ions is concerned.

On the other hand, Na-bentonite shows high adsorption compared to untreated or as received montmorillonite which is assumed to be in the Mg exchanged form. Montmorillonite constitutes about 90% of the bentonite mineral, the other constituents being free silica(quartz) and feldspars.

The adsorption capacity values as determined from the langmuir type plots of the adsorption data of heavy metal ions on some clay minerals like kaolinite, montmorillonite, bentonite, vermiculite etc. are shown in Table 10.7. Most of these values were reported by Bhattacharyya and Sengupta in their different publications and compiled by the authors in their review article [501] which has been referred here instead of individual papers. It must be stated here that the values presented here are highly dependent upon various factors, such as pH, solid concentration, nature and origin of clay minerals, concentration and nature of the background electrolyte solution etc. under which adsorption experiments are carried out. Therefore, the values are merely illustrative in nature and not to be considered as representative of the aluminosilicate minerals.

The effect of pH on the adsorption of heavy metals on montmorillonite is essentially of similar nature to that of kaolinite. However, the initial break or small plateau region in the pH - adsorption curves, signifying ion exchange sorption, occurs at little lower pH value (about 3 - 4) for mont morillonite [483,503]. The rapidity with which subsequent adsorption through complex formation with surface OH groups takes place depends upon the nature of the metal ions [504,505]. The adsorption of Pb, Ni [506], Cu and Cd [507] on bentonite with respect to pH shows essentially similar pattern of behavior. Adsorption increases from pH 2.0, reaches maximum in the range 3 - 4 depending upon the metal ion and then exhibits some decrease to reach a plateau region at pH 5 - 6. However, the authors have not carried out adsorption beyond this pH range and therefore the role of surface OH groups on the adsorption behavior of bentonite remains somewhat unexplored.

The heats of adsorption of different metal ions on montmorillonite and bentonite have been determined from the plots of log (distribution coefficient, k_d) vs 1/T, and the values have also been found to vary with metal ion . While Cd and Cu are found to adsorb with endothermic heat of adsorption [483,496,498], Cr(III) [508], Co [497], Ni [509] and Zn [510] are found to adsorb exothermically on motmorillonite and bentonite. Though Pb is found to adsorb exothermically on montmorillonite [485], but on bentonite it is found to adsorb endothermically [512]. Again, the values quoted in the above references may not be taken as representative of the adsorbate - adsorbent interaction, as kd values are highly dependent upon the experimental conditions [249].

Table 10.7 : Adsorption Capacities of a Few Clay Minerals and their Modified Forms for Some Selected Metal Ions.

Adsorbent Clays	Adsorption capacity (mg/g)							
	Cd(II)	Co(II)	Cr(III)	Cu(II)	Mn(II)	Ni(II)	Pb(II)	Zn(II)
Kaolinite	9.9 a)	11.2 a) 1.7 b)		9.2 a), 4 a) 0.5 b)	10.8 b)	10.4 a)	11.2 a)	
Acid treated kaolinite	11.4 a)	12.1 a)		5.6 a), 10.1 a)			1.9 a)	12.1 a)
Pillared kaolinite-1*	9.7 a)			3.0 a)			8.8 a)	10.7 a)
Pillared kaolinite-2#	9.8 a)			3.2 a)			8.4 a)	9.4 a)
Montmorillonite	32.7 a) 6.3 d)	28.6 a)		25.5 a)	4.8 d)		28.4 a)	33.0 a)
Acid treated montmor.	33.2 a)	29.7 a)		28.0 a)			29.5 a)	34.0 a)
Pillared montmorillonite-1*	27.7 a)			7.1 a)			22.0 a)	26.6 a)
Pillared montmorillonite-2#	30.0 a)			27.3 a)			19.5 a)	22.2 a)
Na-bentonite	26.2 c)		49.8 c)	30.6 c)		24.2 c)		23.1 c)
Ca-bentonite	7.3 c)		44.4 c)	7.7 c)		6.3 c)		5.8 c)
Natural bentonite	9.3 e) 11.2 d)	6.8 f) 15.2 g)		12.4 f)	6.0 d)	4.3 h)	7.4 h)	15.7 f) 53.0 i)
Vermiculite (natural)		44.7 j)		53.3 j)			54.6 j)	62.8 j)

References : a) 501, b) 492, c) 563, d) 561, e) 572, f) 562, g) 570, h) 506, i) 511, j) 514

* ZrO_2 pillared clay , # TBA (tetrabutyl ammonium) pillared clay.

Illite

The adsorption of cations on illite has not been studied as widely as on kaolinite and montmorollonite. Like these two clay minerals , illite has also two types of cation adsorption sites as has been discussed in section 7.5.3 . However, the main difference from the other two clay minerals is the presence of vacant cation sites between two adjacent tetrahedral sheets in the frayed edges of illite. These sites are in addition to those formed by the deprotonation of aluminol and silanol groups at the edges.

Adsorption of metal ions on illite also depends on pH . The effect is most perceptible in higher electrolyte concentration (say 0.1M) than in lower concentration (<0.01M) at which the adsorption of most metal ions is very high even at moderate pH value (≅4 - 5). As an indication of the two stage adsorption process, the initial slow stage in the pH range 3.0 - 4.5 is followed by sharp rise to 100% adsorption at pH above 5 is particularly noticeable in 0.1M $NaNO_3$ solution for metal ions like Cd^{2+}, Ni^{2+}, Zn^{2+}, Cu^{2+} etc [243]. The selectivity sequence of metal ions towards illite can be determined under this condition. But in 0.001M $NaNO_3$ solution the adsorption edges of different metal ions almost merge with one another, thereby making it difficult not only to distinguish two stages of adsorption, but also to determine any selectivity sequence among them. According to Gu and Evans [243] such a sequence observed for the adsorption of five metal ions on permanent negative (surface) sites of Fithian illite is Pb ≅ Cu>Zn ≅Ni >Cd, and on edge sites is Pb>Cu>Zn ≅ Ni>Cd .

As far as the mechanism or best fitting of adsorption model is concerned, all the authors [241,243] proposed outer sphere bidentate $\{(X\text{-})_2\text{-}M^{2+}\}$complex formation at the surface site. However, more than one mechanisms have been proposed on edge site. While for adsorption of Cd [241] and Cu [513] on Muloorina illite the authors proposed the formation of one inner sphere (SO^-M^+) and one outer sphere (SOH^+-- M^{2+}) complexes, others [243] proposed the formation of only SO^-M^+ for the adsorption of heavy metals on the edge sites of illite. However, the authors [243] also proposed the formation of one inner sphere hydroxy complex species (X^-MOH^+) in addition to outer sphere bidentate complex to achieve best model fit for adsorption on surface sites.

Vermiculite

Literature on the adsorption of cations on vermiculite is even scarcer than illite, though the former clay mineral is an important constituent in many soils. Because of extensive substitution in both tetrahedral and octahedral sheets vermiculite has highest permanent negative charge on its surface and consequently, has highest CEC value among all the clay minerals. DeFonseca et al [514] have studied the adsorption of Pb, Cu, Ni, and Co on a Brazilian vermiculite sample without any further treatment. Although the authors have not made any attempt on the determination of surface charge, it appears that unlike other 2:1 layer silicates variable charge on the edges is less than the surface charge. For adsorption at different pH values, the authors have used various constant pH buffer solutions ranging from pH 1.15 to 9.0. It is observed that except Pb, pH variation has small effect on the adsorption of other three metal ions. The nature of pH vs. adsorption (in per unit weight) plots, indeed, do not suggest the presence of two types of surface sites on vermiculite [514]. However, in the case of adsorption isotherm experiments at the natural pH, small breaks are observed at low equilibrium concentration, though the exact pH value and the nature and concentration of supporting electrolyte have not been stated. The selectivity sequence found by the authors [514] from the isotherm experiments is Ni>Cu>Co>Pb, which is in accordance with increasing ionic radii and enthalpy of hydration of the ions. Others [515] have shown that selectivity regarding the uptake of metal ions on vermiculite is sensitive to pH as well as ionic strength of the medium. At low ionic strength ([0.01M) the uptake of heavy metal ions is so high that not only the selectivity order but also the distinction between initial uptake through ion exchange and subsequent adsorption on the edges are almost blurred. But, at higher ionic strength (~ 0.1M) the two stages of adsorption is clearly distinguishable by sharp increase in uptake above pH 5.5. The affinity sequence observed by Abollino et. al. [505] for both batch and column adsorption experiments is Mn>Ni>Zn.>Cd>Cu>Pb.

10.5.2.3 Metal Ion Adsorpion ont Zeolites

A brief description of the structural properties of zeolites has been presented in chapter 7. Zeolites used for the removal of heavy metal ions from waste water are usually of natural origin, keeping in view not only the cost of adsorbent material, but also the possibility of making the process cyclic i..e., regeneration of zeolite after a certain interval of time or cycle of use [516]. For this purpose, the adsorbent must be structurally stable to withstand various treatments involved during the entire process of adsorption and regeneration.

Naturally occurring clinoptilolite, a zeolite mineral belonging to heulandite family, is an important candidate suitable for the adsorption purpose. In fact, most of the studies on cation adsorption on zeolites have been carried out on this mineral which occurs frequently in sedimentary rocks of volcanic origin (zeolitic tuffs) in association with many other secondary minerals like quartz, calcite, aragonite, feldspar, montmorillonite, chlorite besides some other forms of zeolites [207,208]. It differs from heulandite by its high Si / Al ratio(4.5 - 5.3) and high thermal stability (650° - 800°C). Its composition and therefore the chemical formula vary widely depending upon the source. Sprynskyy et. al.[517] quoted the formula as $(Na, K, Ca)_4Al_6Si_{30}O_{72}\cdot 24H_2O$ for a clinoptilolite sample from trans-Carpethian region of Ukrain [518], whereas Erdem et al [519] quoted one formula from another source [520] and is given by: $(Na_{0.1}K_{8.57}Ba_{0.04}), (Al_{9.31}Si_{26.83}O_{72}).\ 19.56H_2O$.

The above formulas suggest that Na, K, and Ca can act as exchangeable cations in clinoptilolite, while in heulandite (low silica variety) alkaline earth metals, like Ca, Sr and Ba act as exchangeable cations. Any way, the dominant excahangeable cation in both the varieties is K, followed by Na and Ca [207]. Generally, Clinoptilolite with void volume fraction of 34% and CEC of 216-220 meq/100g exhibits best selectivity towards the adsorption of different heavy metal ions [517]. Attempts have been made to modify clinoptilolite by chemical treatment to increase its ion exchange capacity [517,521-523]. Chemical, surface and ion exchange properties of clinoptilolite have also been studied by other investigators [524-527].

Adsorption of heavy metal ions on clinoptilolite has been studied more extensively than any of the clay minerals. Besides the earlier literatures cited in well-known monographs on zeolites [207,208,518], there are many papers published in the last two decades on this subject, a few of which may be mentioned here [516,517,523, 528-534]. Like any other inorganic adsorbent, adsorption on zeolites is dependent upon various factors such as pH [516,517,519,532], initial concentration of metal ion [517,519], structure of zeolites [532,533], nature of metal ion [528,535] etc. Some of the important features of metal ion adsorption on zeolites, especially clinoptilolite are discussed below.

Adsorption of metal ions on zeolites decreases with decrease in pH. However, the effect of this parameter is not as sensitive as in the case of clays. Unlike clay minerals, the literature on acid - base equilibria on zeolite surfaces is very scarce. We may, however, assume that pH dependent surface sites on zeolites are very small, though impurity oxides in raw material may also exhibit the presence of such sites. Thus, adsorption on zeolites is mostly through ion exchange and therefore mildly dependent on pH. In general, zeolites are weakly acidic in nature. It is observed [536] that Na- zeolite (e.g., Na-clinoptilolite) is placed in dilute solution ([0.001M) NaCl or $NaNO_3$, pH value increases due to the exchange of H^+ with Na^+.

$$\text{Na - zeolite} + H_2O \longleftrightarrow \text{H - zeolite} + NaOH$$

Therefore, saturated adsorption at higher pH value on clinoptilolite may lead to surface precipitation of metal hydroxide. But, it is not known if K - zeolite or Ca - zeolite can be exchanged with H^+. For example, NH_4^+ can exchange Na^+ or Ca^{2+}, but not K^+, indicating that K^+ stabilizes zeolite structure [535] possibly because of its suitable ionic radius and poor hydration.

Sprynskyy et.al.[517] have studied the kinetics of adsorption of Cd, Cu, Ni, and Pb on clinoptilolite in detail and suggested the existence of three stages during the period of adsorption . The first stage is characterized by rapid adsorption within 25 - 30 min on the crystal surface followed by minor desorption or counter diffusion of exchangeable cations from interior of the micro-crystals for a short period, which the authors termed as "inversion stage". This stage is most noticeable in the case of adsorption of nickel. The third stage is characterized by slow diffusion through micro pores in fine particles and the process continues up to 5- 6 hr depending upon the concentration of metal ion. About 90% of Ni and 40 - 45% of Cu, Cd and Pb are adsorbed in the first stage.

That the behavior of Ni is different from other divalent cations can be demonstrated in different aspects of its adsorption on clinoptilolite. First, while Ni shows about 40% decrease in adsorption in the presence of Pb, Cd, and Cu, the other metal ions do not show much variation (< 3 - 4%) between single and multi ion adsorption. Second, Sprynskyy et. al. [517] have also studied the adsorption of metal ions on clinoptilolite treated with NH_4Cl and HCl. It has been observed that NH_4-zeolite adsorbs almost twice and H- zeolite adsorbs as much as 35% higher amount of Ni than that untreated clinoptilolite. But, the other three metal ions exhibit minor differences between treated and untreated mineral.

As far as adsorption capacities of different metal ions on clinoptilolite are concerned, there are considerable discrepancies in the literature. This has been briefly but adequately reviewed by Sprynskyy et.al.[517]. The discrepancies arise due to various factors mentioned above which influence the extent of adsorption. Therefore any reported sequence or order of adsorption capacity in respect of different metal ions shall be considered as illustrative in nature. A few such examples are cited and discussed below.

Zanzow et. al.[528] have studied the adsorption of a number of metal ions on clinoptilolite and proposed the following order of selectivity: Pb>Cd>Cs>Cu>Co>Cr>Zn>Ni>Hg. Using the maximum adsorption data derived from Langmuir isotherm, Erdem et. al. [519] suggested the order Co>Cu>Zn >Mn which is different from the above as far as the sequence of Co and Cu is concerned. According to the authors, adsorption depends on the charge density of the hydrated ion and since charge of these four cations is same (+2), Mn(II) ion of largest radius (~0.90A°) show minimum adsorption ; Cu being the smallest (r = 0.72A°) among these four metal ions show highest adsorption. But this view has been contradicted by Sprynskyy et. al., according to whom so long as ionic radii of the adsorbing ions are several times smaller than the pore size of the adsorbent, it is hardly possible to correlate different physico-chemical parameters including ionic radii with their corresponding adsorption capacities. Though some authors [535] argued that hexaquo metal ions in aqueous medium can pass through the open channel or be accommodated in the void space of the zeolite structure, others [517] have expressed their doubt in this regard. Assuming that the size of the open channel in most natural zeolites being 3 - 4 A°, it is difficult for fully hydrated aquo ions to pass through without shedding some of their water molecules [208]. Therefore metal ions of high enthalpy of hydration ($-\Delta H_{hyd}$) i.e., strongly hydrated metal ions like Mn(II) and Ni(II) will face greater hindrance than ions of lower enthalpy of hydration i.e., weakly hydrated ions like Pb(II), Hg(II) etc. Thus Sprynskyy et. al. found the order Pb>Cu>Cd>Ni, using the data obtained from initial as well as equilibrium stages of adsorption. It is interesting to note that this order is reverse of that followed for vermiculite [514] wherein strongly hydrated ion is more preferred than weakly hydrated ion.

Another important observation made by several investigators [517,519] is that the distribution coefficient (k_d) of metal ion sorbed on zeolites increases with decrease in initial concentration of metal ion. This aspect is directly related to the selectivity and capacity of metal ion sorption on zeolites. Erdem et. al. [519] have shown that the sensitivity of the variation of kd as function of the initial concentration decreases in the order Co>Cu>Zn>Mn which is exactly similar to the order of selectivity. This suggests that there are more than one type of adsorption sites for divalent metal ions. At low concentration, adsorption takes place at the energetically most favourable site (i.e., site of easily exchangeable ion) which is possibly sensitive to certain properties of the ions like radii, hydration state, electronegativity etc. But, with increase in concentration, metal ions tend to be adsorbed at weakly exchangeable sites, where the dependency of kd on the physico-chemical properties of ions is rather weak. In other words, at higher metal ion concentration there is little or no competition between metal ions for adsorption sites.

Table 10.8 provides a summary of the literature references on the adsorption of metal ions on various aluminosilicate minerals as discussed in the preceding paragraphs. The references have been classified in the same format as presented in Table 10.6.

Table 10.8 : Literature References on the Adsorption of Metal Ions on Some Aluminosilicate Minerals

Aluminosilicate Minerals	Metal ion*	References
Kaolinite	Cd	482, 483, 486, 487, 488, 490, 493, 495, 544, 548, 564, 576
	Co	486, 487, 489, 492, 497, 502, 551
	Cr(III)	566
	Cu	482, 489, 490, 492, 493, 496, 499, 537, 542, 568
	Fe(III)	489, 551
	Hg	571

	Mn	489, 492
	Ni	486, 489, 492, 493, 548, 551
	Pb	485, 486, 489, 490, 493, 498, 542, 548, 564, 568, 576
	Sr	556
	Zn	490, 539, 567
Montmorillonite	Cd	483, 495, 502, 505, 541, 544, 548, 552, 561
	Co	502, 541, 551, 560
	Cr(III)	563
	Cs	565
	Cu	496, 498, 502, 505, 538, 541, 544, 548, 553, 563
	Eu(III)	559
	Fe(III)	551
	Mn	505, 561
	Ni	498, 502, 505, 548, 551, 555, 563
	Pb	502, 503, 505, 548, 552
	Sr	541, 536
	U(VI)	554
Bentonite	Ca	557
	Cd	483, 507, 443, 549, 561
	Co	551, 562, 570, 575
	Cr(III)	508, 563
	Cu	507, 547, 549, 562, 563, 569
	Mn	561
	Ni	506, 548, 563
	ÒtPb	506, 512, 549
	Sr	557
	Zn	511, 543, 562, 563
Vermiculite	Cd	505, 514, 515
	Co	514
	Cu	505, 514, 515, 540
	Mn	505, 515
	Ni	505, 514, 555
	Pb	505, 514, 515
	Zn	505, 515
Illite	Cd	241, 243
	Cu	243, 513, 538, 540
	Eu(III), Nd(III)	558
	Ni, Pb, Zn	243
Zeolite	Cd	517, 521, 528, 530, 533, 534, 550, 572, 578
	Co	519, 528
	Cr(III)	522, 528, 530
	Cu	517, 519, 522, 528, 533, 534, 550, 572, 577, 578
	Eu(III)	573
	Fe(III)	522
	Hg	528

Mn	519
Ni	517, 528, 534, 578
Pb	517, 521, 522, 528, 530, 533, 534, 550, 577, 578
Zn	519, 528, 534, 577

* Unless otherwise stated the metal ions are of oxidation state +2 .

10.5.3 Adsorption on Inorganic Materials other than Oxides and Silicates

Apart from oxides and aluminosilicates, there are also a large variety of inorganic materials which act as ion exchangers from aqueous medium. These include insoluble oxo- or polyoxo salts such as phosphates, vanadates, arsenates, tungstates, molybdates, silicates etc. Earlier literature on these materials can be available in the well-known monograph by Amphlett [579]. Except a few which occur in nature (e.g., apatites) most of the above materials are synthetically prepared in the laboratory. Consequently, their role as natural scavenger in our environment is known to be very limited . As the number of such synthetic ion exchangers is growing reasonably large because of the possibility of a wide variety of combinations, it is not possible to include each and every variety in the present discussion. Besides, the efficiency of many of these materials as adsorbents for a wide range of cations has not yet been established by laboratory investigations. Therefore, a brief discussion is included here on one variety of oxo salt, namely orthophosphates which have been extensively studied as ion exchangers.

Alkaline earth metals form a series of insoluble orthophosphates known as apatites. They are used in the manufacture of super phosphate fertilizer. Sedimentary apatites (e.g., carbonate apatite) in finely ground form are also used as sources of slow release phosphate in soil, especially in acidic soil. Apatites are generally found to contain either F^- or Cl^- or OH^-. Since an additional monovalent anion , usually OH^- , is necessary in their structures for charge balance, these salts are also known as hydroxyapatites (HAP). Fluorapatite($Ca_{10}(PO_4)_6F_2$) is generally found in igneous and metamorphic rocks, while the corresponding chlor- and hydroxyapatites occur rarely in nature, but can be prepared in the laboratory. The methods of preparation of various HAPs and their adsorption properties have been discussed by Mishra [580] . Like oxides and aluminosilicates, adsorption of cations on HAPs is affected by pH of the solution, concentration of adsorbing and competing cations, if any. The dependence on pH suggests the the presence of ionizable OH groups on the surface , besides the OH groups that are already present in their structures. Sugiyama et al [581] observed that ion exchangeability of CaHAP with Pb was strongly influenced by preparation method and particle size fraction. Particle sizes finer than 0.85mm was found to have higher sorption capacity than those in the size range 0.85 - 1.7mm. Sample calcined at 500°C were characterized by XRD and XPS, but they showed lower adsorption than uncalcined sample in the acidic medium. Attempts have been made to convert natural rock phosphate into mesoporous HAP for removal of heavy metal ions [582].

Sugiyama et al [583] developed a novel method of continuously removing cations like, Cd, Co, Cu and simultaneously regenerating the adsorbent CaHAP spread as thin film on a glass plate. This method was, however, found to be unsuitable for the removal of Pb [583,584] as this metal ion is adsorbed irreversibly on CaHAP as insoluble PbHAP through ion exchange with dissolved Ca. The other metal ions such as Cd, Co, Cu adsorbed on CaHAP can , however, be eluated with an acidic solution of $Ca(NO_3)_2$.

The adsorption of Pb on HAPs of Ca, Ba, and Sr or on phosphates of trivalent metals like Fe, Al and Cr has also received much attention by different researchers. The following experimental observations have confirmed that the mechanism of adsorption of Pb on solid phosphate is different from the adsorption mechanism of other divalent cations like Cu, Cd, Sr, and Zn [585]. First, the removal of Pb is almost independent of pH of the aqueous solution and is complete in the pH range 3-12 , whereas the sorption of Cd, Sr and Zn gradually increases with increase in pH [586-588]. Secondly, the contact time required for attaining equilibrium was 30 min for Pb, as against 24h for other cations. Thirdly, the maximum adsorption capacity of Pb on HAP or on phosphates of Al, Fe and Cr is at least 5 - 6 times higher than that of other metal ions like Cd, Cu, Co, Sr, Zn etc. Fourthly, as consequence of the high selectivity for Pb, its removal by HAP is not influenced by the presence of other cations like Ca, Mg, Na and K, whereas the presence of Ca and Mg decreases the adsorption of Cd, Cu, Sr, and Zn etc. All these differences have been attributed to the difference in the mechanism of adsorption of Pb as insoluble PbHAP, while other cations adsorb specifically on CaHAP.

Another important characteristic of the adsorption of Pb on HAP is the observed interdependence of cation and anion when both are present in the adsorption system. Thus, Cl- does not exhibit specific adsorption either on CaHAP or on PbHAP, but when present along with Pb^{2+} the exchange of both takes place on CaHAP and a proportionality between the sorbed quantities of these two ions is found [589]. Similar observation has been made for the adsorption of Pb on BaHAP [590]. The exchange capacity of the latter is greatly improved not only by the use of recycle experiment instead of batch experiment, but also by adding dilute HCl to the system. This is because HCl facilitates surface dissolution of BaHAP to release Ba^{2+} which is then exchanged with Pb^{2+} to form insoluble PbHAP on the surface of the adsorbent. Kinetics, thermodynamics and other aspects of the adsorption of Pb on CaHAP have also been studied [591,592].

Adsorption of Cd from contaminated water and soil has also been studied [593] from single as well as multimetal system containing Cd, Cu, Pb and Zn in the concentration range 0-08 mmole/l. It has been shown that for single metal ion system , the adsorption capacity on CaHAP varies from 0.058 to 1.681 mmole/g of solid depending on the type of metal ion. In the multi metal system, sorption capacity is reduced by 63-83% compared to the sorption capacity of the corresponding single metal ion system. Cadmium is proposed to be adsorbed by a two step mechanism : First step is the rapid surface complexation followed by partial dissolution of CaHAP and ion exchange with Ca^{2+} resulting in the formation of CdHAP. Adsorption / desorption of Cd and Zn on / from HAP in single and binary metal ion system in aqueous solution has also been studied by Peld et al [594].The removal of Cd by HAPs has been reviewed by Zhu et al [595]. Corami et al [596] have also carried out Cu and Zn from binary system and have found that adsorption capacity at 25oC is reduced by 13-76% for Cu and 10-63% for Zn compared to single metal ion system. Adsorption of Zn on Ca and BaHAP has been compared [597]. It is observed that at pH 6-8 adsorption on CaHAP is not only at least three times more rapid than on BaHAP, but also equilibrium adsorption is higher on CaHAP by a similar extent. Sorption properties Cd and other metal ions on certain binary apatites like Mg-Ca-HAP have been studied [598,599]. Sorption of antimony (Sb) on HAP has also been studied [600]. Surface characterization of carbonate containing mineral fluoroapatite and the adsorption of Ni on this material have been carried out [601].

Apart from hydroxyapatites of alkaline earth metals , phosphates of trivalent (Al, Fe and Cr) and tetravalent metals (Sn, Ti and Zr) have been extensively studied for their ion exchange properties vis-à-vis their molecular and crystalline structures. Many of the earlier literature on this subject can be found in the monograph by Clearfeld [602]. Hydrated zirconium phosphate ($Zr(HPO4)_2 .H_2O$) is one of the oldest known phosphates of tetravalent metal whose structure and ion exchange properties have been studied and compared with other tetravalent metal phosphates [603-606]. Preparation, characterization and sorption properties of amorphous titanium phosphate have been studied [607]. The product was assigned the formula $Ti_4O_2(HPO_4)_3(H_2PO_4)_2(OH)_4.3H_2O$. Jia et al [608] also prepared an amorphous variety of Ti-phosphate and studied the adsorption of Pb, Zn and Cd. Sorption properties of several cations (Cu, Cr(III), Co, Fe(III), Mn, Ni etc) on a crystalline α-$Ti(HPO_4)_2$. H_2O have been studied by Sahu et al [609,610]. In the comparative studies of crystalline phosphates of Ti, Sn and Zr by Parida et al [605], it has been observed that α-$Ti(HPO_4)_2$. H_2O exhibits highest sorption capacity over the hydrated phosphates of Zr and Sn.

Mustafa et al [588,611-613] have carried out extensive investigation on the ion exchange sorption of several mono- and divalent cations on the phosphates of trivalent metals like Al, Fe and Cr. At the outset, the authors characterized the adsorbents by various physical and chemical methods which include potentiometric titration by continuous or rapid method in the presence of metal ions and 0.1M KCl as background electrolyte solution. It has been observed that the titration curves shift either to acid or alkaline side depending upon the first hydrolysis constant (pK_{MOH}) of metal ions. In other words, titration curves in the presence of more easily hydrolysable metal ions (lower pK_{MOH}) tend to shift to alkaline side relative to those metal ions which hydrolyze at higher pH values. This behavior is similar to that observed for oxides and silicates as discussed in the preceding chapters, suggesting similar amphoteric dissociation of surface OH groups. It has been proposed that phosphates also undergo hydration in aqueous solution and subsequent dissociation of adsorbed water results in the formation of ionizable OH groups on the surface [588,614]. Consequently, selectivity sequence of metal ion adsorption on these phosphates follows similar pattern of order as in the case of oxides.

As far as the temperature effect on adsorption is concerned, Mustafa et al [615] have observed that while for $AlPO_4$ adsorption tends to increase with increase in temperature (endothermic adsorption), for $CrPO_4$ adsorption tends to decrease with increase in temperature (exothermic adsorption) [588]. The thermodynamic parameters were estimated from the plot of log(distribution coefficient) vs reciprocal temperature (1/T) for different metal ions.

Besides materials of unique and distinct chemical entities of inorganic origin, a large number of naturally occurring materials, wastes from manufacturing and services industries, wastes from mineral and metallurgical processing industries, wastes from agro- horticultural activities have been used as adsorbents for the removal of heavy metals from aqueous solution. They are used mainly from economic consideration and ready availability. As declared in the preface of the book, non-homogeneous materials which cannot be characterized or identified as compounds of inorganic origin will be kept beyond the scope of discussion in the book. We, therefore, refrain from further discussion on the "low cost" adsorbent materials. However, interested readers may consult an exhaustive and interesting review article on this topic by Babel and Kurniawan [616].

CHAPTER 11
ADSORPTION OF ANIONS ON INORGANIC MATERIALS

11.1 Introduction

Like cations, adsorption (and / or desorption) of inorganic and organic anions on inorganic adsorbents, has been extensively studied to understand its manifestations in certain natural phenomena pertaining to agriculture, geochemistry and environment. Since, natural adsorbents are complex and heterogeneous in nature, little progress can be made on the basic understanding of the above phenomena by using such materials in the adsorption studies. Therefore, attempts have been made to study adsorption of anions by using not only well-defined solids as adsorbents, but also specified concentration and nature of background electrolyte solution. Theoretical analysis including modeling of the experimental data is, however, similar to those used for cation adsorption and discussed in detail in the preceding chapters. Unlike the adsorption of cations, the choice of adsorbent for anion adsorption is limited mostly to hydrous oxides of iron and aluminum. This is because the PZC of these adsorbents occur at relatively higher pH values, thereby facilitating anion adsorption over a wide pH range, especially below PZC.

11.2 Characteristic Features of Anion Adsorption

Anions are basically ligands i.e., they possess one or more atoms with a lone pair of electrons which can be donated to form coordinate bonds with metal ions. As it has been stated, the same principle is believed to be involved in the specific adsorption of cations onto an oxide surface which acts as a ligand to form complex with the adsorbing cation. Therefore, adsorption of anionic ligand takes place by replacing OH group from hydrated oxide surface. This is analogous to the formation of anion -metal complex in the solution phase [304]. The equilibrium constant of a surface reaction can therefore be estimated from the corresponding reaction for complex formation in the solution phase. This aspect has been discussed in more details in the next section. Let us consider the solution phase reaction between partially hydrolyzed species of Fe(III) with conjugate acid of an anionic ligand L.

$$FeOH^{2+} + HL \longleftrightarrow FeL^{2+} + H_2O \tag{11.1}$$

$$K_{soln} = \frac{[FeL^{2+}]}{[FeOH^{2+}][HL]} \tag{11.2}$$

The corresponding surface complex formation is represented by

$$FeOH^{o}_{surf} + HL \longleftrightarrow FeL_{surf} + H_2O \tag{11.3}$$

$$K_{surf} = \frac{[FeL_{surf}]}{[FeOH^{o}_{surf}][HL]} \tag{11.4}$$

The electrostatic adsorption of ligand L^- with hydrous oxide surface below pH_{PZC} can also be represented in the following form

$$FeOH_{2\ surf}^{+} + L^{-} \longleftrightarrow FeOH_2^{+}\text{---}L^{-} \tag{11.5}$$

The expression (11.5) is also called outer-sphere complex formation as in the case of cation adsorption. Stoichiometry, however, does not distinguish between reactions (11.3) and (11.5) and even the binding constant values are not much different one another. Such weekly bound adsorption is entropy driven i.e., tend to be exothermic in nature. Generally, carboxylic acid or phenolic OH groups adsorb through hydrogen bonding with singly or triply coordinated surface groups on goethite [59].

However, such complex formation is not envisaged when adsorption occurs through simple exchange with ions of similar charge and type. It is dominated by electrostatic contribution to the overall free energy of adsorption and is therefore influenced by the ionic strength of the medium. This phenomenon is called "ion pair" formation with positively charged ions including H^+ at the solid surface. Examples of non-specifically adsorbed anions include NO_3^- and ClO_4^-. At least one water molecule separates these anions from the adsorbent surface.

Specific adsorption is, by far, the most common phenomenon associated with a wide variety of anions, both inorganic and organic. Here we shall concentrate primarily on ions of inorganic origin such as phosphate, arsenite / arsenate, chromate, selenite / selenate, fluoride etc. Some ions of organic origin such as acetate, oxalate, tartrate, EDTA etc. are frequently encountered in environment and even tend to influence the adsorption of inorganic ions. This aspect of competitive adsorption will be briefly discussed at the end of the chapter.

11.2.1 Linear Free Energy Relationship (LFER) in Anion Adsorption Equilibria : Effect of pH on Anion Adsorption.

In the previous chapter on cation adsorption it has been shown that there exists linear relationship between surface binding constants of divalent metal ions and their corresponding hydrolysis constants. A similar log - log relationship also exists between acidity constants of anions and their corresponding binding constants with hydrous iron oxides. The linear relationship as illustrated in Fig. 11.1 reflects the tendency of divalent anions to be protonated in solution and subsequently reacts with metal ion on the oxide surface. In other words, binding strength of an anion with the oxide surface increases with increase in the degree of second deprotonation of the corresponding diprotic acid. Though sufficient literature data are available for pK_a values of many di- and trivalent anions [78,617], reliable data on the binding constants of the corresponding anions with amphoteric oxides are scarce. Therefore, unlike wide range of predictive capability of the LFER between hydrolysis constants and surface binding constants of divalent metal ions, the same may not be possible with confidence from Fig. 11.1. However, even under such a constraint, Dzombak and Morel [617] have estimated the binding constants of several anions with HFO using a similar relationship. The surface reactions of mono- and divalent anions derived from deprotonation of conjugate acid H_2L are given below

$$FeOH^o_{surf} + HL^- \leftrightarrow FeL^-_{surf} + H_2O \qquad K_2 \qquad (11.6)$$

$$FeOH^o_{surf} + L^{2-} \leftrightarrow FeOHL^{2-}_{surf} \qquad K_3 \qquad (11.7)$$

For monovalent anions like F^-, CN^-, CNS^- etc simply replace L^{2-} with L^- in the above equations. For trivalent anions like orthophosphate (H_3PO_4) the surface binding constants can be defined as follows, corresponding to the three acid dissociation constants of H_3PO_4 to yield $H_2PO_4^-$ ($K_a=10^{-2.15}$), HPO_4^{2-} ($K_a= 10^{-7.20}$) and PO_4^{3-} ($K_a= 10^{-12.35}$)

$$FeOH^o_{surf} + L^{3-} + 3H^+ \leftrightarrow FeH_2L_{surf} + H_2O \qquad K_{1surf} \qquad (11.8)$$

$$FeOH^o_{surf} + L^{3-} + 2H^+ \leftrightarrow FeHL^-_{surf} + H_2O \qquad K_{2surf} \qquad (11.9)$$

$$FeOH^o_{surf} + L^{3-} + H^+ \leftrightarrow FeL^{2-}_{surf} + H_2O \qquad K_{3surf} \qquad (11.10)$$

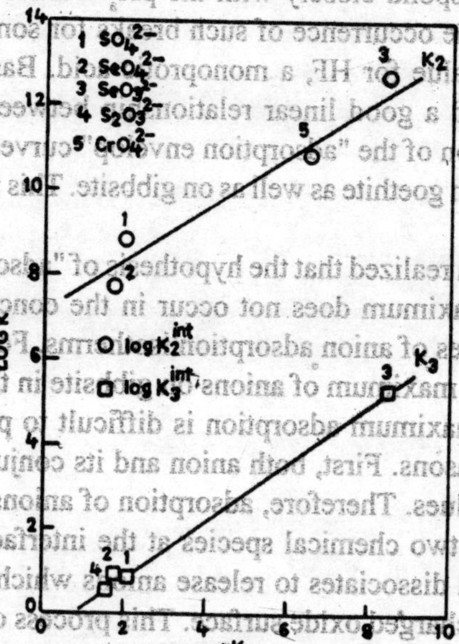

Fig.11.1 Linear free energy relationship between dissociation constants (logK_a) of the conjugate acid of divalent anions and their corresponding binding constants (log K^{int}) with HFO surface. Uncertainty in the log K^{int} values is at least one order of magnitude. Redrawn after ref. 617.

$$FeOH^{o}_{surf} + L^{3-} \Longleftrightarrow FeOHL^{3-}_{surf} \qquad K_{4surf} \qquad (11.11)$$

The linear log - log relationship between K_a and surface binding constant has, however, been dealt in a different way by Hingston et al [618] in their investigation on the effect of pH on anion adsorption on goethite and gibbsite. It has been observed that the adsorption isotherm of anions on goethite at constant pH generally follows Langmuirian behavior i.e., initial increase in adsorption with increase in concentration, gradually reaching a saturation plateau at higher concentration. However, somewhat different behavior is observed in the case of gibbsite. After an initial slow increase in adsorption at very low concentration, adsorption increases rapidly with increase in concentration without reaching any saturation plateau [273]. Therefore, in the case of gibbsite apparent Langmuir maximum is observed only at relatively lower concentration. The plot of Langmuir maxima vs. pH was termed "adsorption envelop" and an attempt was made to correlate this parameter with the properties of both adsorbent and adsorbate [619]. It was observed that apparent Langmuir maxima for monoprotic anions occurred at pK_a of their corresponding conjugate acids. For anions of polyprotic acids, breaks in the slopes of the plots of adsorption envelop vs.

pH were found to correspond closely with the pK_a values of the conjugate acids. Figure 11.2 illustrates the occurrence of such breaks for some polyprotic conjugate acids as well as peak value for HF, a monoprotic acid. Based on this observation Hingston [273] obtained a good linear relationship between the pK_a of conjugate acids and pH of inflection of the "adsorption envelop" curves for the adsorption of a large number of anions on goethite as well as on gibbsite. This is illustrated in Fig. 11.3.

Subsequently it has been realized that the hypothesis of "adsorption envelop" is inadequate, as Langmuir maximum does not occur in the concentration range used in most experimental studies of anion adsorption isotherms. For example, it is difficult to determine adsorption maximum of anions on gibbsite in the pH range 3 - 8. Even for goethite, value of maximum adsorption is difficult to prove without doubt because of at least two reasons. First, both anion and its conjugate acid are present in solution at lower pH values. Therefore, adsorption of anions on oxide is the sum of the adsorption of these two chemical species at the interface. As pH is allowed to increase, conjugate acid dissociates to release anions which tend to adsorb electrostatically on positively charged oxide surface. This process of increase in adsorption with increase in pH continues within a narrow pH range until the pK_a value of the conjugate acid is reached. But this behavior is not universal for all anions and is predominantly followed for monoprotic acids and for some diprotic acids e.g., H_2SO_4,

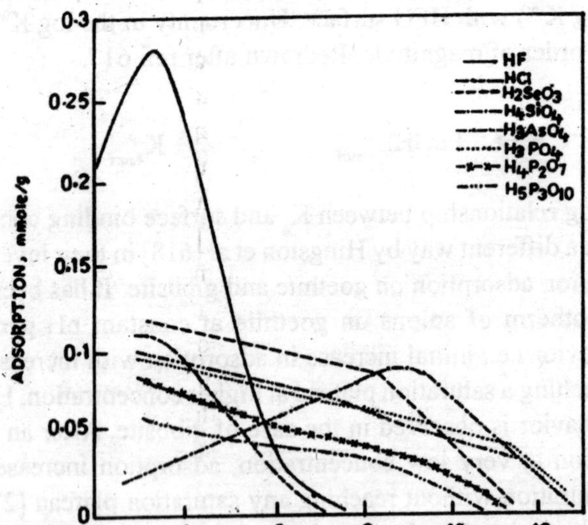

Fig.11.2 Variation of the adsorption capacities of anions on goethite with respect to pH showing the occurrence of inflexion or change in slope values (adsorption envelops). Redrawn after data from refs. 618 and 619.

H_2CrO_4, and their mixtures [249]. Further increase in pH value leads to the decrease in adsorption due to the gradual decrease in the positive surface charge. Thus, when adsorption is non-specific in nature i.e., governed by electrostatic force, the surface must have overall positive charge (SOH_2^+), hence the region of maximum adsorption lies between pK_a of the conjugate acid and pH_{PZC} of the solid. In contrast, an overall positive charge is not essential for specific adsorption and adsorption may take place even at pH above PZC. Polyprotic acids such as phosphate, arsenate, silicate etc. can form more than one species in solution over a wide pH range. It has however been estimated that under the condition of maximum adsorption the number of surface sites occupied by adsorbed anion is only a small fraction of the total number of available surface sites.

Secondly, the adsorption maximum determined from constant pH isotherm by using Langmuir equation and subsequently used in the plot of "adsorption envelop" is itself considered to be an uncertain quantity. This is again due to inadequate range of concentration used in isotherm experiment. Adsorption may strictly continue over a long period of time even after attaining apparent equilibrium state. Therefore, instead of "adsorption envelop" plot , Hingston et al [619] suggested the plot of pH_{50} (pH for 50% adsorption) vs. pK_a for the adsorption of various anions on goethite and gibbsite, since pH_{50} is directly related to the binding strength of anion to the solid surface. However, in such case one set of results cannot be compared with another set unless pH_{50} values for both are determined under similar conditions. Taking these constraints into consideration, more useful LFER than that as illustrated in Fig. 11.1 can be obtained.

In the case of cation adsorption we have observed an excellent LFER between surface binding constant and the corresponding hydrolysis constant of metal ion. Similarly, it is logical to assume that ligand exchange reaction of anion at the oxide surface should be analogous to the corresponding complexation reaction with the same metal ion in solution phase. This hypothesis can serve as yet another LFER, amenable to graphical analysis and predictive capability. In the early eighties attempts were made by Sigg and Stumm [304] and Kummert and Stumm [620] to demonstrate such a relationship for the adsorption of several anions on goethite and gibbsite surfaces respectively. The essential prerequisite in this regard is that both surface and solution phase complexation reactions must be exactly identical. The literature data which can satisfy this pre-requisite are quite scarce. For example, if a reaction between hydrous iron oxide and the conjugate acid of ligand L is expressed as follows

$$FeOH^o_{surf} + HL \leftrightarrow FeL_{surf} + H_2O \tag{11.3}$$

then the solution complexation reaction must also be expressed in the similar

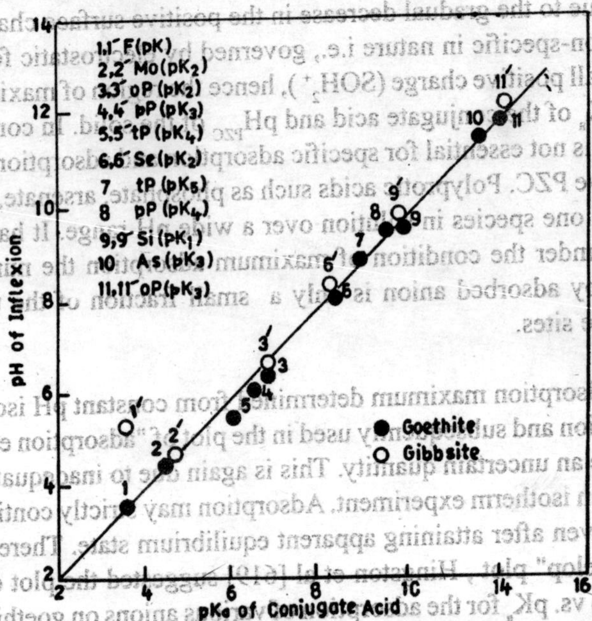

Fig. 11.3 Linear relationship between pK_a of the conjugate acids of anions and their corresponding pH values of the adsorption envelops on goethite and gibbsite. Superscripts without prime denote adsorption on goethite (filled circle) and those with prime denote adsorption on gibbsite (open circles). Anions: fluoride(F), molybdate(Mo), orthophosphate(oP), pyrophosphate(pP), tripoly- phosphate(tP), selenite(Se), silicate(Si), arsenate(As). For significances of the terms pK, pK_2, pK_3, pK_4 and pK_5 see text. Redrawn after ref. 273.

fashion, i.e.,

$$FeOH^{2+} + HL \longleftrightarrow FeL^{2+} + H_2O \qquad (11.1)$$

But, the literature seldom expresses the complexation reaction between Fe^{3+} and L^- in the same form as expressed in (11.1). For example, the reaction between Fe^{3+} and $H_2PO_4^-$ is expressed as follows

$$Fe^{3+} + H_2PO_4^- \longleftrightarrow Fe\,H_2PO_4^{2+} \qquad \log K = 3.65 \qquad (11.12)$$

However, to convert this to the form as expressed in (11.1) the following equilibrium reactions need to be considered

			$\log K_{soln}$
$Fe^{3+} + H_2PO_4^-$	$\leftrightarrow$	$FeH_2PO_4^{2+}$	3.65
$FeOH_2^+$	$\leftrightarrow$	$Fe^{3+} + OH^-$	- 11.81
H_3PO_4	$\leftrightarrow$	$H_2PO_4^- + H^+$	-2.15
$H^+ + OH^-$	$\leftrightarrow$	H_2O	14.0

Adding these four reactions we have

$$FeOH^{2+} + H_3PO_4 \leftrightarrow FeH_2PO_4^{2+} + H_2O \qquad \log K = 3.69 \qquad (11.13)$$

The problem, however, does not end here. The equilibrium constant data of the reaction between H_3PO_4 and hydrous iron oxide as expressed in the form (11.3) is rarely available in the literature. The reaction is normally expressed in the form as given in (11.8) i.e.,

$$FeOH^o_{surf} + PO_4^{3-} + 3H^+ \leftrightarrow FeH_2PO_4{}^o{}_{surf} + H_2O \qquad (11.14)$$

The $\log(K_{1surf})$of the above reaction is 31.3 ± 2.9 [617]. This value does not match with the value calculated for solution phase reaction (11.13). Therefore, we need to have the equilibrium constant value of the solution phase reaction of similar type as given below

$$FeOH^{2+} + PO_4^{3-} + 3H^+ \leftrightarrow FeH_2PO_4^{2+} + H_2O \qquad (11.15)$$

The $\log(K_{soln})$ value of reaction (11.15) is 25.4 [617] which is close to the value obtained for surface reaction (11.14). Similarly, while the complexation constants of the following surface reactions are known [617], the corresponding constants for solution phase complexation reactions are not available.

$$FeOH^o_{surf} + H_3AsO_3 \leftrightarrow FeH_2AsO_{3surf} + H_2O \qquad (11.16)$$

$$\log(K_{1surf}) = 5.41 \pm 0.15$$

$$FeOH^o_{surf} + H_3BO_3 \leftrightarrow FeH_2BO_{3surf} + H_2O \qquad (11.17)$$

$$\log(K_{1surf}) = 0.62 \pm 066$$

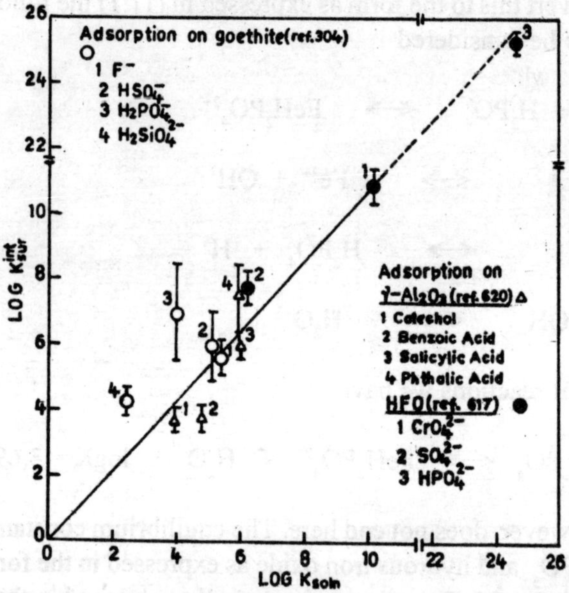

Fig.11.4 Log - log plots of surface binding constants of some anionic ligands with different oxide substrates and their corresponding solution phase complexation constants with similar soluble ionic species. Despite considerable uncertainties in the values of surface binding or complexation constants, good linear relationship between homogeneous and heterogeneous phase reaction constants can be noticed.

Figure 11.4 illustrates the log - log plot of surface complexation or binding constants of some anionic ligands with several oxide substrates and their corresponding solution phase constants, all collected from limited literature sources. It may be noted from the figure that the uncertainty in the values of surface complexation constants is quite high for certain anions. Despite this, the satisfactory linear relationship with slope value around unity suggests similar nature of binding between metal ion and ligand in homogeneous (solution) phase as well as in heterogeneous (solid / solution) phase reactions. According to Dzombak and Morel [617] Fe^{3+} is a hard acid and many of the ligands in Fig. 11.4 are hard bases. The interaction between hard acid and hard base is facilitated on solid surfaces as Fe^{3+} is closer to electron withdrawing oxygen atoms in solid phase than in the solution phase. Using just four sets of surface and solution phase equilibrium constant values for the formation of FeH_2PO_4, $FeHPO_4^-$, $FeCrO_4^-$ and $FeSO_4^-$ species, Dzombak and Morel [617] obtained linear LFER plot from which they derived binding constant values of the formation of $FeHSeO_3^o$,

$FeHSiO_3^o$, $FeS_2O_3^-$, FeF^o and $FeCNS^o$ on HFO surfaces. However, the most important factor on which the accuracy of binding constant values derived from LFER plot is the range of data which should be as wide as possible. It is also desirable that data sets used in the LFER plot should be derived from self consistent models. This means that surface binding constants should not be considered as valid unless they are extracted under conditions which take into account mechanistic variables such as sorbate / sorbent ratio, variation in the strength of binding site etc. In all, this is an area which has still remained largely unexplored.

11.2.2 Modeling of Anion Adsorption

Anion adsorption on oxide surface can be modeled with the help of different mechanistic models in the same manner as followed in the case of cation adsorption. This aspect has been discussed in detail in chapter 8. For satisfactory model fitting, the formation of at least two surface species, namely SA^{2-} and SHA^- have been considered for the adsorptive interaction of a triprotic acid (H_3A) with a hypothetical oxide surface S, as shown in reaction (8.87) and (8.88). Similarly, for a diprotic acid (H_2A) the formation of surface species SA^- and SHA generally need to be considered

$$SOH + A^{2-} + 2H^+ \leftrightarrow SHA + H_2O \qquad K_1 \qquad (11.18)$$

$$SOH + A^{2-} + H^+ \leftrightarrow SA^- + H_2O \qquad K_2 \qquad (11.19)$$

This perhaps, is the main difference of anion adsorption from cation adsorption where the formation of one surface complex species is usually sufficient to fit the data unless adsorption is carried out at pH very close to or higher than the onset of hydrolysis of cation or at high sorbate / sorbent ratio. The mass law equation, the mass balance equations, the coulombic correction factor, the net surface charge equation etc. as required for model fitting, have been presented in section 8.2.2.2. A tableau showing the stoichiometric relationship between components and species for a hypothetical triprotic acid has also been provided in the same section. All these information serve as necessary input for FITEQL or any other suitable computer program which may be used to test the model fitting of anion adsorption data. Tadanier and Eick [312] have also used FITEQL program for fitting the adsorption of phosphate on goethite by CD-MUSIC model. Indeed, this model can describe the adsorption behavior of anions of polyprotic acids on well-crystalline materials like goethite, better than any other models [621,655].

11.3 Adsorption of Some Important Inorganic Anions on Inorganic Materials / Adsorbents

Adsorption of inorganic anions like phosphate, arsenite and arsenate, selenite and selanate, chromate, fluoride etc on common inorganic solids like oxides of iron and aluminum, aluminosilicates has been extensively studied. The volume of literature on this topic does not lag much behind the adsorption of cations, despite the limited variety of anions of inorganic origin. Like cations, most of these studies have either agricultural or environmental or geochemical significances. An attempt has been made in the following sections to review the adsorption behavior of some important and widely investigated inorganic anions.

11.3.1 Phosphate

Phosphorus, in low to appreciable concentration (~ 10^{-3} M) is present in many waste effluents discharged from domestic, industrial and mining sources. It occurs mostly in the form of soluble phosphates, both organic and inorganic (orthophosphate) and also as partially soluble polyphosphates. Phosphorus may also come from agricultural fields as dissolved phosphatic fertilizer. The release of soluble phosphates in natural water systems (lake, river and sea) is of great environmental concern, because phosphorus is an essential nutrient for the growth of micro- and macro organisms in the ecosystem. Its presence beyond certain limit (100mg/l) leads to eutrophication of lakes, streams and even rivers [622] as it stimulates the growth of photosynthetic aquatic macroorganisms like phytoplankton, algae etc. and subsequent destruction of fish and other aquatic animal lives. This is because, the decay of algae and other aquatic plants results in the decrease in dissolved oxygen in water to sustain animal life. Many beautiful lakes and streams have undergone drastic shrinkage and have even faced total death over the years due to eutrophication. As many natural water systems which provide for water human consumption have been shrinking due to rapid urbanization and consequent contamination with various pollutants, phosphorus removal has received much attention as early as 1960s.

Different techniques such as physical, chemical and biological can be employed to remove phosphorus from waste water. Physical techniques like electro-dialysis and reverse osmosis are either too expensive [623] or lack adequate efficiency [624]. Biological treatment, though quite efficient, is very sensitive to operational variables and therefore needs expert handling. In chemical methods, depending upon pH of the solution, precipitation by lime or iron(III) sulfate or aluminum sulfate solution is frequently used when the concentration of phosphorus is high. Thus, lime precipitation is carried out at pH 9.5-10, precipitation by iron and aluminum is carried out at pH 6.5-8.0 [625,626]. Though the phosphates of Ca, Fe(III) and Al are all highly insoluble, certain residual P(1-3 mg/l) tends to remain in solution depending upon the solubility product of the precipitated phosphate as well as pH of the solution. Further, much care is necessary to avoid the addition of excess precipitant.

Adsorption is not only an economical technique, but also simple in operation as the variables are rather flexible and the separation is almost complete within a very short time. Though hydrous oxides of iron(III) and aluminum are the most widely studied adsorbents for phosphate, many other non-conventional solids have also been used as adsorbents. These are: blast furnace slag [627,628], fly ash [629-631], zeolites synthesized from fly ash [632,633], synthetic calcium aluminum silicate [634], calcined or uncalcined dolomite [635,636], red mud [637] etc. It may be noted that all the above adsorbents are rich in either aluminum oxide or iron oxide or calcium oxide or magnesium oxide or combination of more than one oxides. This is because all these oxides individually function as an excellent adsorbent not only for phosphate but also for arsenite and arsenate. A review article by Yeoman et al [638] provides a good collection of literature on the removal of phosphate by various adsorbents and a critical evaluation of their performance.

The presence of positive charge on hydrous oxides of Fe(III) and Al over a wide pH range (from 2 to about 9) in aqueous solution facilitates adsorption of any anions including phosphate. Thus, Georgentus and Grigoropoulou [639] have immobilized relatively high concentration of ortho- and metaphosphates (condensed phosphate) by using alum and $Al(OH)_3$ as function of pH, coagulant dose, temperature and aging of $Al(OH)_3$. Alum is found to be more effective than $Al(OH)_3$, although the latter is more active adsorbent. The difference is attributed to the high adsorption capacity of freshly precipitated $Al(OH)_3$. Adsorption experiment carried out on $Al(OH)_3$ aged for different periods of time. The results confirm that aging decreases adsorption capacity. Orthophosphates are preferably adsorbed than metaphosphates because of favorable orientation of charge per P atom towards surface. Adsorption of both the types of phosphate tend to increase with increase in temperature from 20° to 60°C which is attributed to the break-down of polymerized, in-situ precipitated $Al(OH)_3$, thereby creating new surface for the adsorbent. Besides $Al(OH)_3$, activated Al_2O_3 has also been widely used as adsorbent for phosphate [640-642].

However, hydrous oxides of iron such as goethite, akaganeite and ferrihydrite are the most widely used adsorbents for studying the adsorption behavior of phosphate and arsenate. The initial rate of adsorption on well crystallized goethite and akaganeite is very rapid [643,64644], with about 80% being adsorbed in less than one hour. The time to reach equilibrium adsorption, however, depends upon the degree of crystallinity and texture of the adsorbent. While for highly crystalline solid equilibrium may be attained in 15-20 h, for poorly crystalline or porous solid it may take 3 days to 8 weeks due to slow diffusion of phosphate through meso and micropores [643,644]. While the rate process in the initial stage can be best described by the pseudo second order law, the second stage follows one or two dimensional (cylindrical) intra particle diffusion into porous solid [636,645].

Similarly, the rate of desorption of phosphate from hydrous oxides of iron in alkaline solution (0.1 M NaOH) has also been studied. Desorbability of an anion from porous adsorbent depends upon the nature of adsorbent, the anion, pH, ionic strength of the solution, surface coverage and temperature. For example, desorbability of phosphate increases with pH, concentration of supporting electrolyte and temperature [273 and some earlier references mentioned therein]. As phosphate is strongly adsorbed as inner sphere complex on goethite or on any hydrous oxide of iron, desorption is naturally a very slow process and is partially irreversible [100]. Chitrakar et al [644] carried out desorption of phosphate loaded (23 mg P/g) goethite and akaganeite and observed that about 19 mg P/g (about 83%) could be desorbed from goethite, but subsequent adsorption under similar conditions showed only 7.5 mg P/g i.e., a decrease of about 67% from its original loading capacity, which did not improve in the next two adsorption - desorption cycles. This is attributed by the authors to the increase in crystallinity of the goethite sample upon alkali treatment and partial dissolution of fine particles (about 1%). In contrast, the authors observed that even after ten adsorption - desorption cycles the loading capacity of akaganeite did not show any decrease from the initial value. There was no improvement in the crystallinity of the sample as revealed by XRD and the sample dissolution was practically negligible (less than 0.01%).

Desorption of adsorbed anions has many practical implications. For example, if the phosphate added as fertilizer in the soil gets easily desorbed due to the increase in soil alkalinity, then the anion is likely to be washed out by rain or flood water into river and finally into the sea. This results in the fertility of the soil. Similarly, if arsenite or arsenate and fluoride instead of being immobilized by the underground soil minerals are released in ground water, then they pose serious health hazards to thousands of people. The same problem may also be applicable to the loss of adsorbability or scavenging capability of the sediments of natural water system for toxic ions due to drastic variation in the environment.

11.3.1.1 Mode of coordination of adsorbed phosphate

Ligands are adsorbed on oxide surface by forming a variety of complex species, depending upon the structure of ligand, ionic strength and pH of the solution, temperature etc. In micro-level, the mode of coordination of a ligand with the surface can be determined directly by using modern spectroscopic methods like ATR-FTIR and Extended X-ray Absorption Fine Structure(EXAFS) [646,647]. Persson et al [648] proposed the formation of three types of complexes between oxyanions (PO_4^{3-}, AsO_4^{3-}, SO_4^{2-} etc) and goethite surface with different degrees of protonation depending upon pH of the solution. These are: (1) Non-protonated monodentate, (2) protonated monodentate, (3) non-protonated bidentate and (4) protonated bidentate complexes.

These are schematically illustrated in Fig. 8.6 using, for example, phosphate as ligand.

As far as phosphate adsorption is concerned, the formation of both monodentate and bidentate inner sphere complexes on goethite is proposed, though most of the investigators are in favour of bidentate complex formation. In 1970s and 1980s several authors [304, 649-651] suggested that coordination of phosphate on iron oxides changes from bidentate to monodentate at high surface coverage with decrease in pH. This view was later on supported by other workers [312,652,653]. A binuclear bidentate complex that bridges two OH groups at the corners of two adjacent $[Fe(OH)_6]$ octahedra is proposed and is supported by IR spectroscopy of the adsorbed phosphate complex [100,654]. Such a binding site is termed as double corner linkage (2C) and occurs frequently on the (101) plane of crystalline goethite surface [81]. Mononuclear bidentate complex, on the other hand, is formed through the linkage of two OH groups along the edges of some $[Fe(OH)_6]$ octahedra and is termed double edge linkage (2E). This type of sites occur more frequently on the exposed surfaces of hydrous oxides of iron other than goethite (e.g., ferrihydrite, lepidocrocite and akaganeite) and are considered to be high energy sites which are preferentially occupied at low surface coverage. This possibly explains the differences in the adsorption of oxyanions on goethite from that on ferrihydrite and lepidocrocite [655].

Other group of researchers has approached the problem indirectly at macro level by using best fitting of the experimental adsorption data with those calculated from model equations by assuming the formation of certain most probable surface complex species having different modes of coordination. Hiemstra and Riemsdijk [109] made an attempt to fit data of phosphate adsorption on goethite by their own experiment as well as from literature sources, to CD-MUSIC model. They observed that best fit of the data over a wide range of pH could be obtained from bidentate complex species, monodentate species could possibly exist only at very low surface coverage. Subsequently, more investigators employed CD-MUSIC and other models to demonstrate that adsorption of phosphate on goethite takes place mostly through the formation of binuclear bidentate complex at low to intermediate pH values where the adsorption capacity is highest [621,656,657]. These findings clearly contradict the views expressed by earlier workers [304,649-651] that monodentate complex is formed at high surface coverage.

That two neighboring singly coordinated OH groups combine with a phosphate ion to form binuclear bidentate complex on goethite finds additional support from two important observations. First, many researchers have found that the maximum adsorption of phosphate is about 2.5 mmol P/m^2 [100,621]. This value corresponds to an adsorption density of one phosphate per nm^2 which is very close to the area ($0.68 nm^2$) occupied by two singly coordinated OH groups on the (101) plane and thus supports bidentate complex formation at saturation coverage. Secondly, the sorption capacity of phosphate per unit area of synthetic and natural goethite is essentially similar, irrespective of their morphologies. This is attributed to the fact that predominant crystal face on all goethite samples is (101). For 31 synthetic goethite samples with widely varying surface area, Torrent et al [100] have observed the following correlation between adsorption capacity and surface area ($r^2 = 0.939$)

$$P_{ads}\ (\mu mol/m^2) = 2.50 \times A\ (\text{area in } m^2/g) - 0.37$$

In contrast, no such direct relationship has been found for hematite, because some crystal planes which are commonly exposed in aqueous suspension of hematite are inactive i.e., they do not have pairs of singly coordinated OH groups. The adsorption capacity of hematite, therefore, depends upon crystal morphology. The average Langmuir capacity for 43 hematite samples of different morphologies is found to be $0.98 \pm 0.7\ \mu mol / m^2$ [659]. Another set of 30 hematite samples with varying Al for Fe substitution showed an average capacity of $1.03 \pm 0.53\ \mu mol / m^2$ [660]. Minor increase in adsorption capacity is attributed to increase in the surface area of hematite due to Al for Fe substitution. However, the important point to note here is the large standard deviation of adsorption capacity compared to goethite, suggesting that adsorption of phosphate on hematite is dependent on the morphology of the particles, whereon the favorable adsorption planes like (110), (100), (012) and (113) are exposed.

11.3.2 Arsenite and Arsenate

To provide safe drinking water to millions of people and to meet the ever growing demand of water for irrigation of agricultural land, exploitation of ground water which began in a modest scale in 1960s in the subcontinent, assumed a very large scale in 1980s. As a result of these efforts, thousands of hand-pumped tube wells at a depth of 20-70 meters and deep tube wells at a depth of 110-120 meters were drilled to extract water from underground. Though these efforts were presented as great achievements to ushering agricultural revolution and to provide so-called 'safe' drinking water to the millions of population in rural Bangladesh and its adjoining state of West Bengal in India, little was realized at that time that it also ushered greatest mass poisoning of human history by the way of arsenic contamination. The wide spread occurrence of arsenic at the concentration above 0.05 mg/l in ground water of the southern districts of West Bengal was confirmed by the scientists of Jadavpur University, Kolkata in the early 1990 and the results appeared as a.consolidated report in 1996 [661]. Thus, more than 50 million people of Bangladesh and at least 10-15 million people of West Bengal are already in the great danger of arsenic poisoning. As the geographical area of the survey is expanding, more and more regions are reported to be in the danger

Table 11.1 Extent of Arsenic Poisoning due to Contamination of Ground Water

Country	Population potentially exposed (million)	Concentration of As in ground water in mg/l	Geological conditions favoring the occurrence of As
Bangladesh	40-50	10 to 2500	Natural; alluvial/deltaic sediments with high phosphate, silicate and organics
India (West Bengal)	10-15	10 to 3000	Similar to above
Vietnam	> 1	10 to3050	Natural; alluvial/deltaic sediments.
Taiwan	0.1 to 0.2	10 to1820	Natural: coastal zones, black shales.
Mongolia	0.1 to 0.6	10 to 2400	Natural: alluvial and lake sediments, high alkalinity.
Chile	> 0.4	100 to 1000	Natural and anthropogenic; volcanogenic sediments, closed basins, lakes, thermal springs, mining activities etc.
Mexico	> 0.4	8 to 620	Natural and anthropogenic; volcanogenic sediments, mining
Hungary and Romania	> 0.4	2 to 176	Natural; alluvial sediments and organics

Partially adapted from D. K. Nordstroem, *Science*, **296**(2002), 2143

zones. These include large tracts of Jharkhand and Bihar, districts in the Brahmaputra basin of lower Assam etc. Besides India and Bangladesh, arsenic content in the ground waters of many other countries is reported to be above 0.05 mg/l. These countries are: Vietnam [662], Taiwan [663], Mongolia [664], Argentina [665], Chile [666], Mexico [667], Hungary, Romania and even some parts of the USA [668]. Table 11.1 illustrates how severe is the problem around the world.

In nature, As is found everywhere, in ocean, in soil, in air, in sediments and even in plants and animals. On an average, its concentration in the earth's crust is about 1.5 to 5 mg/kg. The common arsenic containing ores are arsenopyrite (FeAsS) and arsenic rich pyrite ($Fe(S,As)_2$), though there are several less common and even some mineral of rare occurrences. Geological studies [669,670] reveal that in the Gangetic basin of Bengal (including Bangladesh) both arseno-pyrite and arsenic bearing pyrite are the major sources of As in the alluvial sediments at about 50-70 meters below the surface. However, its concentration range of 2-10 mg/kg is not significantly higher than the average value of 2-6 mg/kg found in other parts of the country. Therefore, the question that has not yet been clearly answered is how has the As leached out from its minerals and accumulated to such an alarming level in the ground water.

11.3.2.1 Mobilization of Arsenic in the Underground Sediment

Before dealing with the mobilization of As and its subsequent accumulation in ground water, we may examine the chemical speciation of this element in aqueous solution. Though As occurs in several oxidation states (-3, 0, +3, +5), but in natural waters it occurs mostly in the form of oxyanions of either trivalent As (arsenite, AsO_3^{3-}) or pentavalent As (arsenate, AsO_4^{3-}). A number of factors such as pH, redox potential(Eh), concentrations of iron (both Fe(II) and Fe(III)), sulfide and sulfate, salinity, temperature etc. determine the form of As species in natural waters. However, Eh and pH are the two main factors which determine the stability and dominance of the anion species. Arsenous acid (H_3AsO_3) is stable at pH 0-9, while $H_2AsO_3^-$, $HAsO_3^{2-}$ and AsO_3^{3-} exist as stable species in the pH intervals 10-12, 13 and 14 respectively. For arsenic acid (H_3AsO_4) the corresponding pH values of the stability of the species are: H_3AsO_4 (pH 0-3), $H_2AsO_4^-$ (pH 3-7), $HAsO_4^{2-}$ (pH 7-11) and AsO_4^{3-} (pH >11) [671].

Mobilization of arsenic is a low temperature geochemical process that involves leaching of naturally occurring arsenic bearing minerals under both oxidizing and reducing conditions at relatively high pH. While oxidizing conditions are required for arsenic bearing pyrite, reducing conditions helps where arsenic occurs in association with iron(ferric) or manganese oxides entrapped within the sediment layers. While reducing conditions can be attained through anaerobic decay of organic matters or by bacterial action, oxidizing conditions are possibly achieved either by anthropogenic actions like indiscriminate discharge of chemicals on earth surface or use of fertilizers (e.g., nitrates) which may ultimately seep into the arsenic bearing mineral zones.

Some scientists believe that excessive withdrawal of ground water create relatively low pressure zone in the underground i.e., above water table. The resulting pressure difference between the surface (atmospheric pressure) and the underground causes rapid diffusion of air deep into the porous ground where oxygen is dissolved in the pore water. The aerated water facilitates the oxidative dissolution of pyritic minerals with concomitant release of arsenic acid and sulfate in water through a series of possible reactions as discussed below.

Since the oxidation reaction takes place at neutral or slightly basic pH value, only Fe(II) and sulfuric acid are the main initial products

$$2FeS_2 + 7O_2\,(aq) + 2H_2O \leftrightarrow 2FeSO_4 + 2H_2SO_4 \qquad (11.20)$$

At low pH, dissolved oxygen reacts with pyrite producing $FeSO_4$ and elemental sulfur (S^o) which however oxidizes to sulfuric acid giving the overall reaction

$$2FeS_2 + 7O_2\,(aq) + 8H_2O \leftrightarrow 2FeSO_4 + 2H_2SO_4 + 6H_2O \qquad (11.21)$$

Ferrous sulfate is oxidized to ferric sulfate which, in turn, also oxidize pyrite

$$4Fe^{2+} + O_2 + 4H^+ \leftrightarrow 4Fe^{3+} + 2H_2O \qquad (11.22)$$

$$FeS_2 + 8H_2O + 14Fe^{3+} \leftrightarrow 15Fe^{2+} + 2SO_4^{2-} + 16H^+ \qquad (11.23)$$

The last reaction requires considerable accumulation of ferric ion which is difficult to achieve unless a highly oxidizing condition prevails to oxidize all the Fe^{2+} ions to Fe^{3+} ions. During the process zero valent arsenic is oxidized to either arsenous oxide (As_2O_3) or arsenic oxide (As_2O_5) which combines with water to give the corresponding arsenous (H_3AsO_3) or arsenic (H_3AsO_4) acids. Further, phosphate ions (from phosphatic fertilizer) that seep into the underground displace AsO_4^{3-} ions from the surfaces of hydrous iron oxides. The displaced arsenate ions enter the ground water. However, for this process to take place the pH value of the pore water has to be above 8.0 which is maintained mainly through the introduction of bicarbonate ions, generated by the weathering and dissolution carbonate bearing sedimentary rocks.

It must be mentioned here that the above pathways proposed for the mobilization of As and its accumulation in ground water are confronted with various counter arguments. Therefore, scientists have not yet been able to reach any consensus process(es) in this regard. It is possible that a combination of several processes might have triggered the release of As and its accumulation in ground water.

11.3.2.2 *Extent of Toxicity*

The toxicity of As depends on its oxidation state. Trivalent As is at least hundred times more toxic than its pentavalent state. The effects can be both chronic and acute. There is no universality on the nature of its effects on human body. These appear to be different for different individuals, population groups and geographic regions. There is a considerable debate as to whether there is linear relationship between the toxicity of arsenic and its concentration. The World Health Organization (WHO) has stipulated maximum concentration limit (MCL) of As in drinking water to be 10 μg/l. But, most of the countries whose ground water is highly contaminated with As are still following the MCL of 50 μg/l, because of various social, economic and technological constraints with which these countries are confronted to provide the affected people with arsenic free drinking water.

The poisoning effects of As manifest slowly in several stages. In the first stage, one observes black spots and hyper pigmentation (hyperkeratosis), inflammations of the respiratory tract, eyes, nausea and vomiting. In the second stage, small palpable nodules appear on the palms and soles (skin lesions), the feet and legs swell and the disorder of kidney and lever begins. In the final stage, cancers of liver, kidney and other parts of the body set in [673-675].

11.3.2.3 *Removal of Arsenic*

The technologies developed so far to remove arsenic from ground water may be broadly divided into two categories depending upon whether treatment is done before lifting water from underground aquifer (in-situ treatment) and those after having it extracted from underground. While the methods under the second category have been thoroughly tested and their efficacies established, the methods under the first category are relatively recent and are still under the state of development. The techniques under the second category include ion exchange, reverse osmosis, electrodialysis [676], precipitation-coagulation followed by filtration [18,677,678]. The last technique is still the most popular and widely used where As(III) is the predominant arsenic species and Fe(II) content in the ground water is too low to remove most (>95%) of the dissolved arsenic. The technique involves oxidation of As(III) to As(V) by adding $Fe_2(SO_4)_3$ or $FeCl_3$ or alum along with an oxidizing agent such as $KMnO_4$ or NaOCl at pH value above 7.2. Under this condition arsenite (AsO_3^{3-}) is oxidized to arsenate (AsO_4^{3-}) oxyanion and simultaneous precipitation of $Fe(OH)_3$ or $Al(OH)_3$ takes place. The nascent hydroxides show strong affinity for AsO_4^{3-} ion. Indeed, this is the most critical in this technique as the efficiency of th entire process depends on this step. The precipitates with adsorbed As(V) oxyanion is allowed to coagulate in the presence of Fe(III) or Al(III) ion and finally filtered [679-688]. However, the use of a strong oxidizing agent like H_2O_2, $KMnO_4$, NaOCl etc is not always advisable, since this may lead to the deterioration of water quality through the production of by-products like colloidal MnO_2 or the problem of coloring of water with $KMnO_4$ addition and therefore needs further treatment for its removal. Lime softening by the addition of limestone or slaked lime raises the pH of water to 10.5-11.0 and therefore additional step is required to reduce pH. However, the main disadvantage of precipitation and coagulation technique is the generation of substantial amount of sludge and the problem of its subsequent disposal without contaminating surface and ground water.

In view of the above constraints associated with precipitation-coagulation technique adsorption method is receiving more and more attention for As removal, especially in small scale systems, due to its simple, sludge free operation, easy availability of low cost adsorbents and regeneration capability. Considering the great affinity of arsenate ion towards iron oxides, various type of naturally occurring iron ores [689-692] and materials coated with iron oxides [25,693-696] have been studied as adsorbents for As removal. Others include amorphous hydrous ferric oxide (HFO) or very poorly crystalline (two weak lines) ferrihydrite [18,19,21,697], goethite [621,697-700], metal ion doped goethite [701,702], aluminum oxides/hydroxides [703-707], mixed oxides and hydroxides of Fe(III), Al and /or Mn [708-710], TiO_2 [23], MnO_2 [711-713], SnO_2 [714], zero valent iron [715-717], zeolites [671], clay minerals [718-720], soils [721], bauxite and red mud [22,722-724], fly ash[725] etc. However, the danger of using an industrial waste material is that while trying to remove one toxic element, we may inadvertently introduce more than one no-less toxic elements in the drinking water. We shall now discuss below a few basic aspects of the sorption of As species on oxides of iron and/or aluminum.

11.3.2.4 Kinetics of the Sorption of Arsenic Species

The kinetics of the sorption of phosphate and arsenate have been studied by batch process as a function of concentration, pH, temperature and stirring rate. The rate of adsorption generally follows two step process: Initial fast reaction (5-10 min) is followed by slow process which may take more than 2-3 hr to reach equilibrium [726]. As(III) adsorption on hematite was found to be fast and equilibrium reached within an hour[20]. The rate process followed pseudo first order kinetics. Gupta et al [25] also observed that the initial rate of adsorption of As(III) on iron oxide coated sand was very fast reaching equilibrium in 2h. The adsorption data were found to follow linear relationship between log(% adsorption) and log(t). The slope value of the plot is 0.42. As this value deviates from 0.5, the authors argued that intra-particle pore diffusion was not rate controlling and suggested that it might be controlled by diffusion through liquid film around the solid particles. However, the authors neither studied the influence of the size of iron oxide coated sand particles, nor made any attempt to calculate diffusion coefficient. In the absence of these information, to reach such a conclusion is rather doubtful. Lakshmipathiraj [699,727] also observed very rapid adsorption of both As(III) and As(V) on goethite and manganese substituted iron hydroxides. However, adsorption increases more sharply for As(III) than for As(V) with increase in temperature from 30° to 60°C. By plotting -ln(t) against reciprocal temperature (1/T) the authors determined apparent activation energy for different fractional conversions. It is also doubtful whether such plots are possible in the case of surface reaction like adsorption, where, unlike solid state reaction, temperature dependency of the rate process is much irregular even within10-15°C. Adsorption may change from initial sharp rise to sudden slowing down to attain saturation within a short span of time and temperature as is evident from the plots presented by the authors. Some adsorption reactions are either insensitive to temperature variation or even show decrease in rate with increase in temperature. It is difficult to establish kinetic mechanism for fast reactions unless sophisticated experimental procedure like pressure jump technique [40,259] is used.

Many investigators have shown that adsorption isotherms of As(III) and As(V) follow Langmuir equation over a wide range of equilibrium concentration. But, the experimental conditions such as pH values and sorbate / sorbent ratios employed are so different among different authors that it is difficult to compare the values of Langmuir capacity. For example, at pH 4.6, adsorption capacities of As(III) and As(V) on ferrihydrite, as reported by Raven et al [728], are 0.6 and 0.25 mole per mole of Fe respectively, and at pH 9.2, the corresponding values are 0.58 and 0.16 mole per mole of Fe. Dixit and Hering [697] have reported 0.31, 0.016 and 0.025 mole of As(III) / mole of Fe at pH 8.0 on HFO, goethite and magnetite respectively. Similarly, the authors have also found adsorption capacity values of As(V) on HFO and goethite at pH 4.0 to be 0.24 and 0.016 mole /mole of Fe respectively. In these experiments, the As/Fe molar ratio ($\times 10^{-3}$) for HFO, goethite and magnetite were 30-300, 2-18 and 8-23 respectively. It is interesting to note that despite difference in the pH values, adsorption capacity of the two arsenic species on goethite remains same. The effect of pH on the sorption of arsenic on iron oxides is discussed in more details in the following section.

11.3.2.5 Effect of pH on the Equilibrium Adsorption of Arsenic Species

Like any other inorganic anions, adsorption of both arsenite an arsenate on hydrous iron oxides (and on other oxides also) is highly dependent on pH. However, there are some distinct differences in the influence of pH on the adsorption behavior of the two oxyanions of arsenic. This is illustrated in Fig. 11.5 which shows adsorption of the

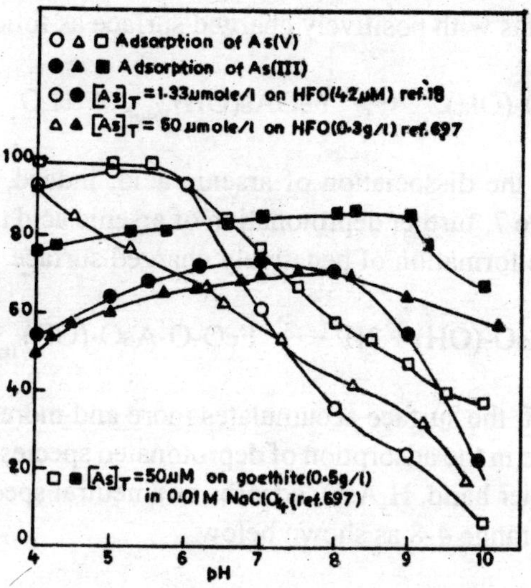

Fig.11.5 Comparison of the adsorption capacities of As(V) and As(III) on HFO and goethite as function of pH, total concentration of As and concentration of solid adsorbents. Redrawn after data from refs.18 and 697.

two species on HFO and goethite as function of pH. For the sake of comparison adsorption data from two different sources are included in the figure which shows that at lower pH values adsorption of As(V) is higher than that of As(III) on both HFO and goethite, but falls rapidly with increase in pH above 5.5. The data obtained from Dixit and Hering [697] show that the fall begins from pH 4 in the case of HFO, possibly due to higher initial concentration of As(V). In contrast, adsorption of As(III) gradually increases from pH 4 to 8.5 and even up to 9 in the case of goethite before showing decreasing trend. This is perhaps the most notable difference between the two oxyanions of As. The reason lies in the acid dissociation of the two oxyacids of As which occurs at widely different pH values (see section 11.3.2.1). While the undissociated H_3AsO_3 is stable in the pH range 0-9, H_3AsO_4 forms at least two charged species ($H_2AsO_4^-$ and $HAsO_4^{-2}$) in the same pH range. The two species occur in almost equimolar proportion at pH 9.2.

While arsenate adsorbs as inner sphere complex on iron oxides, arsenite adsorbs as both inner and outer sphere complexes by ligand exchange with OH_2 and OH^- in the coordination sphere of surface Fe atom (more discussion will follow in the next section). While arsenate adsorption results in the net release of OH^- ion in the entire pH range 4-9, arsenite adsorption results in the net release of H^+ ion at lower pH (~ 4) and OH^- ion at pH around 4-9. At pH around 4 arsenate occurs predominantly as $H_2AsO_4^-$ which reacts with positively charged surface as follows

$$FeOH_{2\ surf}^{+} + O^{-}\text{-}AsO\text{-}(OH)_2 \leftrightarrow FeOAs(OH)_{2(surf)} + H_2O \quad (11.24)$$

Proton released by the dissociation of arsenic acid, indeed, combines with FeOH surface. At pH above 7, further deprotonation of arsenic acid takes place and adsorption occurs with the formation of negatively charged surface

$$FeO^-H^+_{surf} + (O^-)_2\text{-}AsO\text{-}(OH) + 2H^+ \leftrightarrow FeO\text{-}O\text{-}AsO\text{-}(OH)^-_{(surf)} + H_2O + H^+ \quad (11.25)$$

With increase in pH, the surface accumulates more and more negative charges with consequent decrease in the adsorption of deprotonated species ($HAsO_4^{2-}$) as shown in Fig.11.5. On the other hand, H_3AsO_3 adsorbs as a neutral species with the release of H^+ in the entire pH range 4-8 as shown below

$$FeOH_{2\ surf}^{+} + HO\text{-}As\text{-}(OH)_2 \leftrightarrow FeO\text{-}As\text{-}(OH)_{2(surf)} + H_2O + H^+ \quad (11.26)$$

At pH around 9, in addition to the adsorption of neutral species, the following surface reactions also occur through ligand exchange

$$FeO^-H^+_{surf} + O^- As(OH)_2 \leftrightarrow FeO\text{-}As(OH)_{2(surf)} + OH^- \quad (11.27)$$

$$2(FeO^-H^+) + O^-\text{-}As(OH)_2 \leftrightarrow Fe_2O_2As(OH)_{2(surf)} + H_2O \quad (11.28)$$

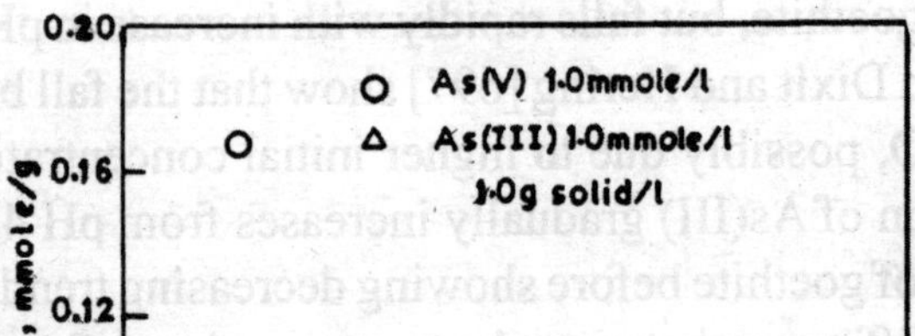

Fig. 11.6 Variation of the concentration of charged surface sites during the adsorption of As(V) and As(III) on $Fe(OH)_3$ coated alumina. Redrawn after data from ref. 696.

H^+ ion released by the deprotonation of arsenous acid is taken up by the negatively charged surface at pH>PZC so as to facilitate the above reaction and thereby increasing the adsorption of As(III) even at pH little above PZC as shown in Fig. 11.5. Spectroscopic studies provide strong evidence for the formation of binuclear bidentate complex during the adsorption of As(III) on iron oxides as discussed in the following section.

The above surface reactions suggest that in contrast to the adsorption of As(V), there is no change in the charge character of iron oxide surface adsorbing As(III), a phenomenon which may explain the increase in adsorption at higher pH. Direct quantitative measurement of surface charge during the adsorption of As(V) and as(III) on iron hydroxide coated aluminum oxide carried out by Hlavay and Polyǎk [696] is shown in Fig. 11.6. The figure clearly demonstrates that the charged surface created by the adsorption of As(III) is negligibly small compared to that created by the adsorption of As(V), especially at pH > 8.

11.3.2.6 Mode of Adsorption of Arsenic Species on Oxide Surface

To understand more precisely the mechanism of adsorption of various arsenic species on oxides of iron and aluminum or any adsorbent containing either of these oxides, it is necessary to determine the mode of binding of the species with the adsorbent surface. Both macroscopic (iep shift, ionic strength effect, model fitting etc) and microscopic (FTIR, Raman, x-ray absorption spectroscopy etc.) methods have been used in this regard. Electrophoretic mobility (EPM) measures movement of the charged particles in response to an applied electric field as function of pH of the aqueous medium. The pH value at which no or zero movement indicates isoelectric point (iep) of the oxide particle, as discussed in more detail in section 6.6.2. Shifts in iep and reversal of the direction of movement of the particles observed with increasing adsorption of sorbate ion gives evidence of the specific adsorption of anion on the surface plane or very close to the surface plane (inner sphere complex). Such shifts have been found for the adsorption of arsenate [621,729-731] and arsenite [18,731] on HFO and arsenate on Al_2O_3 or amorphous Al_2O_3 [731,732]. However, arsenite adsorption on Al_2O_3 does not exhibit any shift in iep indicating the formation of either outer sphere complex or inner sphere complex without any change in the surface charge [731].

As far as the effect of ionic strength is concerned, As(V) adsorption on HFO [729] and on Al_2O_3 [731] and As(III) adsorption on amorphous alumina [733], all as function of pH (adsorption edge), showed little dependence on ionic strength of the medium, suggesting inner sphere complex formation with the adsorbent surfaces. On the other hand, adsorption of As(III) on both HFO and amorphous alumina showed decrease in adsorption with increase in ionic strength, i.e., adsorption edge shifts to acid side indicating the formation of outer sphere complex [731]. This conflicting observation has been attributed to the difference in the range of ionic strength of the medium used by different investigators. While Manning and Goldberg [733] used 0.02-0.1M $NaNO_3$, Goldberg and Johnston[731] used the range 0.1-1.0M $NaNO_3$ which is much higher than the former authors. In view of the apparent ionic strength effect on the adsorption of As(III), it is believed that arsenite ion is more weakly adsorbed than arsenate ion.

Successful model fitting is an important macroscopic technique by which the mode of adsorption may be ascertained to a considerable extent, despite differences in opinion. The main objection arises from the use of certain adjustable parameters, which determine the nature of adsorbed species, are often based on assumed surface speciations. In other words, the species chosen depends upon the model being used and therefore cannot provide unequivocal indication of surface speciation. Therefore, it is widely agreed that an independent method like spectroscopy is necessary for confirmation of the mode of adsorption. However, if assumption on predominant species is made on the basis of precise understanding of the process chemistry, the modeling can also serve as a useful tool in establishing the mechanism of adsorption.

Major contribution in this regard on the adsorption of phosphate, arsenate and arsenite has been made by Goldberg and her co-workers[698,718,731,733]. Using the simple constant capacitance model (CCM) with minimum number of adjustable parameters, the authors successfully predicted adsorption from single anion as well as from binary (competitive) systems comprising phosphate and/or molybdate [698], arsenate and/or arsenite [733] on oxides and clay minerals. The following surface complexation reactions have been proposed by the authors for model fitting.

(A) Adsorption of arsenate, arsenite and other oxyanions on oxides and clay minerals as inner sphere monodentate complex:

$$SOH + H_3AsO_4 \longleftrightarrow SH_2AsO_4 + H_2O \qquad {}^*K^1_{As(V)}(int) \tag{11.29}$$

$$SOH + H_3AsO_4 \longleftrightarrow SHAsO_4^- + H_2O + H^+ \qquad {}^*K^1_{2As(V)}(int) \tag{11.30}$$

$$SOH + H_3AsO_4 \longleftrightarrow SAsO_4^{2-} + H_2O + 2H^+ \qquad {}^*K^1_{3As(V)}(int) \tag{11.31}$$

Table 11.2 Inner Sphere (Monodentate) surface Complexation Constants for the Adsorption of Arsenate and Phosphate on Hydrous Oxides of Iron and Aluminum and Some Clay Minerals.

Log(constant)	Goethite[a)]	Gibbsite[a)]	Amor.Al_2O_3[b)]	HFO[b)]	Kaolinite[b)]	Illite[b)]	Montmorillonite[b)]
$\log {}^*K^1_{As(V)}(int)$	8.40	8.60					
$\log {}^*K^1_{2As(V)}(int)$	3.40	2.20	2.51				-4.52
$\log {}^*K^1_{3As(V)}(int)$	-2.30	-5.70	-2.45	-2.35	-4.69	-5.21	
$\log {}^*K^1_{As(III)}(int)$			2.20	4.68	2.12	1.19	
$\log {}^*K^1_{2As(III)}(int)$			-5.17	-2.27	-5.43	-5.66	3.92
$\log {}^*K^1_{P}(int)$	8.05	9.90					
$\log {}^*K^1_{2P}(int)$	3.40	3.20					
$\log {}^*K^1_{3P}(int)$	-2.20	-4.70					

a) Ref. 698 , b) Ref. 718

Symbol $^*K$ is used to differentiate it from K where polyprotic acid is expressed in its dissociated form (see chapter 8 and elsewhere in this chapter). We may write similar equations for H_3PO_4 giving rise to the corresponding complexation constants viz., ${}^*K^1_P(int)$, ${}^*K^1_{2P}(int)$ and ${}^*K^1_{3P}(int)$. Reactions with arsenous acid (H_3AsO_3) was assumed to form mainly two surface species, namely SH_2AsO_3 and $SHAsO_3^-$ with corresponding intrinsic constants of ${}^*K^1_{As(III)}(int)$ and ${}^*K1_{2As(III)}(int)$ respectively. The formation of $SAsO_3^{2-}$ was not considered, as such complexation reaction takes place only at very high pH. The values of the intrinsic constants determined by the authors [698,718] are shown in Table 11.2.

(B) Adsorption of arsenate and arsenite as inner sphere bidentate surface complex. Along with the formation of above monodentate surface complexes, the authors [698,733] also envisaged the formation of bidentate complexes (two sites model). Such surface sites are designated as $X(OH)_2$, where X represents either Fe or Al. Anionic ligand like arsenate is proposed to be coordinated to two surface OH groups to form mononuclear bidentate complexes as shown below.

$$X(OH)_2 + H_3AsO_4 \longleftrightarrow XO_2AsO(OH) + 2H_2O \qquad {}^*K^2_{1As(V)}(int) \qquad (11.32)$$

$$X(OH)_2 + H_3AsO_4 \leftrightarrow XO_2AsO(O)^- + 2H_2O + H^+ \quad {}^*K^2_{2As(V)}(int) \qquad (11.33)$$

The two surface oxygen atoms form the bridge between anion and solid surface. The adsorption of phosphate has also been represented by similar two sites model. The values of the intrinsic constants are given in Table 11.3.

Table 11.3 Inner Sphere Bidentate (Mononuclear) Surface Complex Constants for the Adsorption of Arsenate and Phosphate on Goethite and Gibbsite

log (constant)	Goethite[a)]	Gibbsite[a)]
$\log {}^*K^2_{1As(V)}(int)$	8.90	7.75
$\log {}^*K^2_{2As(V)}(int)$	2.90	2.90
$\log {}^*K^2_{1P}(int)$	8.90	9.90
$\log {}^*K^2_{2P}(int)$	3.10	3.20

a) Ref. 698

Dixit and Hering [697] have studied the model fitting of the adsorption edges of As(III) and As(V) on goethite, HFO and magnetite at varying mole ratio of total As/Fe, depending upon the adsorption capacity of the adsorbent solid. The authors considered the formation of the following monodentate species in the pH range 4-10.

Arsenate adsorption

$$SOH + AsO_4^{3-} + 3H^+ \leftrightarrow SH_2AsO_4 + H_2O \quad K^1_{As(V)}(int) \qquad (11.34)$$

$$SOH + AsO_4^{3-} + 2H^+ \leftrightarrow SHAsO_4^- + H_2O \quad K^1_{2As(V)}(int) \qquad (11.35)$$

$$SOH + AsO_4^{3-} + H^+ \leftrightarrow SAsO_4^{2-} + H_2O \quad K^1_{3As(V)}(int) \qquad (11.36)$$

Arsenite Adsorption

$$SOH + AsO_3^{3-} + 3H^+ \leftrightarrow SH_2AsO_3 + H_2O \quad K^1_{As(III)}(int) \qquad (11.37)$$

$$SOH + AsO_3^{3-} + 2H^+ \leftrightarrow SHAsO_4 + H_2O \quad K^1_{2As(III)}(int) \qquad (11.38)$$

The above reactions differ from those of (11.29) - (11.33) in the sense that both the acids (H_3AsO_4 and H_3AsO_3) have been expressed here in their dissociated forms to illustrate the adsorption of protons during anion adsorption. Consequently, the values of intrinsic complexation constants as presented in Table 11.4 differ from those given in Table 11.2 by the corresponding value of protonation constants of AsO_4^{3-} and AsO_3^{3-} ions as presented in Table 11.5. The protonation constants of PO_4^{3-} are also included in the table for the sake of comparison of the data among three oxyanions.

Table 11.4 Inner Sphere Mono- and Bidentate (Binuclear) Surface Complex Constants for the Adsorption of Arsenate, Arsenite and Phosphate on Goethite and HFO.

log (Constants)	HFO	Goethite
$\log K^1_{As(V)}$(int)	29.88[a)]	31.00[a)]
$\log K^1_{2As(V)}$(int)	24.43[a)]	26.80[a)]
	-	25.00[b)]
$\log K^1_{3As(V)}$(int)	18.10[a)]	20.22[a)]
$\log K^1_{As(III)}$(int)	38.76[a)]	39.93[a)]
$\log K^1_{2As(III)}$(int)	31.87[a)]	32.40[a)]
$\log K^1_{P}$(int)	31.3 ± 2.9[c)]	-
$\log K^1_{2P}$(int)	25.4 ± 0.2[c)]	-
$\log K^1_{3P}$(int)	17.7 ± 0.5[c)]	20.80[b)]
$\log K^2_{As(V)}$(int)	-	36.30[b)]
$\log K^2_{2As(V)}$(int)	-	28.80[b)]
$\log K^2_{P}$(int)	-	36.40[b)]
$\log K^2_{2P}$(int)	-	30.20[b)]

a) Ref.697 b) Ref. 621 c) Ref. 617

Table 11.5 Protonation Constants of H_3AsO_3, H_3AsO_4 and H_3PO_4

AsO_3^{3-} Protonation Reactions	log K_a
$AsO_3^{3-} + H^+ \leftrightarrow HAsO_3^{2-}$	13.40[a)]
$AsO_3^{3-} + 2H^+ \leftrightarrow H_2AsO_3^-$	25.52[a)]
$AsO_3^{3-} + 3H^+ \leftrightarrow H_3AsO_3$	39.74[a)]
AsO_4^{3-} Protonation Reactions	
$AsO_4^{3-} + H^+ \leftrightarrow HAsO_4^{2-}$	11.60[a)]
$AsO_4^{3-} + 2H^+ \leftrightarrow H_2AsO_4^{2-}$	18.35[a)]
$AsO_4^{3-} + 3H^+ \leftrightarrow H_3AsO_4$	20.60[a)]
PO_4^{3-} Protonation Reactions	
$PO_4^{3-} + H^+ \leftrightarrow HPO_4^{2-}$	12.35[b)]
$PO_4^{3-} + 2H^+ \leftrightarrow H_2PO_4^{2-}$	19.47[b)]
$PO_4^{3-} + 3H^+ \leftrightarrow H_3PO_4$	21.62[b)]

a) Ref. 697 b) Ref. 617

Reactions analogous to (11.28) for the formation of bidentate (binuclear) inner sphere complexes for the adsorption of arsenate, phosphate and arsenite ions have also been proposed by different authors [312, 617,621]. For example,

$$2(FeOH)_{surf} + AsO_4^{3-} + 3H^+ \leftrightarrow Fe_2O_2AsO_2H + 2H_2O \quad K^2_{As(V)}(int) \quad (11.39)$$

$$2(FeOH)_{surf} + AsO_4^{3-} + 2H^+ \leftrightarrow Fe_2O_2AsO_2^- + 2H_2O \quad K^2_{2As(V)}(int) \quad (11.40)$$

While the intrinsic constants values as derived by Tadanier and Eick [312] for the adsorption of PO_4^{3-} on goethite are given in Table 8.5, the values reported by the other workers[617,697] for AsO_4^{3-} and PO_4^{3-} are given Table 11.4.

11.3.2.7 Competitive Adsorption of Arsenic Species in the Presence of Other Anions

The competitive adsorption of the two arsenic anion species in the presence of phosphate, silicate, sulfate and bicarbonate has been a subject of great interest to the environmental chemists. This is because, in some countries including Bangladesh, arsenic contaminated ground water also contains substantial amounts of phosphate, silicate and bicarbonate [734]. All these ions interfere in the removal of arsenic by adsorption process. The structural chemistry of phosphate, arsenate, silicate and sulfate have many common features, because of the close proximity of P, Si, S and As in the periodic table. Though the individual influence of sulfate on the adsorption of arsenate is not significant [735], the presence of phosphate distinctly reduces the adsorption of arsenate on HFO [697,736,740], goethite [697,698, 737,738] and clay minerals [739]. Figure 11.7 clearly demonstrates that adsorption of both arsenite and arsenate on goethite is suppressed in the presence of phosphate. It is interesting to note from the figure that while the adsorption of arsenate continues to decrease from pH 4 in the presence of phosphate, adsorption of arsenite gradually increases up to pH 8.75 even in the presence of phosphate. This suggests that arsenite is predominantly adsorbed as a neutral species in the pH range 4-9, irrespective of the surface charge acquired by the adsorption of phosphate on goethite. The adsorption of arsenite has also been found to be affected by the increasing concentration of phosphate [697,736], although some sites were found to exhibit higher adsorption of As(III) than phosphate [736]. The effect of molybdate on the adsorption of arsenate on goethite and gibbsite has been studied in detail by Manning and Goldberg [698]. It has been observed that the presence of molybdate decreases the adsorption of arsenate, especially at pH value less than 5.0.

The practical aspects of the influence of other anions, either added from outside or produced in situ on the removal of arsenic have been studied using Bangladesh ground water which contain substantial quantities of phosphate and silicate besides arsenic [740]. Meng et al [734] have also studied the removal of arsenic from contaminated ground water (both actual and simulated) of Bangladesh, containing phosphate, silicate and bicarbonate. Like phosphate, the presence of silicate adversely affects the adsorption of arsenic by iron oxyhydroxide. Indeed, both the groups of investigators referred above have reported that the removal of arsenic from Bangladesh ground water becomes more difficult due to the presence of soluble silicate and phosphate [741]. The salient findings of the work of Meng et al [734] are as follows.

(1) The apparent binding constants with $Fe(OH)_3$ surface indicate that the affinity of the anions for the solid decreases in the following order:

Arsenate > Phosphate >Arsenite > Silicate > Bicarbonate

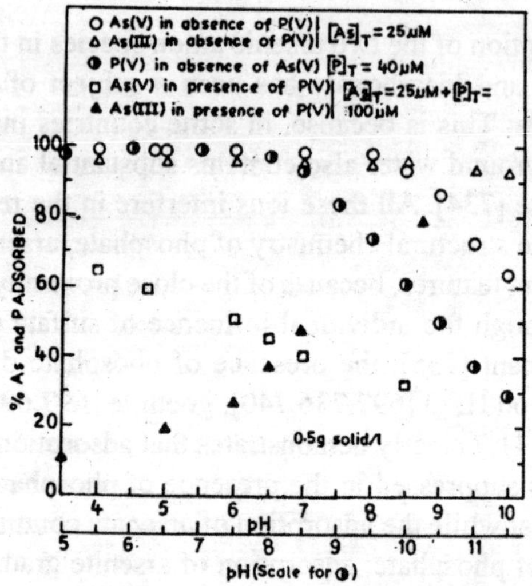

Fig.11.7 Adsorption of As(V) and As(III) on goethite (0.5 g/l) in the absence and in the presence of phosphate. Redrawn using the data from refs. 696 and 697.

(2) The adverse effect of anions on the removal of As(III) decreases in the following order: Phosphate(P)>Silicate(Si) > Bicarbonate(HCO_3) >Chloride(Cl)

(3) The combined effect of these anions on the removal of As(III) from a solution containing 300 μg/l of total As is of the following order:
$P + Si + HCO_3 > P + Si > P + HCO_3 > Si + HCO_3$

(4) The removal of As(V) is drastically reduced in a multi anion system except Si + HCO_3 However, none of the above anions individually affect the removal of As(V) by $Fe(OH)_3$ to a significant extent.

11.3.3 Selenite and Selenate

Selenium, belonging to Group VI p element {$(Ar)s^2 4p^4$}, form compounds of different oxidation states such as -2, 0, +2, +4 and +6. Though it occurs in widely scattered state in nature as a mixture of several isotopes, selenium bearing minerals are seldom encountered in nature. It occurs in elemental state along with free sulfur deposits and also in very small quantities or in traces as selenides in sulfide ores. The oxidized products of selenium formed during the roasting of sulfide ores are discharged along with waste water unless recovered for other uses. Selenium dioxide is also formed during the combustion of coal and ultimately finds its way into the soil as soluble selenious acid through seepage from ash pond or from fly ash used as land fills. Selenium contamination may also occur from the effluents of various manufacturing industries like photo-electric solar cells, photocopiers, stainless steel, ceramic products etc.

Selenium is an essential micronutrient for both animals and plants and possibly for human beings also [742]. At the same time, it is also potentially toxic beyond certain concentration. However, the difference between toxicity and requirement levels is not only narrow, but is also ill-defined [743]. Total selenium content in soil varies from 0.1 to 20 mg/l depending upon the industrial and mining activities in the area. This limit is considered high as far as safe environment is concerned, because selenium also enters human body through the consumption of plant products. It is comparable to arsenic in toxicity giving rise to similar symptoms as observed with arsenic. Fortunately, however, selenium content in most ground water is less than 10 μg/l which is within the permissible limit. This is because most of the predominant oxyanions of selenium is immobilized by hydrous iron oxide in soil.

The bioavailability of selenium depends on the nature of its chemical form. Like arsenic, selenium forms two oxyanions, namely selenite (SeO_3^{2-}) and selenate (SeO_4^{2-}), wherein the oxidation states of Se are +4 and +6 respectively. Both the anions occur in natural as well as contaminated water. Like arsenic the oxidation states of selenium in nature is governed by the environmental conditions, namely redox potential, pH, temperature and the presence of other cations and anions. However, unlike arsenic, the species of higher oxidation state i.e., Se(VI) is found to be more mobile than Se(IV) in natural waters and in soils. This is because the former (selenate) shows much weaker or lower affinity towards iron oxides than the latter (selenite) [30,32,744-746] thus rendering selenate easily transportable in pore waters of soil and making it available for uptake by plants. This is mainly attributed to two reasons. First, the acid dissociation constant (K_a) of selenious acid (H_2SeO_3) and selenic acids (H_2SeO_4) differ very widely as shown in Table 11.6. The data presented in the table suggest that while H_2SeO_4 dissociates at lower pH (< 5.0), H_2SeO_3 remains in un-dissociated state even at pH 8.0 - 8.5, so that electrostatic interaction between anion and positively charged hydrous oxides of Fe and Al can still occur, though for specific adsorption such interaction is not an essential requirement.

Among the oxyhydroxides of iron, adsorption of selenium species has been studied mainly on goethite [30,32,314,745,747,748] and HFO [31,32,745,748,749]. The other adsorbents used for the adsorption of selenium species are: Natural iron oxides

[746,750], activated Al_2O_3 [705,751-753], TiO_2 [754], layered double hydroxides [755,756], leached residue of manganese nodules [757], kaolinite [758], different types of soil [759,760], cement minerals [761] etc.

Table 11.6 logK_a Values of the Two Selenium Oxyacids in Aqueous Solution (I=0).

Reactions	-log K_a
$SeO_3^{2-} + H^+ \leftrightarrow HSeO_3^-$	8.46[a], 8.5[b]
$HSeO_3^- + H^+ \leftrightarrow H_2SeO_3$	2.61[a]
$SeO_3^{2-} + 2H^+ \leftrightarrow H_2SeO_3$	11.07[c]
$SeO_4^{2-} + H^+ \leftrightarrow HSeO_4^-$	1.70[a], 1.92[b]
$HSeO_4^- + H^+ \leftrightarrow H_2SeO_4$	-3.0[b]
$SeO_4^{2-} + 2H^+ \leftrightarrow H_2SeO_4$	-1.08[c]

a) Ref. No. 617 b) Akhmetov, N. S. *General & Inorganic Chemistry* (Engl. Transln.), Mir Publishers, Moscow, 1983, p346. c) calculated

For all the above adsorbents, adsorption of selenate was found to be much weaker than that of selenite which was found to adsorb on oxides of iron and aluminum even at pH above 8.0, whereas the adsorption of selenate was practically negligible at pH above 6.0. This is typically illustrated in Fig. 11.8 which shows the adsorption behavior of these two selenium species on goethite as function of pH in $NaNO_3$ solution of different ionic strength. From this important work of Hayes et al [32] it can seen that ionic strength has little influence on the adsorption of selenite, which suggests that this species adsorbs as inner sphere complex. On the other hand, adsorption of selenate is influenced by the changes in the ionic strength of $NaNO_3$ solution, indicating that selenate adsorbs as outer sphere complex onto goethite surface.

As far as kinetics of adsorption is concerned, the initial rate of adsorption of selenite on hematite at pH 3.5 for $[Se]_o = 4 \times 10^{-4}$ M was very fast, reaching about 95% adsorption within 5-10 min [750]. The remaining 5% adsorption continued in a span of 4-65h, indicating that the adsorption process in this stage is more complex than simple surface reaction. The rate of reaction was slower at pH 7.0 than at pH 3.5, which was attributed to possible decrease in the concentration of positively charged

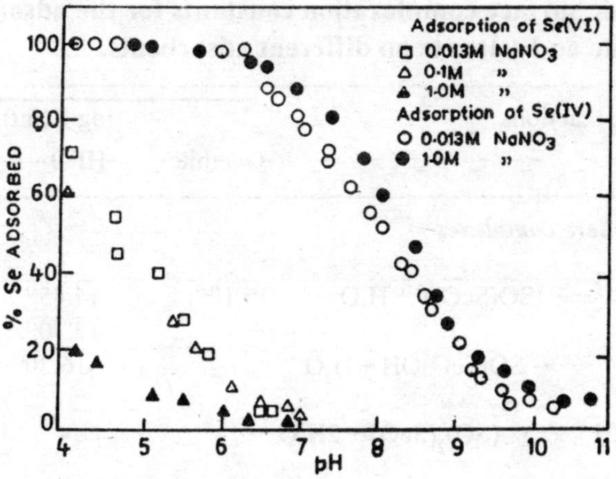

Fig.11.8 Effect of pH and ionic strength of the background electrolyte solution ($NaNO_3$) on the adsorption of Se(IV) and Se(VI) on HFO(10^{-3}M Fe).

sites. Increase in temperature from 25° to 50°C had no appreciable effect on sorption kinetics. Selenite adsorbed at pH 3.0 on hematite could be desorbed at pH 12.0, though the rate of desorption was much slower than the rate of adsorption. The equilibrium desorption was attained after 5h [750]. Similar slow rate of desorption of selenite from goethite was observed by others also [744,747]. However, Suarez et al [730] have failed to desorb selenite from ferrihydrite or HFO. Clearly, the mode of adsorption of selenite depends upon the nature of arrangement of atoms on the crystal plane of the adsorbent.

As far as the mode of adsorption of selenite is concerned, it is widely recognized that the anion is adsorbed as inner sphere complex on the oxides of Fe and Al. Macroscopically, this view has been supported by model fitting of the adsorption data as mono- or bidentate inner sphere complexes on iron oxides [32,746,750], on TiO_2 [754] and on Al_2O_3 [752]. Electrophoretic mobility measurement shows that the adsorption of Se(IV) decreases zeta potential as well as pH_{iep} of Al_2O_3 [752]. Microscopically, the formation of inner sphere surface complex was also correlated as early as in 1987 by EXAFS studies which suggested binuclear, bidentate complex formation [762]. This finding also received further support from similar spectroscopic studies [31,752]. While the three Se - O bond distances are at 1.70A°, the two Se - Fe distances are at 3.38A° for best fitting of the spectral data. On the other hand, SeO_4^{2-} does not shed its hydration sheath when it approaches the adsorbent surface and therefore adsorbs as outer

Table 11.7 Intrinsic surface complexation constants for the adsorption of selenite and selenate on different adsorbents.

Sl. No.	Surface reactions	log K(int)		
		Goethite	HFO	Hematite
	Inner sphere complexes			
1	$SOH + SeO_3^{2-} + H^+ \leftrightarrow SO(SeO_2)^- + H_2O$	15.10[a)] -	14.45[a)] 12.70[c)]	16.7 - 17.3[b)] -
2	$SOH + SeO_3^{2-} + 2H^+ \leftrightarrow SO(SeO)OH + H_2O$	-	16.20[c)]	25.0 - 25.9[b)]
3	$2SOH + SeO_3^{2-} + 2H^+ \leftrightarrow (SO)_2(SeO) + 2H_2O$	-	-	30.4 - 31.0[b)]
4	$SOH + SeO_3^{2-} + H^+ + Na^+ \leftrightarrow SO(SeO_2)^- + Na^+ + H_2O$	14.10[a)]	-	-
5	$SOH + SeO_4^{2-} + H^+ \leftrightarrow SO(SeO_3)^- + H_2O$	8.50[a)] -	7.70[a)] 7.73[c)]	- -
6	$SOH + SeO_3^{2-} \leftrightarrow SOH(SeO_3)^{2-}$	-	5.17[c)]	-
7	$SOH + SeO_4^{2-} \leftrightarrow SOH(SeO_4)^{2-}$	-	0.80[c)]	-
	Outer sphere complexes			
8	$SOH + SeO_3^{2-} + 2H^+ \leftrightarrow SOH_2^+\text{---}(HSeO_3)^-$	2.18[a)]	2.12[a)]	-
9	$SOH + SeO_4^{2-} + 2H^+ \leftrightarrow SOH_2^+\text{---}(HSeO_4)^-$	15.70[a)]	14.50[a)]	-
10	$SOH + SeO_4^{2-} + H^+ \leftrightarrow SOH_2^+\text{---}(SeO_4)^{2-}$	8.90[a)]	9.60[a)]	-

a) Ref. 32 , b) Ref. 750 and c) Ref. 617

sphere complex, possibly via hydrogen bonding with surface oxygen atoms. But, some controversy still persists regarding the adsorption of selenate as EXAFS data appear to support the formation of both outer sphere [752,762] and inner sphere [31] complexes. Infra-red spectroscopy also appears to suggest the formation of some inner sphere complex for selenate adsorption [763]. However, this controversy has been attributed to the pH value at which the adsorption of selenate is carried out; lower pH value favours the formation of outer sphere complex and relatively higher pH value (>4.5) favoring inner sphere complex formation. Typical values of the intrinsic surface complex formation constants (log K(int)) are quoted in Table 11.7 for

benefit of the readers.
Selenite exhibits strong competition with silicic acid (H_4SiO_4) (at concentration below the solubility limit of amor. SiO_2) for magnetite surface with the resultant decrease of its concentration in solution [764]. Das and Parida [757] have observed that the adsorption of selenite on a sample of leached residue of manganese nodule from binary mixture containing equimolar concentration of either PO_4^{3-}, or ClO_4^-, or SO_4^{2-} or NO_3^-, decreases from 91% (in the absence of any of these anions) to 63%, 67%, 74% and 80% respectively. The degree of influence of these anions on the adsorption of SeO_3^{2-} therefore decreases in the same order i.e., $PO_4^{3-} > ClO_4^- > SO_4^{2-} > NO_3^-$.

11.3.4 Cr(VI) or Chromate

Like selenium, chromium is also an essential micronutrient for plant as well as animal metabolism up to a certain tolerable concentration (50μg/l). Beyond this concentration limit it can also turn out to be quite toxic and may causes diseases like nauseas, skin ulceration, lung cancer etc.[765]. When its gradual accumulation (in human body) reaches the level of 0.1g / kg of body weight it becomes almost lethal. Cr(VI) is much more toxic than Cr(III). Sources of Cr(VI) in the environment are mostly from human activities such as wastes from steel industries, chrome / electroplating works, drainage from chrome ore mining areas etc. However, the most notorious sources of Cr(VI) pollution are leather tanning industries where $Na_2Cr_2O_7$ is extensively used.

Unlike phosphate, arsenate and selenite, chromate is a weakly adsorbing anion. Like other anions, chromate is also effectively adsorbed by hydrous oxides of iron. Its adsorption is highly dependent on pH. The adsorption is highest at pH 5 - 6, above which adsorption decreases sharply and becomes almost negligible at pH 8.0 - 8.5 [749]. Figure 11.9 shows percent adsorption of Cr(VI) on HFO as function of pH (adsorption edge) as well as concentration of electrolyte solution . The dissociation and protonation constants of chromic acid as given below in Table 11.8 are wide apart, indicating that mono protonated species ($HCrO_4^-$) is the predominant adsorbing species at lower pH values (< 6.0) and deprotonated species (CrO_4^-) at pH values higher than 6.5 [621].

Like sulfate and selenate , the bonding of chromate ion with iron oxide surface is rather weak and adsorption decreases with increase in ionic strength of electrolyte solution [749,766] indicating that adsorption takes place through outer sphere complex formation [752,767-769]. However, IR spectroscopy as well as shifts in electrophoretic mobility (or zeta potential) and pH_{iep} values tend to suggest also the formation of inner sphere complex [29,770]. Some authors [771], using CD-MUSIC model

Table 11.8 log Ka values* of chromic acid in aqueous solution (I = 0)

Reactions	log K_a
$CrO_4^{2-} + H^+ \leftrightarrow HCrO_4^-$	- 0.20
$HCrO_4^- + H^+ \leftrightarrow H_2CrO_4$	6.50
$CrO_4^{2-} + 2H^+ \leftrightarrow H_2CrO_4$	6.30 (calc.)

* Ref. 617

fitting studies, have proposed that the adsorption of chromate on goethite occurs through inner sphere complex formation. Satisfactory fitting of chromate adsorption data by TLM using electrostatic interaction or outer sphere complex formation has been shown by Davies and Leckie [749]. Indeed, placing a large anion like chromate on o-plane is spatially not a feasible proposition [772]. Some typical intrinsic surface binding constants obtained by different workers for the adsorption of chromate ion

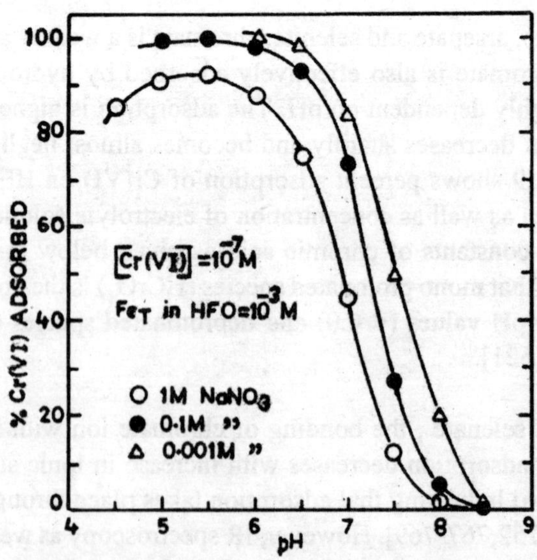

Fig.11.9 Adsorption of Cr(VI) on HFO(10^{-3}M Fe) as function of pH and ionic strength of the medium ($NaNO_3$). Redrawn using the data from ref. 617.

on HFO are given in Table 11.9. Attempts have been made to adsorb Cr(VI) on clay minerals as well as on alumina surface [773]. Adsorption capacities of these adsorbents obey the following sequence:

Table 11.9: Intrinsic Binding Constants for the Adsorption of Chromate on HFO.

Sl. No.	Reaction	log $K_{Cr(VI)}$(int)	Reference
1.	$FeOH^{o}_{surf} + H^{+} + CrO_4^{2-} \leftrightarrow FeCrO^{-}_{4surf} + H_2O$	10.1	767
		10.8	617
		12.8	774
2.	$FeOH^{o}_{surf} + 2H^{+} + CrO_4^{2-} \leftrightarrow FeOH_2^{+}\text{---}HCrO_4$	20.4	774
3.	$FeOH^{o}_{surf} + 2H^{+} + CrO_4^{2-} \leftrightarrow FeHCrO_4 + H_2O$	17.6	767

$$Al_2O_3 > \text{kaolinite} > \text{montmorillonite} > SiO_2$$

Speciation studies suggest that $HCrO_4^{-}$ is the major adsorbed species. However, at concentration above 10^{-3} M, adsorption of dichromate ion ($Cr_2O_7^{2-}$) has also been found to occur. $HCrO_4^{-}$ ion adsorbs on alumina mainly as outer sphere complex. Rai et al [766], who have investigated the adsorption behavior of a number of adsorbents for chromate, have observed the following sequence in respect of their adsorption capacities.

$$\text{Hematite} > \text{goethite} = HFO > Al_2O_3 > \text{kaolinite} > \text{montmorillonite}$$

Activated alumina has also been used by other investigators [774] and their findings more or less corroborate with those of Fritzen et al [773] and Rai et al [766]. Weng et al [775] have studied the adsorption of Cr(VI) on anatase, a variety of TiO_2 and found that the contribution of specific chemical interaction to the total free energy of adsorption is much more higher than those of electrostatic and salvation energies. Besides the above adsorbents of distinct chemical identities, several unconventional or low cost solid materials have also been used for the adsorption of chromate [776].

Adsorption of Cr(VI) on HFO is considerably reduced in the presence of SO_4^{2-} [249,749] and CO_3^{2-} ions [772]. These weakly adsorbing ions compete with Cr(VI)

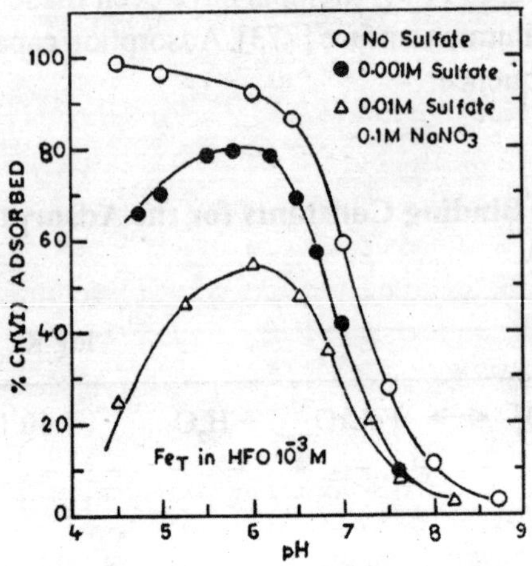

Fig.11.10 Effect of the presence of SO_4^{2-} ion on the adsorption of Cr(VI) on HFO (10^{-3}M Fe). Solid lines represent data fitted to TLM according to ref.749. Redrawn using the data from ref.249.

adsorption sites, because of their similar mode of complexation with the surface sites. Indeed, by incorporating sulfate and carbonate complexation constants with the surface sites, a considerable improvement in the model fitting of the experimental data was noticed. Figure11.10 demonstrates the good fitting of the experimental data by considering the appropriate SO_4^{2-} and CO_3^{2-} complexation in the triple layer model [249]. The following values of surface complexation constants were used for model fitting: log ($K_{CrO_4^{2-}}$) =10.1, log ($K_{HCrO_4^-}$) = 17.6, log ($K_{SO_4^{2-}}$) = 9.4 and log ($K_{HSO_4^-}$) = 15.5.

11.3.5 Fluoride (F⁻)

Like oxyanions of P, As, Se and Cr(VI) , fluorine in its ionized form is also an environmentally hazardous element. The severity of the environmental scourge of this ionic species afflicting millions of people in many countries of the world including India, through contaminated drinking water and various food items, is no way less than that of arsenic. Fluoride, when taken in very small quantity (< 0.75 mg/l) is, indeed, beneficial for healthy growth of teeth and bones of young people. But, when its concentration in water or in food items exceeds beyond a certain limit (~1.5 mg/l or per g) it causes a disease called "fluorosis". This is a slow, progressive, crippling malady that affects young and old, rich and poor, rural and urban population alike.

The disease was first detected in India in the early 1930s and the first report of endemic fluorosis appeared in 1937 when it was prevalent in areas which now come under the states of Andhra Pradesh, Tamilnadu, Punjab and Uttar Pradesh. During the period 1950 - 1980, nine more states were identified as endemic for fluorosis. Since 1990 at least four more states have been added to the list of those endemic for the disease. Thus, the total number of states that have been engulfed by this environmental menace are seventeen, which include almost the northern, western, central and eastern India, and most of the southern India excluding the state of Karnataka. More than 30 million people in rural India consume water with fluoride content above the prescribed limit of 1.5 mg/l [777-779]. Besides India, fluoride contaminated ground water occurs in many countries of Africa, West Asia, America and in some countries of Europe [780-783].

11.3.5.1 Toxic Effects of Fluoride

Long term consumption of water (or food items) containing more than 1.3- 1.5 ppm F^- may lead to dental fluorosis. Yellow patches develop on teeth which then turn brown and finally black due to destruction of natural enamel, affected teeth may get pitted, perforated and chipped off in the final stage.

Long term consumption of water containing 3-6 ppm F^- may lead to skeletal fluorosis. Fluoride can damage foetus if mother takes highly contaminated water or food during pregnancy. Infant death may occur due to calcification of blood vessels. The other symptoms of skeletal fluorosis [777] are:

1. Severe pain in the backbone and in the joints.
2. Severe pain in the hip region.
3. Stiffness of the backbone and the joints.
4. Calcification of ligaments and increased bone density.
5. Constriction of vertebral canal and invertebral foramen resulting in the pressure on nerves.
6. Ultimate paralysis of limbs.

Besides fluorosis, high intake of fluoride leads to many other manifestations (or signs) of ailments such as:

1. <u>Neurological manifestations</u> like nervousness, tingling sensation of fingers and toes, excessive thirsts and tendency to urinate.
2. <u>Muscular manifestations</u> like muscular weakness, stiffness, pain in the muscles.
3. <u>Allergy manifestations</u> like inflammatory, painful skin rushes which may

appear either as red or bluish or pinkish, disappearing in 7-10 days.

4. Gastro-intestinal manifestations like acute abdominal pain, diarrhea, constipation, blood in stools etc.

11.3.5.2 Sources of Fluoride in Ground Water

Like other toxic pollutants, there are mainly three sources from which fluoride is incorporated in the ground water, namely (a) geological (b) industrial and (c) agricultural. Fluoride occurs in different types of rocks which include high and low calcium granites, ultramafic rocks, slates, sand stones etc. The main fluoride containing minerals in these rocks are fluorspar (CaF_2), cryolite ($NaAlF_6$), fluorapatite ($3Ca_3(PO_4)_2 \cdot CaF_2$), sellaite ($MgF_2$) etc. All these minerals are insoluble in water. However, under acidic condition these minerals may undergo slow dissolution leading to the accumulation of fluoride in pore waters of subsoil aquifer. It also occurs widely in the dissolved state in river and in sea water.

The geochemistry of the dissolution of fluoride-bearing minerals and the accumulation of F^- ion in ground water is rather complex and is not yet fully understood. Rain water percolating through underground subsoil takes up H^+ from soil organic acid and when such water percolates through rock bearing strata, dissolution of fluoride-bearing minerals takes place. While most of the dissolved F^- combines with Ca^{2+} or even $CaCO_3$, a part of it enters into clay minerals which remain in suspended state in ground water. Subsequent ion exchange process releases F^- in the water from clay minerals. Therefore, the concentration of F^- in ground water depends upon a) pH of the percolating water, b) Ca^{2+} present in the water i.e., water with high Ca^{2+} content is low in F^- content, c) presence of dissolved CO_2 or HCO_3^- ion. Increase in the concentration of HCO_3^- ion favors more dissolution of fluoride. d) Residence time of percolating water in the solid i.e., longer residence time favors more dissolution of fluoride bearing minerals.

Fluorine and its compounds are used extensively in many industries such as electronics (semiconductors, printed circuit board etc.), manufacture of phosphatic fertilizer using fluorapatite, preparation of high purity graphite using HF to remove traces of silica, electrolysis of molten alumina and in nuclear energy application. Toxic wastes from these industries may spill into the environment, if proper care is not taken to separate it from the wastes.

As far as the agricultural sources are concerned, F^- comes mainly from phosphatic fertilizer and the use of fluoride contaminated river water for irrigation purpose. Certain pesticides / insecticides also contain fluoride.

11.3.5.3 Remedial Measures

Like arsenic, there are broadly two types of remedial measures to control fluoride concentration in ground water: i) In- situ control of the contamination in aquifer itself. ii) Removal of fluoride from contaminated ground water using various physico-chemical methods like reverse osmosis, dialysis / electro-dialysis, electro-coagulation, electro-flocculation, ion exchange, precipitation - coagulation, adsorption etc.[784]. In the first type of approach various artificial recharge techniques may be employed to reduce the concentration of F^- by diluting with water of no or low F^- contamination. One such technique is called Aquifer Storage Recovery (ASR) wherein water is stored underground through deep bore well and is utilized in times of emergency needs. The dilution process can also be done by transferring or transporting potable water from nearby areas to the problem areas. Recharging of contaminated ground water may also be done with rain water harvested in large tanks on the over ground.

The second type of remedial measures which involve treatment of post withdrawal contaminated ground water may be further subdivided broadly into two categories, namely physical and chemical methods. The physical methods are of two types, namely those employing membrane technique and those based on non-membrane technique. The methods employing membrane technique include reverse osmosis, nanofiltation, dialysis and electro-dialysis. These techniques are mostly useful when the concentrations of soluble salts including fluoride are very high such as brackish water or ground water of high salinity, industrial effluents etc. Interested readers may consult the review articles [11,12,15,784] for more information with relevant references. Electro-coagulation and electro-flocculation belong to non-membrane physical methods [785-787]. These techniques employ electrochemical principles to separate charged species. The above cited review articles may also be consulted for more details.

Depending upon the nature of treatment processes, the chemical methods of defluoridation are of two types. The first involves insolubilization of fluoride by adding either lime to precipitate calcium fluoride or adding sequentially alkali(lime), chlorine (for oxidation of iron) and alum or aluminum chloride to form mixed precipitates of aluminum and calcium fluoride. The presence of aluminum salt helps to coagulate the precipitates. The method is known as "Nalgonda Technique", in commemoration of the place, a district in the state of Andhra Pradesh where the endemic fluorosis was first detected in 1937 [788,789]. But, both lime and alum treatments raise the pH of the water to a high level (~10) and therefore further treatment with acid is necessary to reduce the pH of water to potable level which involves additional process cost [790,791]. Some more disadvantages of this technique have been identified and listed by Meenakshi and Maheswari [12]. Despite such constraints, the technique has been adopted in many places as small community defluoridation units.

The other chemical method of defluoridation is based on adsorption on suitable solid materials or adsorbents. Among these materials, activated alumina has been found to be the most used one, especially when the initial fluoride content is not very high (<5-8 mg/l) [11,12,15,784, 792,793]. Activated alumina is a granular, highly porous material containing hydrated Al_2O_3. It is generally prepared by the pyrolysis of gibbsite or materials containing gibbsite, either slowly at relatively lower temperature (<400°C) or rapidly at higher temperature [794,795]. Slow heating at lower temperature results in a product which is more crystalline with lower surface area and porosity than that obtained by rapid heating (flash pyrolysis) at higher temperature. Rapid decomposition is carried out in flue gases at 600° - 800°C and exposed time of 1-3 seconds. Alumina activated in this way is called "transitional alumina", because its composition is somewhat intermediate between $Al(OH)_3$ and Al_2O_3 with 0.2 - 1.0 mole of water per mole of Al_2O_3. Because of high surface area (~ 200 m^2/g), this poorly crystalline product exhibits rapid and high sorption of F-ion. Activated alumina can be regenerated by treating with 4-5% NaOH solution followed by washing with dilute HCl to neutralize excess alkali on the surface.

Besides activated alumina, a wide variety of materials have been used as adsorbents for the removal of fluoride from water. These are: i) Bone char prepared by calcining raw bones at 1100°C and consists mainly of carbonate apatite [789,796]. ii) Alum sludge [797,798]. iii) Clay minerals and soils which have long been considered as potential low cost adsorbents for fluoride. Kaolinite type clays exhibit pH dependent variable charge at their edges (see section 7.2.1). At lower pH (< 4.0) the positive charge at the edges can adsorb F^- ion. However, the overall adsorption efficiency is low [799-802]. Montmorillonite and bentonite, on the other hand, have loosely bound structural (interlayer) OH^- ions which can exchange with F^-, in addition to the pH dependent variable charge at the edges. Consequently, these clay minerals show more adsorption of F^- ion than kaolinite [801,803- 806]. Considerable amount of work has also been carried out on the adsorption of F^- ion on spent bleaching earth, a waste product from edible oil industry and consists mainly of bentonite clay [807-809]. Attempt has also been made to improve the adsorption capacity of clay minerals to a certain extent by alkali treatment [801]. However, considerable success has been achieved in this regard by using metal oxide coated clays, especially bentonite, as adsorbents [810,811]. Kamble et al [812] have shown that bentonites when coated with La_2O_3 give the best results, followed by MgO and Mn_2O_3 coated clays. iv) Soils: Adsorption of F^- ion on soils depends, however, not only on their clay mineral contents, but also on their hydrated oxide, especially amor. aluminum hydroxide contents. The presence of high exchangeable sodium content has been found to inhibit the adsorption of fluoride [813-816]. v) Fly ash [817-818]. vi) Bauxite: This is an important hydrated aluminum and iron oxides based mineral and is therefore considered as one of the low cost materials having potential applicability for removal of fluoride from water. Bauxite is abundantly available in some states of India where ground water is also contaminated with fluoride. Therefore attempts have been made to study the adsorption behavior of both raw (untreated) [13,14,819] as well as heat treated bauxite [14]. It has been observed [13] that high alumina and low iron refractory grade bauxite exhibits much higher adsorption than common low alumina(~ 55% Al_2O_3) feed bauxite. It has also been shown [14] that preheating titanium rich bauxite at 300° - 450°C considerably increases its adsorption capacity for F^- ion. vii) Red mud: This waste product generated after alkaline leaching of bauxite ore (Bayer's Process) contains residual aluminum, iron and titanium oxides. Since this material is strongly alkaline, it has to be treated with strong acid solution, washed, dried and if necessary, reactivated [820]. All these, however, add to the process cost. viii) Lateritic ores: Goethite rich lateritic iron ores have been used as adsorbents for F^- ion from drinking water [821,822]. But, the efficacy of iron oxide rich geomaterials for fluoride removal is generally found to be lower than those of aluminum oxide rich materials. ix) Zeolites: The efficacy of zeolites for the removal of fluoride from water depends on its Si / Al ratio i.e., zeolites of high aluminum content are most suitable as adsorbents for F^- ion [823]. Drost [824] has wrongly attributed active alumina to certain naturally occurring zeolites of high aluminum content which show considerable affinity for anions at pH 6.5. Perhaps, he got this view from earlier literature wherein attempts were made to adsorb various anions on such materials.

The main purpose of the extensive literature cited above on defluoridation of drinking water was to develop low cost inorganic based solid materials with high adsorption capacity for F^- ion. In this regard the investigators have studied several parameters which are common to almost all the above references. These are: (a) Effect of pH which generally shows an optimum or a range of optimum values, below and above of which adsorption tends to decrease. (b) Amount of adsorbent (g/l) which depends on the degree of affinity of the solid (adsorption capacity per unit mass) for F^- ion. (c) Initial concentration of F^- ion in the system which is obviously above 2.0 mg/l, but some authors used concentration as high as 50-80 mg/l. (d) Contact time i.e., time to reach equilibrium adsorption at the optimum pH value. This parameter depends upon the rate of adsorption, which is invariably very fast in the initial 10-15 min and then proceed slowly until the equilibrium is reached. Generally, 95% of adsorption is achieved in 2-4h depending upon the nature of the solid. (e) Adsorption capacity is determined either from Freundlich or from Langmuir equation and its value is generally compared with activated alumina for which the capacity varies from 1.8 to 2.4 mg/g. Any solid material showing capacity higher than this range is considered to be a good adsorbent.

Besides, the search for low cost adsorbent materials attempts have been made to develop adsorbents of distinct chemical entities such as oxides and hydr(oxides) of one or more than one metals. Thus, attempts were made to enhance the adsorption capacity of activated alumina by coating it with either manganese oxide [825,826] or with magnesium oxide [827] or with copper oxide [828]. The last preparation has been found to be specially effective to increase the capacity of F^- from 2.32 mg/g to 7.7 mg/g.

Mixed hydroxides of iron and aluminum coprecipitated from their aqueous chloride or nitrate solution at pH 7.0-7.5 in equimolar proportion, aged, filtered and air dried exhibit higher adsorption of fluoride than their corresponding individual hydroxides [829]. Similar observation has also been made by Sujana et al [830,831]. The authors claim to obtain an adsorption capacity of 91.7 mg/g at pH 4.0. The optimum pH value, however, increases with increase in Al content in the mixture.

Though aluminum oxide based adsorbents are mostly preferred for the removal of fluoride from drinking water, some researchers have found that granulated ferric hydroxide (GFH) can also be used as efficiently as activated alumina for the same purpose [832]. The adsorption behavior of this material depends upon the pH of the aqueous medium, but not on its ionic strength. Maximum adsorption occurs in the pH range 3.0-6.5. Spectroscopic data suggest that F^- ion is adsorbed by hydrogen bonding between F^- and OH groups. The presence of anions like $H_2PO_4^-$, HCO_3^-, SO_4^{2-} and Cl^- in many water samples tend to reduce the adsorption capacity in the following order: $H_2PO_4^- > HCO_3^- > SO_4^{2-} > Cl^-$. A similar sequence of the inhibiting effect of coexisting anions has been observed for the sample prepared by Kumar et al [833].

Hydroxy-apatite (HAP) is another distinct inorganic material that has also been studied as a potential adsorbent for the removal of fluoride from drinking water. The adsorption capacity of this material for F^- ion was found to be almost comparable to that of activated alumina [834]. The uptake of F^- ion by HAP is mainly through ion exchange process and partly through chemisorption. Some other inorganic materials tested for fluoride adsorption exhibit lower capacity in the following decreasing sequence [835]

HAP > fluorspar > quartz > quartz-Fe^{3+} > calcite

11.3.5 .4 Effect of pH Variation

Adsorption of fluoride on inorganic adsorbents is highly sensitive to the variation of pH, almost irrespective of the nature of adsorbent. The phenomenon is specially pronounced for oxidic materials. Figure 11.11 illustrates the effect of pH on the adsorption of F^- on pure Al_2O_3 and on a heterogeneous material like titanium - rich bauxite and red mud. For all the three materials, the optimum pH value lies between 5 and 6, as also observed by other workers [13,836]. The decrease in F^- adsorption below pH 5 is generally attributed to two reasons. First, fluoride promotes dissolution of even highly insoluble metal oxide like Al_2O_3 forming various soluble fluoro complexes such as AlF^{2+}, AlF_2^+, AlF_3, AlF_4^- etc. [837,838]. Increase in temperature facilitates the process of dissolution and therefore, adsorption decreases. More intensive investigation is necessary on this aspect of F^- ion adsorption.

At pH >6 , adsorption of fluoride decreases sharply because of the less favourable electrostatic condition of the surface i.e., decreasing concentration of SOH_2^+ and simultaneously increasing concentration of SO^-. Further, as pH increases competition with OH^- ion for surface sites also increases. At the optimum pH, fluoride adsorption has also been found to be affected by temperature [839]. Increase in temperature tends to decrease adsorption density indicating exothermic nature of the adsorption process. Valdivieso et al [839] attributed the decrease in F^- adsorption on α-Al_2O_3 with increase in temperature from 25° to 40°C to the overall decrease in the positive surface charge caused by the desorption of protons from the surface. This results in the shifting of zeta potential to acid side. This is illustrated in Fig. 11.12, reconstructed using the data obtained by the authors [839] for α-Al_2O_3 in 0.01M $NaNO_3$ solution. The figure shows that higher is the concentration of F^- ion in the

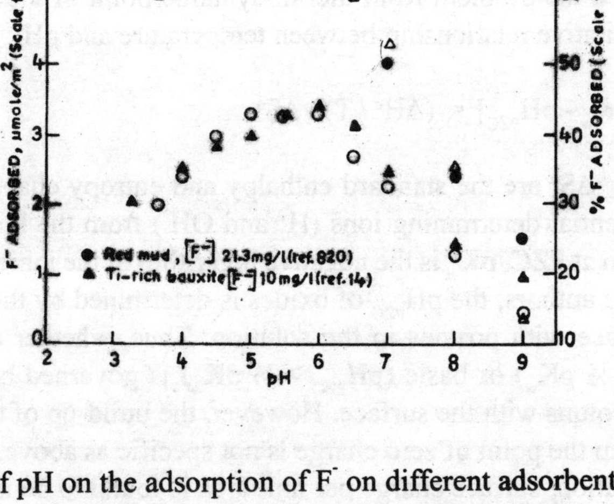

Fig.11.11 Effect of pH on the adsorption of F^- on different adsorbents. Redrawn using the data derived from references 14, 820 and 839.

system the smaller is the pH range in which zeta potential exhibits negative value. This is because protons are accumulated in the outer Helmholtz plane (OHP) as counter ions of negative surface charge and this results in the larger shift of pH_{iep} towards acid side, facilitated by the increase in temperature. This also suggests that F^- is adsorbed specifically even if the surface is negatively charged.

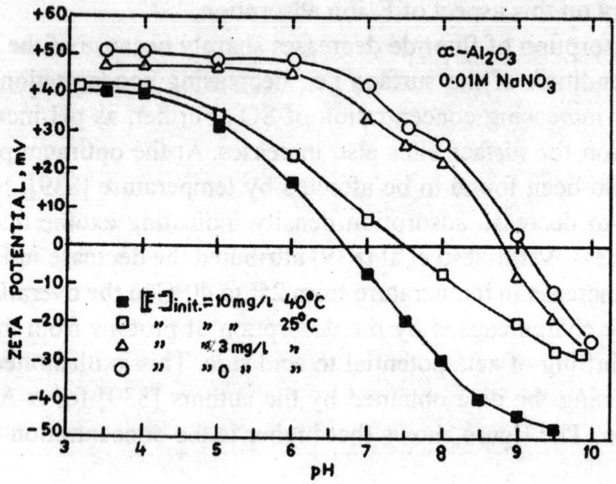

Fig.11.12 Effect of temperature and F^- adsorption on the variation of zeta potential of α-Al_2O_3 with respect to pH. Redrawn after data from ref. 839.

The phenomenon of the shift in the pH_{iep} of alumina with increase in temperature was observed as early as in 1972 [840] and later on by Halter [841]. Berube and DeBruyn [842] approached the problem from thermodynamic point of view and derived the following quantitative relationship between temperature and pH_{PZC}

$$4.6\,R\,[1/2pK_w - pH_{PZC}] = (\Delta H^o / T) - \Delta S^o \qquad (11.41)$$

where, ΔH^o and ΔS^o are the standard enthalpy and entropy changes that occur for transferring potential-determining ions (H^+ and OH^-) from the bulk solution to the interfacial region at PZC, pK_w is the negative logarithm of the ionic product of water. According to the authors, the pH_{PZC} of oxides is determined by the specific interaction of the surface with protons in the solution. Thus, whether an oxide behaves acidic ($pH_{PZC} < ½\ pK_w$) or basic ($pH_{PZC} > ½\ pK_w$) is governed by the enthalpy of interaction of protons with the surface. However, the build-up of charge on the surface starting from the point of zero charge is not specific as above. At constant electrolyte concentration, surface charge per unit area is found to be independent of the nature of oxides of similar group, i.e., closer pH - pH_{PZC} values [843]. In truly indifferent electrolyte solution i.e., in the absence of specific interaction the authors have derived the following equation which they claim more rigorous and general than Eq (11.41)

$$\left(\frac{\partial pH_{pzc}}{\partial T}\right)_{Co} = \frac{\Delta_{ads} H^{o}_{H+}}{2.303RT^{2}} \tag{11.42}$$

where, $\Delta_{ads}H^{o}_{H+}$ is equivalent to ΔH^{o} in Eq.(11.41)

Besides the above mentioned work [839] on the zeta potential of α-Al_2O_3, not much basic investigation on the adsorption of F^-, has been carried out. Even the value of the binding constant of F^- with alumina surface is not available in literature. Fluoride is a strong complexing agent, its stability is only after CN^- and $H_2PO_4^-$. It adsorbs on goethite as inner sphere complex [315]. Similar mechanism has also been observed by Tang et al [832] for fluoride ion adsorption on granular ferric hydroxide. However, the binding constants for the adsorption of F^- ion on HFO as determined from linear free energy relationship are as follows [617]

$$FeOH_{(surf)} + F^- + H^+ \leftrightarrow FeF_{(surf)} + H_2O \tag{11.43}$$

$$\text{Log } K_{Fo}(\text{int}) = 8.7$$

$$FeOH_{(surf)} + F^- \leftrightarrow FeOHF^-_{(surf)} \tag{11.44}$$

$$\text{Log } K_{F^-}(\text{int}) = 1.6$$

Since the surface cannot remain in the charged state without the corresponding counter ion in solution, anion exchange reaction takes place with the release of OH^- in solution.

$$FeOHF^-_{(surf)} \leftrightarrow FeF_{(surf)} + OH^- \tag{11.45}$$

While reaction (11.43) takes place when the surface is positively charged, reaction (11.44) takes place on neutral or uncharged surface.

11.3.6 Adsorption of Other Inorganic Anions

In addition to the anions discussed above there are a few more inorganic anions such a borate, silicate, nitrate, nitrite etc. which are also considered as undesirable contaminants in drinking water. Various chemical methods other than adsorption are employed for their removal. However, silicate and borate can also be removed by adsorption, usually on hydrous iron oxides at higher pH value (8.5-9.0). Silicate can ccur in various polymeric forms which can complicate the adsorption process. However, for the purpose of basic investigation, it is believed to occur primarily as $H_3SiO_4^-$ at pH 9.0 which corresponds to the pK_{a1} of H_4SiO_4. The binding constants $H_3SiO_4^-$ with goethite and HFO have been found to be 4.1 and 3.62 respectively [304, 844]. However, the nature of the surface species has not been fully established because of the variable state of polymerization , though Hansen et al [844] assumed it to be $FeOSi(OH)_3$.

Boron is an important micronutrient for plants. Its mobility in the pore water of subsoil depends on its affinity for hydrous oxides of iron and aluminum. Boron is adsorbed as $B(OH)_3$ and its negatively charged anionic form $B(OH)_4^-$. The pK_a value of $B(OH)_3$ is 9.24 and it shows maximum adsorption at pH 8-9 on goethite [845,846], whereas on ferrihydrite maximum adsorption occurs at 6.5- 8.5 [847]. Adsorption is found to be reversible with respect to pH, concentration and increase in temperature [848].

11.3.7 Quantitative Analysis of Some Selected Anions at Low Concentrations

While cations at low concentrations or even in trace quantities (in ppb level) can be estimated using modern sensitive instruments like flame or flameless (graphite furnace) Atomic Absorption Spectrophotometer (AAS), Inductively coupled Plasma - Optical Emission Spectrometer (ICP-OES), ICP-Atomic Emission Spectrometer (ICP-AES), ICP-Mass Spectrometer (ICP-MS) etc., such direct instrumental methods are not always feasible for anions. In fact, chemical pretreatment followed by instrumental method of analysis is usually adopted for the determination of trace quantities of anions. In the following pages, an attempt has been made to present briefly the methods of determination of trace quantities of anions, whose adsorption characteristics have been discussed in this chapter, for the benefit of the readers.

11.3.7.1 Phosphate

Colorimetric /spectrophotometric method is considered to be most suitable for the estimation of phosphate in water. There are two such methods: One is molybdenum blue method which is based on the reduction of molybdo-phosphate complex by either stannous chloride or amino naphthol disulphonic acid (ANSA) in strong acid medium. The intensity of the blue color measured against reagent blank (all the reagents except phosphate added in the same quantities as in test samples) either at 690nm or at 880nm. Soluble silica, nitrite, Cr(VI) and arsenic are the main interfering agents. The other method is the formation of yellow phosphovanadomolybdate complex which is measured at 460nm. The second method is not only simpler to carry out, but is also free from interference of some common co-existing ions in water or in waste water. However, if any organic matter is supposed to be present, the sample may be heated with a mixture of conc. H_2SO_4 (2 ml) and conc. HNO_3 (5 ml) over sand bath to SO_3 fumes at least twice before the determination is carried out. The following is the step-wise procedure for the determination of phosphate.

1. The residue obtained after fuming with mixed acids is digested in 5 ml of 5M HNO_3 and 10 ml water. The digested solution, if clear, may be directly trans ferred to 50 ml volumetric flask. If suspended particles are present, the solution may be filtered directly into the volumetric flask along with the washings (15-20 ml).

2. Add 5ml of 10% ammonium molybdate solution and then 5 ml of 25 % ammonium vanadate solution prepared in 6N HCl with warming, if necessary.

3. Dilute with distilled water to 50 ml mark and then shake the flask end over end for 10 - 15 times.

5. Prepare a blank solution in a similar manner except the addition of any phosphate bearing solution.

6. Wait for 15 min. for color to be stabilized and measure the absorbance of yellow phosphovanadomolybdate complex at 460nm against blank.

7. Prepare a calibration curve using a series of standard solutions of phosphate prepared by dissolving 0.7165g dried and powdered KH_2PO_4 in 1000 ml distilled water. 1ml of this solution ≡ 0.5 mg PO_4.

11.3.7.2 Arsenic (Arsenite and Arsenate)

Inorganic arsenic which occurs in ground water either as arsenite or as arsenate, is reduced to gaseous arsine (AsH_3) by nascent hydrogen generated in situ by adding zinc metal to acid solution. Sulfate, if present in the solution, is also reduced producing H_2S which is selectively absorbed in a scrubber containing glass wool soaked in lead acetate solution. The generated arsine is the absorbed in a solution containing silver diethyl dithiocarbamate (SDDC) dissolved in pyridine producing red colored arsenic dithiocarbamate complex, the intensity of which is measured spectrophotometrically at 525 nm. The method is suitable for total arsenic concentration in the range 2-100 μg/l . In the earlier literature AsH_3 was used to be absorbed in sodium hypobromite solution so as to be oxidized to As(V) state. Arsenate was then allowed to react with acidic solution of ammonium molybdate in the presence of hydrazine to produce blue colored complex. The method is, however, now-a-days seldom used as SDDC is found to give more accurate results.

Reagents

A) Hydrochloric acid : Concentrated (32%) AR/GR grade.

B) Potassium iodide solution: Dissolve 15g AR/GR grade KI in 100ml distilled water.

C) Stannous chloride solution: Dissolve 40g (As free) $SnCl_2$. H_2O in 100ml of 6N HCl under constant stirring, if necessary heat the solution.

D) Lead acetate solution: Dissolve 10g $Pb(CH_3COO)_2.3H_2O$ in 100ml distilled water.

E) SDDC reagent: Dissolve 1.0g silver diethyl dithiocarbamate in 200ml pyridine and store on a dark brown bottle. However, E. Merck in their pamphlet (EN 2695) recommended the dissolution of 0.25g SDDC along with 0.02g brucine in 100ml 1methyl-2 pyrollidone (extra pure) and storing in dark brown bottle.

F) Zinc Powder: Pure Zn metal (AR/GR), 20-30 mesh.

G) Stock arsenic solution: Dissolve 1.32g As_2O_3 in 20ml distilled water containing 4g NaOH and dilute the solution to 1000ml with double distilled water. 1ml of this stock solution ≡ 1μg As.

H) Working As standard solution: Dilute 5ml stock As solution to 500ml with DD water. 1ml of this solution ≡ 10μg As.

If organic matter or organically complexed As is suspected to be present, then pre digestion with acid is necessary. This is carried out as follows: Transfer a suitable aliquot of sample containing 2-30mg As into a 250ml beaker, add 2ml conc H_2SO_4 and 5ml conc HNO_3. Heat on sand bath to SO_3 fumes, cool, add 10-15ml water and heat again to SO_3 fumes to expel completely the oxides of nitrogen. Cool, add 25ml water and transfer the whole solution along with washings (10-15ml) to arsine generator.

There are various designs of arsine generator as adopted by different groups of workers. Arsenic estimation apparatus essentially consists of three parts: i) Arsine generator wherein arsenate and arsenite are reduced to elemental As which then combines with nascent hydrogen to form A_sH_3. ii) Scrubber containing glass wool soaked in lead acetate solution. iii) Arsine absorber containing SDDC reagent. The absorber may be either a separate system (a capillary tube that leads the gas into a test tube containing a known volume of reagent) or attached to the main apparatus as a fixed volume pipette. The two types of apparatus are shown in Figs. 11.13 and 11.14 respectively.

Procedure : The following are the steps involved in the estimation of total arsenic.

1. Place a suitable aliquot of the original water sample or the entire acid digested sample to the arsine generator flask.

2. Add successively 2ml KI solution (B), 5ml conc. HCl (A) and 10-12 drops $SnCl_2$ solution (C), mixing adequately after each addition. Allow 15 min for complete reduction of As(V) to As(III).

3. In the mean time fill the arsine absorption tube or the pipette with an exact volume of SDDC solution (E) and fit it to the generator via gas scrubber.

4. Add 2-3 g Zn powder to the arsine generator flask and close it immediately, connect it to the gas absorber as shown in Figs 11.13 and 11.14.

5. Allow at least 45 min for reaction to complete.

6. Transfer a portion of the solution from the gas absorber to spectrophotometer couvett with cover and measure the absorbance at 535nm against the reagent blank.

7. In order to prepare calibration curve carry out absorbance measurement in the same manner as above by transferring known volumes of working standard solutions of As (H) to arsine generator.

The above method, though proved to be reasonably accurate, is rather time consuming. Further, the method gives the estimate of total As and not the values of individual species such as As(III) and AS(V) when present together. For more rapid estimation of total As it is necessary to use instrumental methods such as HGAAS, ICP-OES, ICP-AES, ICP-MS etc. Arsenic concentration as low as 1 μg /l, can be estimated by using these instruments. Since AAS is much less expensive compared to the ICP based instruments, the former is most widely for its reasonable accuracy. This method is based on the conversion of As(III) to AsH_3 which is subsequently determined by flame or flameless AAS. Therefore, all As(V) in the sample must be reduced to As(III) by adding KI and HCl before the generation of hydride. The quantities of KI and HCl have to be optimized depending upon the As(V) content in the sample. Generally, few drops of 3N HCl followed by 1ml of 10% KI solution and a waiting time of 1-2 h are sufficient for this purpose. Therefore, reduction step for all the samples should be completed before the start of AAS measurement. The entire sample containing only As(III) is then transferred to the special hydride generator supplied as an accessory along with the main instrument. This is followed by the addition of 1-2ml of 5N HCl and 5ml of freshly prepared 0.5% $NaBH_4$ in about 1% NaOH. Volatile AsH_3 is sucked into the heated zone of the flame or furnace and measured at 193.7nm. A calibration curve is prepared by using suitable aliquots of standard arsenite solution having arsenic concentration in the range 2-30μg/l.

Arsenic speciation can be determined by the cartridge method. A suitable aliquot containing 2-30μg/l total As is allowed to flow through a strong cation exchanger (SCX) and an anion exchanger (SAX) which are already preconditioned with ethanol in deionized or distilled water (1:1 v/v). The cation and anion exchangers retain the organically complexed arsenic and arsenate(V) leaving the arsenite (III) in the solution [849,850]. The arsenate retained by the anion exchange cartridge is eluated with 1M HCl to separate it from organically complexed As. As(III) and As(V) are deter-

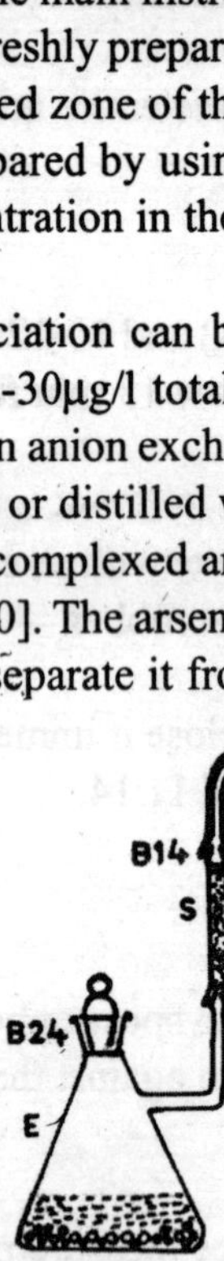

Fig.11.13

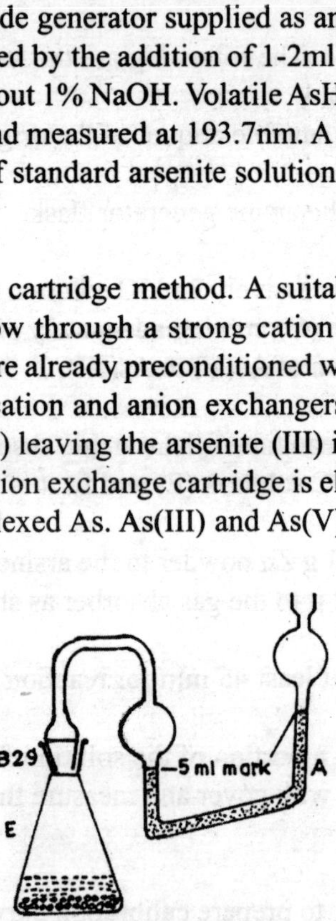

Fig. 11.14

Fig.11.13 A typical glass apparatus with separate arsine absorber used for the determination of total arsenic.

Fig.11.14 Apparatus with attached fixed volume piopett for the absorption of arsine gas (adapted after MERCK)
E = Erlenmayer flask for generation of arsine
S = Scrubber for absorbing hydrogen sulphide
C = Capillary tube
A = Arsine absorber

mined separately by HGAAS method as described above. Subtracting the combined values of As(III) and As(V) from the total As one gets organically complexated As.

Roberts et al [740] have reported that when disodium citrate buffer of pH 4.8 is used instead of HCl for generation of arsine, As(III) can be selectively detected and estimated in an Atomic Fluorescence Spectrometer(AFS), because at pH around 5.0 only As(III) is converted to A_sH_3 [851]. Roberts et al [740] also measured total As after reducing all the As to As(III) using ascorbic acid / KI as reducing agent which is prepared by dissolving 2.5g KI and 0.5g ascorbic acid in 5ml water. The detection limit is 0.1μg/l.

Different species of As when present together in a sample of water have been separated by Ion Chromatographic (IC) technique (e.g., Dionex) using IonPac AG7 and IonPac AS7 as guard and separator column respectively [852]. The mobile phase was nitric acid with ion pairing modifier benzene 1,2-disulphonic acid as proposed by Londesborough et al [853]. The separated species can be detected using ICP-MS technique. The chromatograms can be quantitatively evaluated by integrating the peaks corresponding to As(III), As(V), dimethyl arsenic acid (DMAA) and phenyl arsonic acid (PAA) with the help of peak fitting software module MICROCAL Origin 4.2.

11.3.7.3 Selenium (Selenite and selenate)

Selenium can also be determined by hydride generation atomic absorption spectrophotometry (HGAAS) in he same way as in the case of arsenic. Thus, selenium is determined by converting Se(IV) to volatile SeH_2 which is measured at 195nm using either flame or flameless AAS. Therefore Se(VI), if present, has to be reduced to Se(IV) before hydride generation. It can also be determined by ICP-OES or ICP-AES. Therefore, instrumental methods provide only the total Se content in the sample.

Se(IV) forms highly colored and strongly fluorescent complexes (piazselenol) at pH 1.5-1.8 with different diamino compounds such as 2,3-diaminonaphthalene (DAN), 3,3'-diaminobenzidene hydrochloride (DABH) etc. Organosulfur compound like 2-mercaptoethanol also gives similar colored complex at pH 1.5. All these colored complexes can be extracted or separated by shaking with either cyclohexane or toluene and measured directly in a spectrophotometer or in a fluorimeter. However, mercaptoethanol can also act as a reducing agent for Se(VI). In all cases the pH value should be less than 2.0, because at pH above 2.0 the rate of formation of piazselenol is critically dependent on pH. The minimum detectable concentration limit is 10mg/l. Except the presence of any oxidizing agent, no other inorganic compounds interfere in the formation of the complex. Colored organic compounds which are also extractable by cyclohexane or toluene interfere and therefore such compounds are destroyed either by boiling with few drops of 0.1N $KMnO_4$ and 2-3 ml of 3% $CaCl_2.2H_2O$ solution or by heating with a mixture of nitric and sulfuric acids as is followed in the case of arsenic. Such pretreatment leads to the oxidation of Se(IV) to Se(VI) which again has to be reduced by heating with 4-5 ml conc. HCl over steam bath.

Reagents

1. 2,3-diamino naphthalene (DAN) solution: Place 200 ml of 0.1N HCl in a 250 ml beaker and dissolve 0.2g DAN adding little solid at a time under constant stirring. The solution may be purified by shaking once or twice with small quantity (~ 10ml) of cyclohexane, discarding the organic phase and preserving the aqueous phase in a dark amber colored bottle.
   However, it is recommended that 3,3'- diaminobenzidine hydrochloride (DABH) solution should be prepared on daily basis. Dissolve 0.1g DABH in 20 ml water with intermittent stirring and if necessary, warming over steam bath until a clear solution is obtained.

2. EDTA solution: Dissolve 3.723g disodium salt (AR/GR grade) in 50-60 ml water and transfer to 100 ml volumetric flask, add 3 ml NH_4OH and dilute to the mark with distilled water to obtain 0.1M solution.

3. Hydroxylamine hydrochloride solution: Dissolve 10g salt (AR/GR grade) in 100 ml distilled water.

4. Hydrochloric acid: Concentrated (32% , 11N) or 0.1N.

5. Standard Se(IV) solution: Dissolve 2.190g sodium selenite (Na_2SeO_3) in 100 ml water containing 10 ml conc. HCl and dilute to 1 litre. Dilute 5 ml of this solution to 500 ml. 1 ml of diluted solution ≡ 10μg Se.

Selective estimation of Se(IV)

If the sample is free from any organic matter, then proceed directly to determine selectively Se(IV) as follows.

1. Transfer the sample containing 20-100 mg Se to 150 ml beaker. Adjust the pH value to 1.5 ± 0.3 with 0.1N HCl using a pH meter. If the sample volume is large use 1:3 (v/v) HCl and NH_4OH.

2. Add 5 ml DAN solution or 2 ml DABH solution, heat on a steam bath for

20-30 min.

3. Cool and adjust pH to 8.0 with NH_4OH. Transfer the sample to 50 ml volumetric flask and make up the volume with washings fropm the beaker. Shake well and transfer the entire solution to a 250 ml separating funnel. Add 10 ml cyclohexane or toluene and shake for 30 sec.

4. Remove the organic extract containing piazselenol to a 20 ml centrifuge tube. Repeat extraction, if necessary, with 5ml more of organic solvent. Centrifuge the entire organic phase to separate fine droplets of water present in the organic phase. Alternatively, filter through dry filter paper containing 100-150 mg anhydrous Na_2SO_4.

5. Measure absorbance of the complex against a blank at 420-480 nm depending upon the absorbance maximum, scanned separately. Use cells of 1 cm path length.

6. Construct a calibration curve using standard Se(IV) solutions of suitable Se contents and follow the above procedure for each sample.

<u>Total Selenium content</u>

To determine total selenium content transfer a suitable volume of water sample to a 150 ml beaker. After destroying and oxidizing any organic matter either by $KMnO_4$ treatment or by SO_3 fuming with mixed acids, cool and dilute with 20 ml water. Add 5ml of conc. HCl and 5ml of 25% NH_4Cl to the solution, , cover the beaker and warm the solution for 20-30 min over steam bath. Allow the solution to cool and add 5ml of 0.1M EDTA and 2ml of 10% hydroxylamine hydrochloride solutions. Then follow the procedure given above for Se(IV) estimation. Subtracting Se(IV) content from total Se one can get Se(VI) content in the sample.

11.3.7.4 Chromate [Cr(VI)]

Both Cr(III) and Cr(VI) can be determined by AAS method at 358nm using air - acetylene flame. Chromium(VI) can also be estimated with the help of ICP-AES or ICP-OES. Wherever such facilities are not available, spectrophotometric method can be used to Cr(VI) at concentration level as low as 3 μg/l. The most widely used method is based on the measurement of the intensity of reddish violet colored complex formed by the reaction between Cr(VI) and diphenyl semicarbazide. The elements which interfere in the above reaction are Mo, V, Fe(III) and Cu. The influence of these elements may be eliminated by treating a suitable quantity (10-15 ml) of sample containing 50-100 μg/l of Cr(VI) with cupferron and subsequently separating the cupferron complex with chloroform. The steps are as follows:

1. Add 5 ml of 6N H_2SO_4 to the sample. Dilute with another10-15 ml of distilled water, cool the sample in an ice-cold bath.

2. Transfer the ice-cold sample to a 125 ml separating funnel. Add 5 ml ice-cold cupferron solution (5% w/v), mix by gentle shaking and add 5 ml chloroform. Shake the mixture for 1 min and separate the organic phase. Depending upon the contents of the interfering elements, more than one such extraction steps may be necessary using each time 5 ml portion of cold cupferron solution and 5 ml chloroform.

3. The aqueous portion is transferred to a 200 ml conical flask, covering its mouth with a funnel. Boil for 5 min to expel all chloroform and cool. Add 5 ml conc. HNO_3 and 2ml conc. H_2SO_4, heat on a sand bath to SO_3 fumes. When nearly dry, remove the flask, cool, add 5ml HNO_3, repeat fuming until dry, cool.

4. Wash the sides of the funnel as well as the flask with 25-30 ml distilled water. Neutralize excess acidity with 1:3 NH_4OH using methyl orange as an indicator.

5. Add 2-3 ml of 18N sulfuric acid and 5-6 drops of phosphoric acid. Add 4% $KMnO_4$ solution drop wise until pink color persists. Boil the solution for 2 min.

6. Add 0.5% sodium azide solution drop wise until color fades completely, cool, filter if necessary.

7. Transfer to a 50 ml volumetric flask, add 2 ml diphenyl semicarbazide solution (0.5% in acetone) and mix thoroughly. Make up the volume with water, wait for 10-15 min. Measure absorbance at 540 nm against reagent blank.

8. Prepare a calibration curve using standard $K_2Cr_2O_7$ solution. Transfer 0.283 g dry $K_2Cr_2O_7$ in 1 litre volumetric flask, dissolve in about 200 ml of water and finally make up the volume with distilled water. Dilute 5 ml of this stock solution to 250 ml.1 ml of the diluted solution ≡ 2 μg of Cr(VI).

If the interfering elements are present in quantities less than ten times that of Cr(VI), then proceed from step 5.

11.3.7.5 Fluoride

Quantitative analysis of fluoride has been the subject of interest to many investigators because of the health hazards posed by its presence in the ground waters of many countries. As a result, a number of methods have been developed of which colorimetric and ion selective methods are the most widely used ones. Colorimetric / spectrophotometric methods are based on the reaction between fluoride and Zr-dye lake. Fluoride (F^-) ion reacts with dye lake, dissociating a portion of it into a colorless complex anion (ZrF_6^{2-}) and the dye. As the amount of F^- is increased in the system the color produced becomes progressively lighter depending on the reagent used. The interfering ions in this method are Fe^{3+}, Cl^-, PO_4^{3-}, hexametaphosphate ($[NaPO_3]_6$), SO_4^{2-} etc. When one of them is present in substantial quantity (>0.1-0.2 μg/l) or whenever there exists doubt about their total quantity, the sample should be distilled as follows.

Both direct and steam distillation can be used. For direct distillation, add 200 ml of conc. H_2SO_4 slowly with stirring to 400 ml ice cold distilled water. Transfer the acid solution into a round bottomed distillation flask, heat slowly until the temperature reaches 180°C. Discard the initial distillate, Cool to 120°C and add 300 ml of sample water slowly into the distillation flask followed by several glass beads. Add little Ag_2SO_4 (approx. at the rate o 5 mg/mg of Cl^- when the sample contains high chloride). Distill at temperature not exceeding 180°C. When the volume of water reduces to half of its original volume add more distilled water slowly into the distillation flask. Continue distillation at 180°C until sufficient i.e., not less than 300 ml of distillate is collected.

Steam distillation is carried out essentially in the same manner as direct distillation, except that steam from a superheated steam generator is allowed to enter into the distillation flask through a narrow tube reaching up to about ½ inch from the bottom of the flask. Carry out distillation at 130-135°C until the quantity of the distillate is no less than the volume of sample added in the flask.

11.3.7.5.1 Spectrophotometric method using SPADNS

<u>Reagents</u>

1. Standard fluoride solution: Dissolve 0.2210g NaF (AR/GR grade) in distilled water and dilute to 1000 ml. 1 ml of this solution ≡ 100 mg of F^-.
2. SPADNS solution: Dissolve 0.9580g SPADNS (extra pure grade) in distilled water and dilute to 500 ml. The solution is stable indefinitely if protected from sunlight.

3. Zr-acid solution: Dissolve 0.1330g zirconium oxychloride ($ZrOCl_2 . 2H_2O$) in about 25 ml distilled water. Add 350 ml conc. HCl, dilute to 500 ml with water in a volumetric flask. Equal volumes of SPADNS solution and Zr-acid solution may be mixed together to give a single (or mixed) reagent which is stable for at least 6 months.
4. Sodium arsenite solution: Dissolve 2.5g AR/GR grade $NaAsO_2$ in water and dilute to 1000 ml.
5. Reference solution: Add 10 ml of mixed reagent to 100 ml water. Dilute separately 7 ml of conc. HCl to 10 ml and add to the diluted mixed reagent solution. The resulting solution is used as reference for spectrophotometric measurement.

Procedure

Calibration curve : Prepare a series of solutions containing 0.1 - 1.4 mg F^-/l by diluting appropriate quantities of standard fluoride solution to 50 ml with distilled water transfer 10 ml of mixed reagent to each solution and mix well. Measure the absorbance with respect to the reference solution as zero absorbance.

Fluoride in test sample : Transfer a known quantity of raw or distilled water sample in a beaker and dilute to 50 ml. Add 10 ml of mixed reagent mix and read absorbance after 15 min. If the sample contains residual chlorine, remove it by adding one drop of arsenite solution for each 0.1mg Cl^-. Accuracy of the method is ± 0.05mg F^- / l.

11.3.7.5.2 Ion-Selective system

Now-a-days highly sensitive and precision ion-selective electrodes are available in the market. These electrodes sense the EMF of the relevant ion with respect to a reference electrode (usually calomel). Since the EMF is directly related to the activity or concentration of the ion, the instrument can be calibrated using the appropriate range of its known concentrations. However, the electrode does not respond if the relevant ion (i.e., F^-) is in the complexated state with any cation, e.g., Al^{3+}. To overcome this problem a buffer solution of suitable pH value together with a strong complexing agent dissolved in an aqueous solution containing high concentration of indifferent electrolyte is used. This is called Total Ionic Strength Adjustment Buffer (TISAB). For the determination of fluoride TISAB is prepared as follows:

Place approximately 500 ml distilled water in an 1 liter beaker, add 57 ml glacial acetic acid, 58g NaCl and 4g 1,2-cyclohexylene diaminetetraacetic acid (CDTA) under constant stirring. Place the beaker in a cold water bath containing several ice cubes. Add slowly 125-150 ml of 6N NaOH until the pH is between 5.0 and 5.5. Transfer to one liter volumetric flask and make up the volume with water. All standard solutions as well as test samples must be diluted with TISAB in 1:1 proportion. Use plastic beakers in all measurements. For procedure and other information on the care of electrodes, consult the manual / instructions provided by the manufacturer.

BIBLIOGRAPHY

1. Gouy, G., J. *Phys.,* **9** (1910), 457
2. Chapman, D. L., *Phil. Mag.,* **25** (1913), 475
3. Stern, O., Z. *Electrochem.,* **30** (1924), 508
4. Grahame, D. C., *Chem. Rev.,* **44** (1947), 441
5. Davis, J. A. and Hayes, K. F. (eds), *Geochemical Processes at Mineral Surfaces,* ACS Symp series 323, Amer. Chem. Soc., Washington D C, 1986.
6. Garrels, R. M. and Christ, C. L., *Solutions, Minerals and Equilibria,* Harper and Row, New York, 1965.
7. Hochella, M. F. (Jr) and White, A. F. (eds), *Mineral - Water Interface Geochemistry* Reviews in Mineralogy, Vol. 23, Mineralogical Society of America, Washington D C,1990.
8. Riley, J. P. and Skirrow, G. (eds), *Chemical Oceanography,* Vol.1, 2nd edn., Academic Press, New York, 1975.
9. Riley, J. P. and Chester, R., *Introduction to Marine Chemistry,* Academic Press, London, 1971.
10. Crerar, D. L. and Barnes, H. L., *Geochim. Cosmochim. Acta,* **38**(1974), 279
11. Onyrange, M. S. and Matsuda, H., *Chapter 1: Fluoride Removal from Water using Adsorption Technique. Advances in Fluorine Chemistry,* **2**(2006), pp 1-48.
12. Meenakshi, and Maheswari, R. C., *J. Hazard. Mater.,* **137**(2006), 456.
13. Mohapatra, D.; Misra, D.; Mishra, S. P.; Roychowdhury, G. and Das, R. P., *J. Colloid Interface Sci.,* **275**(2004), 355.
14. Das, N.N.; Pattanaik, P. and Das, R., *J . Colloid Interface Sci.,* **292**(2005), 1
15. Mohapatra, M.; Anand, S.; Mishra, B. K.; Giles, D. E. and Singh, P., *J. Environ. Management,* **91**(2009), 67.
16. Holm, T. R.; Anderson, M. A .; Iverson, D. G. and Stanforth, R. S. in *Chemical Modelling of Aqueous systems : Speciation, Sorption, Solubility and Kinetics* (E. A. Jenne, ed.) ACS Symp. Ser. 93, Am. Chem. Soc., Washington DC, 1979, pp 711 - 736.
17. Mohan, D. and Pittman (Jr), C. U., *J. Hazard. Mater.,* **142**(2007), 1-53.
18. Pierce, M. L. and Moore, C. B., *Environ. Sci. Technol.,* **14**(1980), 214.
19. Pierce, M. L. and Moore, C. B., *Water Res.,* **16**(1982), 1247.

20. Singh, D.B.; Prasad, G.; Rupainwar, D.C. and Singh, V. N., *Water, Air, Soil Pollution,* **42**(1988), 373.
21. Jain, A.; Raven, K. P. and Loeppert, R. H., *Environ. Sci. Technol.,* **33**(1999),1179.
22. Fuhrman, H.G-; Tjell, J. C. and McConchie, D., *J. Colloid Interface Sci.,* **271**(2004), 313.
23. Dutta, P. K.; Roy, A. K.; Sharma, V. K. and Millero, F. J., *J. Colloid Interface Sci.,* **278**(2004), 270.
24. Lenobo, V.; Chabroullet, C.; Shukry, R.; Serpand, B.; Deluchat, V. and Bollinger, J. C., *J. Colloid Interface Sci.,* **280**(2004), 62.
25. Gupta, V. K.; Saini, V. K. and Jain, N., *J. Colloid Interface Sci.,* **288**(2005), 55
26. Vithange, M.; Chandrajit, R.; Bhandara, A. and Weerasooriya, R., *J. Colloid Interface Sci.,* **294**(2006), 265
27. Dzombak, D. and Morel, F. M. M., *Surface Complexation Modeling: Hydrous Ferric Oxide,* John Wiley & Sons, New York, 1990.
28. Mesuere, K. and Fish, W., *Environ. Sci. Technol.,* **26**(1992),2357, 2365.
29. Fendorf, S. E., Eick, M. J., Grossil, P. and Sparks, D. L., *Environ. Sci. Technol.,* **31**(1997), 315.
30. Balistrieri, L. S. and Chao, T. T., *Soil Sci. Soc. Am. J.,* **51**(1987), 1145.
31. Manceau, A. and Charlet, L., *J. Colloid Interface Sci.,* **168**(1994), 87.
32. Hayes, K.F.; Papelis, C.and Leckie, J.O., *J.Colloid Interface Sci.,***125**(1988), 717.
33. Cowan, C. E.; Zachara, J. M . and Resch, C. T., *Environ. Sci. Technol.,* **25**(1991), 437
34. Hoins, U.; Charlet, L. and Sticher, H., *Water, Air, Soil Pollution,* **68**(1993), 241.
35. Peterson, W.; Walkman, K.; Schroer, S. and Schroeder, F., *Anal. Chim. Acta,* **273**(1993), 323.
36. Gunneriusson, L., *J. Colloid Interface Sci.,* **163**(1994), 484.
37. Venema, P.; Hiemstra, T. and van Riemsdijk, W. H., *J. Colloid Interface Sci.,* **83**(1996), 515.
38. Randall, S. R.; Sherman, D. M.; Rognarsdottier, K. V. and Collins, V., *Geochim. Cosmochim. Acta,* **63**(1999), 2971.
39. Hohl, H. and Stumm, W., *J. Colloid Interface Sci.,* **55**(1976), 281.
40. Hayes, K. F. and Leckie, J. O. in *Geochemical Processes at Mineral Surfaces*(J. A. Davis and K. F. Hayes, eds) ACS Symp. Series 323, Am. Chem. Soc., Washington D C, 1986, pp 114 - 41.

41. Swallow, K. C.; Hume, D. N. and Morel, F. M. M., *Environ. Sci. Technol.*, **14**(1980), 1326.

42. Gupta, V. K. and Ali. I., *J. Colloid Interface Sci.* **271**(2004), 32.

43. Weng, C. -H., *J. Colloid Interface Sci.*, **272**(2004), 262.

44. Feng, -Q.; Lin, -Q; Gong, -F.; Sugita, S. and Shoya, M., *J. Colloid Interface Sci.*, **278**(2004), 1

45. Tokimoto, T.; Kawasaki, N. and Nakamura, T., *J. Colloid Interface Sci.*, **281**(2005), 56.

46. Forbes, E. A.; Posner, A. M. and Quirk, J. P., *J. Colloid Interface Sci.*, **49**(1974), 403.

47. Barrow, N. and Cox, V. C., *J. Soil Sci.*, **43**(1992), 437.

48. Gunneriusson, L. and Sjoeberg, S., *J. Colloid Interface Sci.*, **156**(1993), 121.

49. Mishra, S. P.; Dubey, S. S. and Tiwary, D., *J. Colloid Interface Sci.*, **279**(2004), 61.

50. Vogel, A.I. *A Textbook of Quantitative Inorganic Analysis*, 3rd Edn. ELBS and Longmans Green & Co Ltd. 1961, pp 460 - 480.

51. Fuerstenau, D. W. (ed), F*roth Flotation*, AIME, New York, 1962.

52. Fuerstenau, D. W. and Osseo-Assare, K. *J. Colloid Interface Sci.*, **118**(1987), 524.

53. Habashi, F., *Principles of Extractive Metallurgy*, Vol.2, Gordon Breach, New York, 1968.

54. Giovanoli, R. and St‹hli, E., *Chimia*, **24**(1970), 49.

55. Giovanoli, R.; St‹hli, E. and Feitknecht, W., *Helv. Chim. Acta*, **53**(1970), 209, 453.

56. Giovanoli, R.; Oswald, H. R. and Feitknecht, W., *J. Microscopie*, **4**(1965), 711.

57. -Idem- , *Helv. Chim. Acta*, **49**(1966), 1971.

58. Giovanoli, R.; Feitknecht, W. and Fahrer, W., *J. Microscopie*, **7**(1968), 177.

59. Cornell, R. M. and Schwertmann,U., *The Iron Oxides: Structures, Properties, Reactions, Occurrences and Uses*, 2nd edn., Wiley-VCH GmbH & Co., KgaA, Weinheim, 2000.

60. James, R. O. and Parks, G. A., in *Surface and Colloid Science* (E. Matijevic, ed.), Vol.12, Plenum Press, New York, 1972, pp 119 - 216.

61. Lyklema, J. *J. Electroanal. Chem.*, **18**(1968), 341.

62. Perram, J. W.; Hunter, R. J.and Wright, H. J. L., *Aust.J.Chem.*, **27**(1974), 461.

63. Perram, J. W., *J. Chem. Soc. Farad. Trans. II,* **69**(1973), 993.
64. Brunauer, S.; Emmett, P. H. and Teller, E., *J. Am. Chem. Soc.,* **60**(1938), 309.
65. Gregg, S. J. and Sing, K. S. W., *Adsorption, Surface Area and Porosity,* Academic Press, London, 1982.
66. Thomas, J. M. and Thomas, W. J., *Principles of Heterogeneous Catalysis,* Academic Press, London & New York, 1967.
67. Orr, C. and Dallavale, J. M., *Fine Particle Measurement,* McMillan & Co, London, 1965
68. Adamson, A. W., *Physical Chemistry of Surfaces,* 3 rd edn. Wiley -Interscience, New York, 1976.
69. Greenland, D. J. and Mott, C. J. B. *in The Chemistry of Soil Constituents,* (Greenland, D. J. and Hayes, M. H. B. Eds.), chapter 4, John Wiley & sons, Chichester, 1978.
70. Kanungo, S. B. and De, P. K., *Indian J. Technol.,* **8**(1970), 305.
71. van den Hul, H. J. and Lyklema, J., *J. Colloid Interface Sci.,* **23**(1967), 500; *J. Am. Chem. Soc.,* **90**(1968), 3010.
72. Kanungo, S .B.; Tripathy, S S. and Rajeev, *J. Colloid Interface Sci.,* **269**(2004), 1.
73. Siever, R. S., *Geochim. Cosmochim. Acta,* **37**(1973), 1851.
74. Bolt, G. H. and van Riemsdijk, *W. H.* in *Aqueous Surface Chemistry* (W. Stumm, ed.), John Wiley & Sons, New York, 1987, pp127-163.
75. Yates, D. E., *The Structure of Oxide - Aqueous Electrolyte Interface,* Ph. D. Thesis, University of Melbourne, Australia, 1975.
76. Sille'n, L. G. and Martell, A. E. *Stability Constants of Metal ion Complexes,* Special Publication 17 & 25 and subsequent supplements, Chemical Society, London, 1968.
77. Seidell, A. and W. F. Linke, *Solubilities of Inorganic and Metal-Organic Compounds,* D. Van Nostrand, New York, 1958.
78. Smith, R. M. and Martell, A. E., *Critical Stability constants,* Vol. 4 Inorganic Complexes, Plenum Press, New York, 1976.
79. Baes, C. F. (Jr) and Mesmer, R. E., *The Hydrolysis of Cations,* Wiley-Interscience, New York, 1976.
80. Boehm, P., *Discuss. Farad. Soc.,* **52**(1971), 264.
81. Barron, V. and Torrent, J., *J. Colloid Interface Sci.,* **177**(1996), 407.
82. Roy Morrison, S., *Electrochemistry at Semicondutor and Oxidized Metal Electrodes,* Plenum Press, New York.
83. Farmer, V. C. (ed.), *Infrared Spectra of Minerals,* Mineralogical Society, London, 1974.

84. Russel, J. D., *Clay Minerals,* **14** (1979), 109.
85. Parfitt, G. D. and Rochester, C. H. in *Characterization of Powder Surfaces* (G. D. Parfitt and K.S.W. Sing, eds), Academic Press, San Francisco, 1976.
86. Hair, M. L., *Infrared Spectroscopy in Surface Chemistry,* Marcel Dekkar, New York, 1967
87. Anderson, J. H. and Parks, *G. A., J. Phys. Chem.,* **72**(1968), 3662.
88. Kanungo, S. B., *J. Colloid Interface Sci.,* **162**(1994), 86.
89. Tripathy, S.S.; Kanungo, S. B. and Mishra, S. K., *J. Colloid Interface Sci.,* **241**(2001), 112.
90. Kanungo, S. B. and Mohapatra, D. M., *Colloids & Surfaces,* **42**(1989), 173.
91. Tanabe, K., *Solid Acids and Bases,* Academic Press, New York & London, 1970.
92. Frenkel, M., *Clays Clay Minerals,* **22**(1974), 435.
93. Soloman, D. H. and Murray, H. H., *Clays Clay Minerals,* **20**(1972), 135.
94. Karickhoff, S. W. and Baily, G. H., *Clays Clay Minerals,* **24**(1976), 170.
95. Tamura, H.; Tanaka, A.; Mita, K.Y. and Furuichi, R., *J. Colloid Interface Sci.,* **209**(1999), 225.
96. Badlani, M. and Wachs, I. E., *Catalysis Lett.,* **75**(2001), 137.
97. Laragne, J. J. and Brenet, J. P., *Bull. Soc. Chim. (France),* **9**(1968), 3499.
98. Kanungo, S. B. and Parida, K. M., *J. Colloid Interface Sci.,* **98**(1984), 252.
99. Hiemstra, T. and van Riemsdijk, W. H., *J. Colloid Interface Sci.,* **225**(2000), 94.
100. Torrent, J.; Barron, V. and Schwertmann, U., *Soil Sci. Soc. Am. J.,* **54**(1990), 1007.
101. Onada, G. Y. and DeBruyn, P. L., *Surface Sci.,* **4**(1966), 48.
102. Berube, Y. G.; Onada, G. Y. and DeBruyn, P. L., *Surface Sci.,* **7**(1967), 448.
103. Yates, D. E.; Greiser, F.; Cooper, R. and Healy, T. W., *Aust. J. Chem.,* **30**(1977), 1655.
104. Yates, D. E.; James, R. O. and Healy, T. W., *J. Chem. Soc. Farad. Trans. I,* **76**(1980), 1.
105. Jones, P. and Hockey, J. A., *Trans. Farad. Soc.,* **67**(1971), 2679.
106. Russel, J. D.; Parfitt, R. L.; Fraser, A. R. and Farmer, *V. C., Nature,* **248**(1974), 220.
107. Lewis, D. G. and Farmer, V. C., *Clay Minerals,* 21(1986), 93.
108. Hiemstra, T.; DeWit, J. C. M. and van Riemsdijk, W. H., *J. Colloid Interface Sci.,* **133**(1989), 105.
109. Hiemstra, T. and van Riemsdijk, W. H., *J. Colloid Interface Sci.,* **179**(1996), 488.
110. Hiemstra, T. and van Riemsdijk, W. H., *Langmuir,* **15**(1999), 8045.
111. Schindler, P. W. and Stumm, W. in *Aquatic Surface Chemistry* (W. Stumm, ed.), John Wiley & Sons, New York, 1987, pp 83 - 108.

112. Schindler, P. W. and Gamsjaeger, H., *Kolloid Z. u Z. Polymere,* **250**(1970), 759.

113. Stumm, W.; Hohl, H. and Dalang, F., *Croat. Chem. Acta,* **48**(1976), 491.

114. vanRiemsdijk, W. H.; Bolt, G. H.; Koopal, L. K. and Blackmeer, J. *J. Colloid Interface Sci.,* **109**(1986), 219.

115. vanRiemsdijk, W. H.; DeWit, J. C. M.; Koopal, L. K. and Bolt, G. H., *J. Colloid Interface Sci.,* **116**(1987), 511.

116. Bolt, G. H. and van Riemsdijk, W. H. in *Soil Science B: Physico-Chemical Models,* 2nd Edn.(G. H. Bolt, Ed.), Elsevier, Amsterdam, 1982, p 459.

117. Koopal, L. K., *Electrochim. Acta,* **41**(1996), 2293.

118. Hiemstra, T. and van Riemsdijk, W. H., *Colloids & Surfaces,* **59**(1991), 7.

119. Hiemstra, T. and van Riemsdijk, W. H. and Bolt, G H., *J. Colloid Interface Sci.,* **133**(1989), 91.

120. Pauling, L., *The Nature of Chemical Bond,* 3rd Edn. Chapter 13, Cornell Univ. Press, Ithaca, New York, 1967.

121. Bleam, W. F., *J. Colloid Interface Sci.,* **159**(1993), 312.

122. Berube, Y. G. and DeBruyn, P. L., *J. Colloid Interface Sci.,* **27**(1968), 305.

123. Lyklema, J. and Overbeek, J. Th., *J. Colloid Interface Sci.,* **16**(1961), 501, 509.

124. Lyklema, J. and Overbeek, J. Th., *Trans. Farad. Soc.,* **59**(1963), 418.

125. Stumm, W.; Huang, C. P. and Jenkins, S. R., *Croat. Chem. Acta,* **42**(1970), 223.

126. Bockris, J. O'M. and Reddy, A. M., *Modern Electrochemistry,* vol.1, Plenum Press, New York, 1970.

127. Antropov, L. I., *Theoretical Electrochemistry,* Engl. Transln. by A. Beknazarov, Mir Publishers, Moscow, 1977, pp 264 - 272.

128. Levine. S and Smith, A. L., *Discussion Farad. Soc.,* **52**(1971), 290.

129. Parks, G. A. and DeBruyn, P. L., *J. Phys. Chem.* **66**(1962), 967.

130. Atkinson, R. J.; Posner, A. M. and Quirk, J. P., *J. Phys. Chem.* **71**(1967), 550.

131. Yates, D. E. and Healy, T. W., *J. Chem. Soc. Farad. Trans. 1,* **76**(1980), 9.

132. Huang, C. P. in *Adsorption of Inorganics at Solid - Liquid Interface* (Anderson, M. A. and Rubin, A. J. eds.) Ann Arbor Science, Ann Arbor, Michigan, 1981, pp 183-217.

133. Ref. 27, pp 49 - 53.

134. Kanungo, S. B. and Mahapatra, D. M., *J. Colloid Interface Sci.,* **131**(1989), 103.

135. Mular, A. L. and Roberts, R. B., *Trans. Canad. Inst. Mining Metall.,* **69**(1966), 438.

136. Balistrieri, L. S. and Murray, J. W., *Am. J. Sci.,* **281**(1981), 788.

137. Tripathy, S. S. and Kanungo, S. B., *J. Colloid Interface Sci.,* **284**(2005), 30.

138. Pyman, M. A.; Bowden, J. W. and Posner, A. M., *Aust. J. Soil Res.,* **17**(1979), 191.

139. Parkar, J. C.; Zelazny, W, Sampath, S. and Harris, W. G., *Soil Sci. Soc. Am. J.,* **43**(1979), 668.

140. Sposito, G., *The Surface Chemistry of Soils,* Oxford University Press, New York, 1984.

141. Parks, G. A., *Chem. Rev.,* **65**(1965), 177.

142. Kosmulski, M., *J. Colloid Interface Sci.,* **253**(2002), 77; **275**(2004), 214; **298**(2006), 730.

143. Smith, A. L. in *Dispersion of Powders in Liquids* (Parfitt, G. D. ed.) Elsevier, Amsterdam, 1969.

144. Overbeek, J.Th.G., in *Colloid Science,* Vol. 1 (Kruyt, H. R. ed.) Elsevier, Amsterdam, 1952.

145. Hayden, D. A. in *Recent Progress in Surface Sci.,* Vol 1 (Danielli, J. F., Pankhurst, K. G. A. and Riddiford, A. C. eds.) Academic Press, New York, 1964, chap.3.

146. Westall, J. and Hohl, H., *Adv. Colloid Interface Sci.,* **12**(1980), 265.

147. Wise, G. R.; James, R.O.; Yates, D. E. and Healy, T. W. in *International Review of Science, Physical Chemistry Series 2, Vol.6 - Electrochemistry* (Bockris, J. O'M ed.), Butterworths, London, 1976, pp 54-102.

148. Yates; D. E.; Levine, S. and Healy, T. W., *J. Chem. Soc. Farad. Trans. 1,* **70**(1974), 1807.

149. Breeuwsma, A. and Lyklema, J., *Discuss. Farad. Soc.* 52(1971) 324; *J. Colloid Interface Sci.* **43**(1973), 437.

150. Davis; J. A.; James, J. O. and Leckie, *J. O., J. Colloid Interface Sci.,* **63**(1978), 480.

151. Davis, J. A. and Leckie, J. O., *J. Colloid Interface Sci.,* **67**(1978), 90.

152. Regazzoni, A. E.; Blesa, M. A. and Maroto, A. J. G., *J. Colloid Interface Sci.,* **91**(1983), 560.

153. Milonjic, S. K.; Ilic, Z. E. and Kopeconi, M. M., *Colloids Surfaces,* **6**(1983), 167.

154. Houchin, M. R. and Warren, L. J., *J. Colloid Interface Sci.,* **100**(1983), 278.

155. Janssen, M. J. G. and Stein, H. N., *J. Colloid Interface Sci.,* **111**(1986), 112.

156. Koopal, L. K.; van Reinsdijk, and Roffey, M. G., *J. Colloid Interface Sci.,* **118**(1987), 117.

157. Healy, T. W. and White, L. R., Adv. *Colloid Interface Sci.,* **9**(1978), 303.

158. Sposito, G., *J. Colloid Interface Sci.,* **91**(1983), 329.

159. Johnson, R. E. (Jr), *J. Colloid Interface Sci.,* **100**(1984), 540.

160. Hayes, K. F. and Leckie, J. O., *J. Colloid Interface Sci.,* **115**(1987), 564.

161. Chan, D.; Perram, J. W.; White, L. R. and Healy, T. W., *J. Chem. Soc.Farad. Trans.* 1, **71**(1975), 1046.

162. Smit, W., *J. Colloid Interface Sci.,* **109**(1986), 295.

163. Smit, W., *J. Colloid Interface Sci.,* **113**(1986), 288.

164. Kanungo, S. B., Unpublished results

165. Esin, O. A. and Shikov, V. M., *Zh. Fiz. Khim.,* **17**(1943), 236

166. Esin, O. A. and Markov, B. F., *Acta Physicochim. USSR,* **10**(1939), 353

167. Levine, S.; Bell., G. M. and Calvert, D., *Canad. J. Chem.,* **40**(1962), 518

168. Levine, S.; Mingins, J. and Bell, G. M., *J. Electroanal. Chem.,* **13**(1967), 280

169. Drost-Hansen, W., *Ind. Engg. Chem.,* **51**(1969), 10; *J. Colloid Interface Sci.,* **58**(1977), 251.

170. Gast, P. G.; Landa, E. R. and Meyer, G. W., *Clay Clay Mineral,* **22**(1974), 31.

171. Drost - Hansen, W. and Hunter, R. J. *J. Colloid Interface Sci.,* **22**(1974), 23.

172. Schindler, P. W. and Kamber, H. R., *Helv. Chim. Acta,* **51**(1972), 1781.

173. Hingston, F. J.; Posner, A. M. and Quirk, J. P., *Adv. Chem. Ser.* **79**(1968), 82.

174. Blesa, M. A., Nestor, M. F.; Maroto, A. J. G. and Regazzoni, A., *J. Colloid Interface Sci.* **101**(1984), 410.

175. James, J. O.; Davis, J. A. and Leckie, J. O., *J. Colloid Interface Sci.* **65**(1978), 331.

176. Sprycha, R. and Szczypa, J. *J. Colloid Interface Sci.,* **102**(1984), 288.

177. Sprycha, R., *J. Colloid Interface Sci.,* **127**(1989), 1.

178. Sprycha, R., *J. Colloid Interface Sci.,* **102**(1984), 173.

179. Morel, F .M. M. and Morgan, J. J., *Environ. Sci. Technol.,* **6**(1972), 58.

180. McDuff, R. E. and Morel, F. M. M., *Description and Use of Chemical Equilibrium Program, REDEQL - 2.* Techl. Rept. EQ. 73 - 02, W. M. Keck Laboratory, California Institute of Technology, Pasadena, CA - 1973.

181. Truesdel, D. H. and Jones, B. F., *WATEQ: A Computer Program for the calculation of Chemical Equilibrium of Natural Waters.* National Technical Information Service, NTIS Techl. Rept. PB2-20464, Washington DC., 1973.

182. Westall, J. C.; Zachary, J. L. and Morel, F. M. M., *MINEQL: A Computer Program for the Calculation of Chemical Equilibrium Composition of Aqueous Systems,* MIT Technical Note 18, Cambridge, MA, 1976.

183. Westall, J. C., *MICROEQL: A chemical Equilibrium Program in BASIC. Rept. 86-02.* Oregon State University, Corvallis, OR, 1986.

184. Sposito, G. and Mattigold, S. V., *GEOCHEM: A Computer Program for the Calculation of Chemical Equilibrium in Soil Solutions and Other Natural water Systems.* Techl Rept., Dept. of Soil Science, University of California, 1980.

185. Felmy, A. R., Girvin, D. C. and Jenne, E. A., *MINTEQ : A Computer Program for Calculating Aqueous Geochemical Equilibrium.* EPA - 600 /3 - 84 - 032, Office of R& D, US - EPA, Athens, GA, 1984.

186. Westall, J. C., *FITEQL : A Computer Program for the Determination of Equilibrium Constants from Experimental Data.* Techl. Rept. Chemistry Dept., Oregon State University, Corvallis, OR, 1982.

187. Herbelin, A. L. and Westall, J. C., *FITEQL : A Computer Program for the Determination of Equilibrium Constants from Experimental Data, Version 3.2.* Techl. Rept. Chemistry Dept., Oregon State University, Corvallis, OR 97331, 1996.

188. Bousse, L. and Meindle, J. D., *in Geochemical Processes at Mineral Surfaces* (J. A. Davis and K. F. Hayes, eds) ACS Symp. Series 323, Amer. Chem. Soc., Washington D C, 1986, pp 80-98.

189. Healy, T. W.; Yates, D. W., White, L. R. and Chan, D., *J. Electroanal. Chem.*, **80** (1977), 57.

190. Hunter, R. J., *Introduction to Modern Colloid Science,* Oxford University Press, New York, 2002.

191. Hunter, R. J., *Zeta Potential in Colloid Science,* Academic Press, New York, 1981.

192. Hiemenz, P. C., *Principles of Colloid and Surface Chemistry,* Marcel and Dekkar, New York, 1977.

193. Lyklema, *J., Croat. Chem. Acta,* **43**(1971), 249.

194. Bowden, J. W.; Posner, A. M. and Quirk, J. P., *Austr. J. Soil Res.,* **15**(1977), 121.

195. Derjaguin, B. V., *Discuss. Farad. Soc.,* **42**(1942), 109.

196. Bohn, H. L.; McNeal, B. L. and O'Connor, G. A., *Soil Chemistry,* John Wiley & Sons, New York, 1979.

197. Velde, B., *Origin and Mineralogy of Clays : Clay and the Environment,* Springer Verlag, New York, 1995.

198. Bailey, S. W.; Brindley, G. W.; Johns, W. D.; Martin, R. T. and Ross, M., *Clays Clay Miner.,* **19**(1971), 129, 132.

199. Jackson, M. L. in *Chemistry of the Soil* (Bear, F. E. ed.), Am Chem Soc. Monogr. Series, Reinhold Publishing Corp., New York, 1964, pp 71-141.

200. Brown, G.; Newmann, A. C. D., Raymer, J. H., and Weir, A. H. in *The Chemistry of Soil Constituents* (Greenland, D. J. and Hayes, M. H. B., eds.) John Wiley & Sons, New York, 1978, pp 29-178.

201. Jara, A. A., Goldberg, S. and Mora, M. L., *J. Colloid Interface Sci.,* **292**(2005), 160.

202. Schwertmann, U. and Fitzpatrick, R. W., in *Biomeralization : Processes of Iron and Manganese* (Skinner, H. C. W. and Fitzpatrick R. W. eds.) Catena Verlag, 1992, pp 7 - **30**

203. Schwertmann, U. and Taylor, R. M., *in Minerals in Soil Environment*, 2nd edn., Soil Sci Soc. Am. 1989, pp 379 - 438.

204. Nye, W. F.; Levin, S. B. and Kedesdy, H. H., *Proc. Annual Power Sources Conf.*, 3(1959), 125.

205. Malati, M. A., *Chemistry and Industry,* 1971, pp 440 - 452 and similar paper by Kanungo, S. B. and Sant, B. R., *J. Sci. Indust. Res.* (India), **31**(1972), 264.

206. Burns, R. G. and Burns, V. M., in *Marine Manganese Deposits* (Glasby, G. P. ed.), Elsevier, Amsterdam, 1973.

207. Barer, R. M., *Zeolites and Clay Minerals as Sorbents and Molecular Sieves,* Academic Press, New York, 1987.

208. Breck, D. K., *Zeolites Molecular Sieves, Structure, Chemistry and Uses,* John Wiley and Sons, New York, 1974.

209. Schofield, R. K., *J. Soil Sci.,* **1**(1949), 1.

210. Greenland, D. J., *Trans. 10th Int'l. Congr. Soil Sci.* Moscow, vol 2 1974, pp 278-285.

211. Jackson, M. L., *Trans. 7th Int'l. Congr. Soil Sci.* Madison, vol 2, 1960, pp 444 - 455.

212. Barshad, I., *Trans. 7th Int'l. Congr. Soil Sci.*, Madison, vol 2, 1960, pp 435-443.

213. Kerr, G. T.; Zimmermann, R. H., Fox, H. A.(Jr) and Wells, F. H., *Clays Clay Miner.* **4**(1956), 322.

214. Kinniburgh, D. G. and Jackson, M. L. *in Adsorption of Inorganics at Soid/ Liquid Interfaces* (Anderson, M. A. and Rubin, A. J. eds.), chapter 3, Ann Arbor Science Publishers, Inc. 1981.

215. Barshad, I. and Foscolos, A. E., *Soil Sci.,* **110**(1970), 52.

216. van Olphen, H., *An Introduction to Clay Colloid Chemitry,* Wiley Interscience, New York, 1963.

217. Ferris, A. P. and Jepson, W. B., *J. Colloid Interface Sci.,* **51**(1975), 245.

218. Jepson, W. B.; Jeffs, D. S. and Ferris, A. P., *J. Colloid Interface Sci.,* **55**(1976), 454.

219. Schofield, R.K. and Samson, H. R., *Discuss. Faraday Soc.,* **18** (1954), 135.

220. Greenland, D. J.; Oader, J. M. and Sherwin, T. W., *J. Soil Sci.,* **19**(1968), 123, 126.

221. Bradle, H. B., *J. Colloid Interface Sci.,* **227**(2004), 1.

222. Mehra, O. P. and Jackson, M. L., *Proc.7th Nat'l. Conf. Clays, Clay Minerals,* Washington DC, 1960, p137.

223. Beene, G. M.; Bryant, R. and Williams, D. J. A., *J. Colloid Interface Sci,* **147**(1991), 358.

224. Baeyens, B. and Bradbury, M. H., *J. Contam. Hydrol.* **27**(1997), 199.
225. Du, Q.; Sun, Z.; Forsling, W.; and Tang, H., *J. Colloid Interface Sci.*, **187**(1997), 221.
226. Lu, W., *Water Res.*, **35**(2001), 4111.
227. Lu, W. and Smith, E. H., *Geochim. Cosmochim. Acta,* **60**(1996), 3363.
228. Biswas, T. D. and Mukherjee, S. K., *Textbook of Soil Science,* 2nd edn., Tata McGraw-Hill Publishing Co. New Delhi, 1995, p 158.
229. Marshall, C. E. and Bergmann, W. E., *J. Phys. Chem.*, **46**(1942), 52, 325.
230. Marshall, C. E. and Krinbill, C. A,. *J. Phys. Chem.*, **46**(1942), 1077.
231. Low, P. F., *Soil Sci. Soc. Am. Proc.* **19**(1955), 135.
232. Garrels, R. M. and Christ, C. L., *Am. J. Sci.,* **254**(1956), 372.
233. Schindler, P.; W.; Leichti, P. and Westall, J. C., *Nether. J. Agric. Sci.,* **35**(1987), 219.
234. Duc, M., Gaboriaud, F. and Thomas, F., *J. Colloid Interface Sci.*, **289**(2005), 139.
235. Avena, M. J. and De Pauli, C. P., *J. Colloid Interface Sci.* **202**(1998), 195.
236. Avena, M. J.; Cabrol, R. and De Pauli, C. P., *Clays Clay Miner.*, **38** (1990), 35.
237. Missana, T. and Adell, A., *J. Colloid Interface Sci.*, **230**(2000), 150.
238. Bleam, W. F.; Welhouse, G. J. and Janowink, M. A., *Clays Clay Miner.*, **41**(1993), 305.
239. Cao, E.; Bryant, R. and Williams, D. J. A., *J. Colloid Interface Sci.*, **179**(1996), 143.
240. Avena, M.; Mariscal, M. M. and De Pauli, C. P., *Appl. Clay Sci.* **24**(2003), 3.
241. Lackovic, K.; Angove, M. J.; Wells, J. D. and Johnson, B. B., *J. Colloid Interface Sci.,* **257**(2003), 31.
242. Kanungo, S. B. and Mahapatra, D. M., *Proc.3rd Int'l Symp.on Beneficiation and Agglomeration,* R. R. L., Bhubaneswar (India), 1991, pp 37 - 53.
243. Gu, X. and Evans, L. J., *J. Colloid Interface Sci.*, **307**(2007), 317.
244. Swartzen - Allen, S. L. and Matijevic, E., *J. Colloid Interface Sci.*, **50**(1975), 143; 56(1976), 159.
245. Dzombak, D. A. and Morel, F. M. M., *J. Colloid Interface Sci.,* **112**(1986), 588.
246. Davis, J . A.; Fuller, C. C. and Cook , A . D., *Geochim. Cosmochim. Acta,* **51**(1987), 1477.
247. Fuller, C. C. and Davis, J. A., *Geochim . Cosmochim. Acta,* **51**(1987), 1491.
248. McKenzie, R. M. *Austr. J. Soil Sci.,* **5**(1967), 235; **8**(1970), 97.
249. Davis, J. A. and Kent, D. B, in *Mineral-Water Interface Geochemistry* (Hochelle (Jr), M. F. and White, A, F. eds.), Reviews in Mineralogy Series vol. 23, Minerological Society of America, Washington D. C., 1990, pp 177-258.

250. Helffereich, F., *J. Phys. Chem.,* **69**(1965), 1178.
251. Helffereich, F. *Ion - Exchange,* McGraw Hill Book Co. Inc., New York, 1962.
252. Michelson, L. D.; Gideon, P. G. and Kutal, L. H., *Office of Water Research and Technology,* Bull. No. 74, 1975.
253. Smit, W. Holten, C. L. M.; Stein, H. N.; deGoeij, J. J. M. and Theleen, H. M. J., J. *Colloid Interface Sci.,* **63**(1978), 120 ; **67**(1978), 397.
254. Simon, J.; Schulz, W. and Voltz, M., *Z. Anorg. Allgem. Chem.,* **394**(1972), 233.
255. James, R. O.; Stiglich, P. J. and Healy, T. W., *Disc. Faraday Soc.* **59**(1975), 142.
256. Kurbatov, M. H.; Wood, G. B. and Kurbatov, J. D., *J. Phys. Chem.,* **55**(1951), 1170.
257. Bunzle, K.; Schmidt, W. and Sansoni, B., *J. Soil Sci.,* **27**(1976), 32.
258. Sasaki, M.; Moriya, M.; Yasunaga, T. and Astumian, R. D., *J. Phys. Chem.,* **87**(1983), 1449, 1454 and 5254.
259. Hachiya, K.; Sasaki, M., Ikeda, T.; Mikami, N. and Yasunaga, T., *J. Phys. Chem.,* **88**(1984), 27.
260. Chenug, C. W.; Porter, J. F. and McKay, G., *Sepn. Purif. Technol.,* **19**(2000), 55.
261. Reddad, Z.; Gerente, C.; Andres, Y. and Cloiree, P. L., *Environ. Sci. Technol.,* **36**(2002), 2067.
262. Ho, Y. S. and McKay, G., *Trans. IChemE.,* **76**(1998), 332 ; **77**(1999), 165.
263. Banerjee, K.; Cheremisinoff, P. N. and Cheng, S. L., *Water Res.,* **31**(1997), 249.
264. Ho, Y. S.; Ng, J. C. Y. and McKay, G., *Spn. Sci. Technol.* **36**(2001), 115.
265. Ho, Y. S. and McKay, G., *Process Biochem.,* **34**(1999), 45.
266. Ho, Y. S. and McKay, G., *Chem. Engg. J.,* **70**(1998), 115.
267. Ho, Y. S. and McKay, G., *Canad. J. Chem. Engg.,* **76**(1998), 822.
268. Azizian, S., *J. Colloid Interface Sci.,* **276**(2004), 47.
269. Ho, Y. S. and McKay, G., *Water Res.,* **34**(2000), 735.
270. Tien, C. T. and Huang, C. P., in *Heavy Metals in the Environment* (Vernet, J. P. ed.), Elsevier, Amsterdam, 1991, pp 315–316.
271. Hayword, D. O. and Trapnell, B. M.W., *Chemisorption, Butterworths, London,* 1964, p 93.
272. Allen, C. and Scaife, P. H., *Austr. J. Chem.,* **15**(1966), 2023.
273. Hingston, F. J. in *Adsorption of Inorganics at Solid - Liquid Interface* (Anderson, M. A. and Rubin, A. J. eds.) Ann Arbor Science, Ann Arbor, Michigan, 1981, pp 69 - 70.
274. Leckie, J. O. and Tripathy, V. S., *Proc. 5th Int'nl. Conf. of Heavy Metals in Environment.* vol. 2, 1985, pp 369 - 371.
275. Li, Y-Hui, *Geochim. Cosmochim. Acta,* **45**(1981), 1659.

276. Balistrieri, L. S. and Murray, J. W., *Geochim. Cosmochim. Acta,* **47**(1983), 1091.

277. Balistrieri, L. S. and Murray, J. W., *Geochim. Cosmochim. Acta,* **48**(1984), 921; 50(1986), 2235.

278. Tessier, A.; Carignan, R.; Dubreuil, B. and Rapin, F., *Geochim. Cosmochim.cta,* **53** (1989), 1511.

279. Higgo, J. I. and Rees, L. V. C., *Environ. Sci. Technol.,* **20**(1986), 483.

280. Honeyman, B. D. and Santschi, P. H., *Environ. Sci. Technol.,* **22**(1988), 862.

281. Johnson, C. A., *Geochim. Cosmochim. Acta,* **50**(1986), 2433.

282. Kinniburg, D. G., *J. Soil Sci.,* **34**(1983), 759.

283. Kanungo, S. B., *J. Colloid Interface Sci.,* **162**(1994), 93, 103.

284. Kanungo, S. B.; Tripathy, S. S.; Mishra, S. K. Sahoo, B. N. and Rajeev, *J. Colloid Interface Sci.,* **269**(2004), 11.

285. Honeyman, B. D. and Leckie, J. O. in *Geochemical Processes at Mineral Surfaces* (J. A. Davis and K. F. Hayes, eds) ACS Symp. Series 323, Amer. Chem. Soc., Washington D C, 1986, pp 162 - 190.

286. Perona, M. J. and Leckie, J. O., *J. Colloid Interface Sci.,* **106**(1985), 64.

287. Kinniburg, D. G., *Environ. Sci. Technol.,* **20**(1986), 895.

288. Dowd, J. E. and Riggs, D. S. *J. Biol. Chem.,* **240**(1965), 863.

289. Colquhoun, D. *J. Royal Stat. Soc. Ser.* C, **18**(1969), 130.

290. Harter, R. D. and Baker, D. E., *Soil Sci. Soc. Am. J.,* **41**(1977), 1077.

291. Veith, J. A. and Sposito, G., *Soil Sci. Soc. Am. J.,* **41**(1977), 697, 870.

292. Elprince, A. M. and Sposito, G., *Soil Sci. Soc. Am. J.,* **45**(1981), 277.

293. Sposito, G., *Soil Sci. Soc. Am. J.,* **43**(1979), 197; **46**(1982), 1147.

294. Kanungo, S. B. and Parida, K. M., *J. Colloid Interface Sci.,* **98**(1984), 245.

295. Gabano, J. P.; Etienne, P. and Laurent, J. F., *Electrochim. Acta,* **10**(1965), 947.

296. Loganathan, P. and Burau, R. G., *Geochim. Cosmochim. Acta,* **37**(1973), 1277.

297. Rubin, A. J. and Mercer, D. L. in *Adsorption of Inorganics at Solid-Liquid Interface* (Anderson, M. A. and Rubin, A. J. eds.) Ann Arbor Science, Ann Arbor, Michigan, 1981, pp 295-325.

298. Papini, M. P. and Majone, M. in *Encyclopaedia of Surface and Colloid Science* (Hubbard, A. ed.), Marcel Dekkar, New York, 2002, p 3483.

299. Kinniburg, D. G; Baker, J. A. and Whitefield, M. A., *J. Colloid Interface Sci.,* **95**(1983), 370.0

300. Sips, R., *J. Chem. Phys.,* **16**(1948), 490; **18**(1950), 1024.

301. To'th, J., *Acta Chem. Hung.,* **69**(1971), 311.

302. Schindler, P. W.; Furst, B.; Dick, R. and Wolf, P. W., *J. Colloid Interface Sci.*, **55**(1976), 469.

303. Stumm, W.; Kummert, R. and Sigg, L., *Croat. Chem. Acta,* **53**(1980), 291.

304. Sigg, L. and Stumm, W., *Colloids Surfaces,* **2**(1981), 101.

305. Goldberg, S. and Sposito, G., *Soil Sci. Soc. Am. J.,* **48**(1984), 772, 779.

306. Goldberg, S., *Soil Sci. Soc. Am. J.,* **49**(1985), 851.

307. Huang, C. P. and Stumm, W., *J. Colloid Interface Sci.,* **22**(1973), 231.

308. Foissy, A. M'Pandou, A, and Lamarche, J. M., *Colloids Surfaces,* **5**(1982), 363.

309. Smit, W. and Holten, C. L. M., *J. Colloid Interface Sci.,* **78**(1980), 1.

310. Bousse, L. deRooij, N. F. and Bergveld, P., *Surface Sci.,* **135**(1983), 479.

311. Hiemstra, T.; Venema, P. and van Riemsdijk, W. H., *J. Colloid Interface Sci.,* **184**(1996), 680.

312. Tadanier, C. J., and Eick, M. J., *Soil Sci. Soc. Am. J.,* **66**(2002), 1505.

313. Hiemstra, T. and van Riemsdijk, W. H., *J. Colloid Interface Sci.,* **301**(2006), 1.

314. Hiemstra, T. and van Riemsdijk, W. H., *J. Colloid Interface Sci.,* **210**(1999), 182.

315. Hiemstra, T. and van Riemsdijk, W. H., in *Encyclopaedia of Surface and Colloid Science* 1st ed. (Hubbard, A. ed.), Marcel Dekkar, New York, 2002, p 3773

316. Rietra, R. P. J. J; Hiemstra, T. and van Riemsdijk, W. H., *Geochim. Cosmochim. Acta,* **63**(1999), 3009.

317. Jenny, H. and Reitemier, R. F., *J. Phys. Chem.,* **39**(1935), 593.

318. Arnold, P. W. in *The Chemistry of Soil Constituents* (Greenland, D. J. and Hayes, M. H. B., eds.) John Wiley & Sons, New York, 1978, pp. 355-403.

319. Eriksson, E., *Soil Sci.,* **74**(1952), 103.

320. Bolt, G. H., *Soil Sci.,* **79**(1955), 267.

321. Okajaki, R., Smith, H. W. and Moodie, C. D., *Soil Sci.,* **96**(1963) 205.

322. Corey, R. B.; in *Adsorption of Inorganics at Solid - Liquid Interface* (Anderson, M. A. and Rubin, A. J. eds.) Ann Arbor Science, Ann Arbor, Michigan, 1981, Chapter 4, pp.161-182.

323. Farley, K. J.; Dzombak, D. A. and Morel, F. M., *J. Colloid Interface Sci.* **106**(1985), 226

324. Reed, B. E. and Matsumoto, M. R., *J. Environ. Eng.,* **119**(1993), 332.

325. James, R. O. and Healy, T. W., *J. Colloid Interface Sci.,* **40**(1972), 42.

326. Kinniburgh, D. G.; Jackson, M. L. and Sayers, J. K., *Soil Sci. Soc. Am. J.* **40**(1976), 796.

327. Tewari, P. H.; Campbell, A. B. and Lee, W., *Canad. J.Chem.*, **50**(1972), 1642.

328. James, R. O. and Healy, T. W., *J. Colloid Interface Sci.*, **40**(1972), 53, 65.

329. Ref. 27, pp 301 - 306

330. Pearson, R. G., J. *Chem. Educ.*, **45**(1968), 581.

331. Miessler, G. L. and Tarr, D. A., *Inorganic Chemistry,* 3rd edn., Pearson Education (Singapore) Pte. Ltd , Indian Reprint, 2005, New Delhi, pp 197- 204.

332. Burgess, J., *Metal Ions in Solution,* Wiley - Interscience, New York, 1978.

333. Criscenti, L. J. and Sverjensky, D. A., *Am. J. Sci.,* **299**(1999), 828.

334. Brown, G. E. (Jr), Parks, G. A. and O'Day, P. A ., in *Mineral Surfaces,* (Vaughan, D. J. and Pattrick, R. A. D. eds.) Chapman & Hall, London, 1995, pp 129-185.

335. Bargar, J. R.; Brown, G. E. (Jr) and Parks, G. A., *Geochim. Cosmochim. Acta,* **62**(1998), 193.

336. Lützenkirchen, J., *J. Colloid Interface Sci.,* **195**(1997), 149.

337. Sahai, N. and Sverjensky, D. A., *Geochim. Cosmochim. Acta,* **61**(1997), 2801, 2827.

338. Davis, J. A. and Leckie, J. O., *Environ. Sci. Technol.,* **12**(1978), 1309.

339. Elliot, H. A. and Huang, C. P., *J. Colloid Interface Sci.,* **70**(1979), 29.

340. Benjamin, M. M. and Leckie, *J. O., Environ. Sci. Technol.,* **15**(1981), 1050.

341. Elliot, H. A. and Huang, C. P., *Environ. Sci. Technol.,* **14**(1980), 87.

342. Bowers, A. R. and Huang, C. P., *J. Colloid Interface Sci.,* **110**(1986), 575.

343. Schindler, P. W., in *Mineral -Water Interface Geochemistry* (Hochelle (Jr), M. F. and White, A, F. eds.), *Reviews in Mineralogy Series* vol. 23, Minerological Society of America, Washington D. C., 1990, Chapter 7 pp 281 - 306.

344. Benjamin, M. M. and Hayes, K. F., *J. Water Pollution Control Federation,* **54**(1982), 1472

345. Balistrieri, L. S. and Murray, J.W., *Geochim.Cosmochim. Acta,* **46**(1982), 1253.

346. Hoins, U.; Charlet, L. and Sticher, H., *Water, Air, Soil Pollut.,* **60**(1993), 241.

347. Lamy, I.; Djofar, M. and Terce, M., *Water, Air, Soil Pollut.,* **57-58**(1991), 457.

348. Ali, M. A. and Dzombak, M. A., *Geochim. Cosmochim. Acta,* **60**(1996), 5045.

349. Ali, M. A. and Dzombak, M. A., *Geochim. Cosmochim. Acta,* **60**(1996), 291.

350. Nowack, B. and Sigg, L., *J. Colloid Interface Sci.,* **177**(1996), 106.

351. Yang, J. K. and Davis, A. P., *J. Colloid Interface Sci.,* **213**(1999), 77.

352. Davis, A. P. and Upadhyaya, M., *Water Res.,* **30**(1996), 1894.

353. Pimental, G. C. and McClellan, A. L., *The Hydrogen Bond,* Freeman, San Francisco, 1980.

354. Bowers, A. R. and Huang, C. P., *J. Colloid Interface Sci.*, **105**(1985), 197.

355. Girvin, D. C.; Gassman, P. L. and Bolton H.(Jr), *Clays, Clay Miner.* **44**(1996), 757.

356. Bargar, J. R.; Persson, P. and Brown, G. E. (Jr), *Geochim. Cosmochim. Acta,* **63**(1999), 2975.

357. Zachara, J. M.; Smith, S. C. and Kuzel, L. S., *Geochim. Cosmochim. Acta,* **59**(1995), 4825.

358. Girvin, D. C.; Gassman, P. L. and Bolton H. (Jr), *Soil Sci. Soc. Am. J.,* **57**(1993), 47.

359. Szecsody, J. E.; Zachara, J.M.and Bruckhart, P. L., *Environ. Sci. Technol.,* **28**(1994), 1706.

360. Nagy, N. M. and Ko'ny a, J., *Colloids Surfaces A,* **137**(1998), 231.

361. Nagy, N. M. and Ko'ny a, J. Ko'nya, I., *Colloids Surfaces A,* **137**(1998), 243.

362. Huang, C. P.; Rhoads, E. A. and Hao, O. J., *Water Res.,* **22**(1988), 1001.

363. Güclu, K. and Apak, R, *J. Colloid Interface Sci.,* **228**(2000), 252.

364. van Benschoten, J. E.; Young, W. H.; Matsumoto, M. R. and Reed, B. E. *J. Environ. Qual.,* **27**(1998), 24.

365. Harvey, D. T. and Linton, R. W., *Colloids Surfaces,* **11**(1984), 81.

366. Sposito, G. in *Geochemical Processes at Mineral Surfaces* (Davis, J. A. and Hayes, K. F. eds.) ACS Symp. Series 323, Am. Chem. Soc., Washington DC, **1986,** pp 217 - 278.

367. Sposito, G., *Soil Sci. Soc. Am. J.,* **48**(1984), 531.

368. James, R. O.; Wiese, G. R. and Healy, T. W., *J. Colloid Interface Sci.* **39**(1977), 381.

369. Schindler, P. W., in *Adsorption of Inorganics at Solid - Liquid Interface* (Anderson, M. A. and Rubin, A. J. eds.) Ann Arbor Science, Ann Arbor, Michigan, 1981, pp 1 - 50.

370. Chen, Y.-S.R.; Butler, J. H. and Stumm, W., *Environ.Sci.Technol.,* **7**(1973), 327.

371. Bolar, N.S.; Barros, N. J. and Posner, A. M., *J. Soil Sci.,* **36**(1985), 187.

372. Fokkink, L.G. J.; DeKeizer, A. and Lyklema, J., *J. Colloid Interface Sci.,* **118**(1987), 454.

373. Ref. 27, pp 287 -290.

374. Gadde, R. R. and Laitinen, H. A., *Anal. Chem.,* **46**(1974), 2022.

375. Morgan, J. J. and Stumm, W., *J. Colloid Sci.,* **19**(1964), 347.

376. Crawford, R. J.; Harding, I. H.and Mainwaring, D.E., *Langmuir,* **9**(1993), 3050

377. Huang, C. P. and Stumm, W., *J. Colloid Interface Sci.,* **43**(1973), 409.

378. MacCarthy, P. and Smith, G. C., in *Chemical Modeling in Aqueous System: Speciation, Sorption , Solubility and Kinetics* (Jenne, E. A. Ed.) Am. Chem. Soc., Washington D C, 1979.

379. Clauss, C. R. A. and Weiss, K. *J. Colloid Interface Sci.,* **61**(1977), 577.

380. Gadde, R. R. and Laitinen, H. A., *Environ. Letts.* **5**(1973), 223.

381. Vydra, F. and Galba, J., *Coll. Czech. Chem. Commn.,* **34**(1969), 3471.

382. Murray, J. W., *Geochim. Cosmochim. Acta,* **39**(1975), 505, 635.

383. Ainsworth, C. C.; Pilon, J. L.; Gassman, P. L. and van der Slays, W. G., *Soil Sci. Soc. Am. J.,* **58**(1994), 1615.

384. Barrow, N. J.; Bowden, J. W.; Posner, A. M. and Quirk, J. P., *Aust. J. Soil Res.,* **19**(1981), 309.

385. Benjamin, M. M.and Leckie, J. O., *J. Colloid Interface Sci.,* **79**(1981), 209.

386. Bolland, M. D.; Posner, A. M. and Quirk, J. P., *Aust. J. Soil Res.,* **15**(1977), 279,

387. Bruemmer, G. W.; Gerth, J. and Tiller, T. G., *J. Soil Sci.,* **39**(1988), 37.

388. Charlet, L. and Manceau, A. A., *J. Colloid Interface Sci.,* **148**(1992), 443.

389. Collins, C. R.; Regnersdottier, K. V. and Serman, D. M., *Geochim. Cosmochim. Acta,* **63**(1999), 2989.

390. Coughlin, B. R. and Stone, A. T., *Environ. Sci. Technol.,* **29**(1995), 2445.

391. Fendorf, S. and Fendorf, M., *Clays Clay Minerals,* **44**(1996), 220.

392. Forbes E. A.; Posner, A.M. and Quirk, J. P. *J. Soil Sci.,* **27**(1976),154.

393. Gerth, J. and Bruemmer, G. W., *Z. Anal. Chem.,* **316**(1983),

394. Girvin, D.C.; Amos, L. L.; Schwab, A.I.and McGarrah, J.E., *J. Colloid Interface Sci.,* **141**(1991), 67.

395. Grimme, H., *Z. Pflanzenern‹hrung Bodenk.,* **121**(1968), 58.

396. Hildebrand, E. S. and Blum, W. E., *Naturwissenschaften,* **61**(1974), 160.

397. Hsi, C. D. and Langmuir, D., *Geochim. Cosmochim. Acta,* **49**(1985), 1931.

398. Johnson, B. B., *Environ. Sci. Technol.,* **24**(1990), 112.

399. Kinniburg, D.G.; Sayers, J.K. and Jackson, M.L., *Soil Sci. Soc. Am. J.***39**(1975), 464.

400. Kooner, Z. S.; Cox, C, D. and Smoot, L., *Environ. Toxicol. Chem.,* **14**(1995), 2077.

401. Machesky, M.; Andrade, W. and Rose, A.,*Geochim. Cosmochim. Acta,* **55**(1991),769.

402. Christle, I. and Kretzschmar, R., *Geochim. Cosmochim. Acta,* **63**(1999),2929.

403. Marmier, N.; Delise-, A. and Fromage, P., *J. Colloid Interface Sci.,* **211**(1999), 54.

404. Mueller, B. and Sigg, L., *J. Colloid Interface Sci.,* **148**(1992), 517.

405. Music, S. and Wolf, R. H. H., *Mikrochim. Acta,* **1**(1979), 87.
406. Music, S.; Gessner, M. and Wolf, R. H. H., *Mikrochim. Acta,* **1**(1979), 105.
407. Music, S.; Gessner, M. and Wolf, R. H. H., *J. Radioanal. Chem.,* **50**(1979), 91.
408. Music, S.; Gessner, M. and Wolf, R. H. H., *Radiochim. Acta,* **26**(1979), 51.
409. Nechaev, E. A. and Shein, V. H., *Kolloid Zeits.,* **41**(1979), 1028.
410. Padmanabhan, M., *Aust. J. Soil Res.,* **21**(1983), 309, 515.
411. Peterson, W.; Walmann, K.; Schroer, S. and Schroeder, F., *Anal. Chim. Acta,* **273**(1993), 323.
412. Rose, A. W. and Bianchi- Mosquera, G. C., *Econ. Geol.,* **88**(1993), 1226.
413. Schenck, C.V.; Dillard, J. G. and Murray, J. W., *J. Colloid Interface Sci.,* **95**(1983), 398.
414. Sung, W. and Morgan, J. J., *Geochim. Cosmochim. Acta,* **45**(1981), 2377.
415. Tamura, H.; Matijevic, E. and Meites, L., *J. Colloid Interface Sci.,* **92**(1983), 303.
416. Trivedi, P. and Axe, L., *J. Colloid Interface Sci.,* **218**(1999), 554.
417. Waite, T. D.; Davis, S. A.; Payne, T. E.; Waychunas, G. A. and Xu, N., *Geochim. Cosmochim. Acta,* **58**(1994), 546.
418. Xue, J. and Huang, P. M., *Geoderma,* **64**(1995), 343.
419. Music, S. and Ristic, M., J., *Radioanal. Nucl. Chem.,* **120**(1988), 289.
420. Ardizzone, S.; Formaro, L. and Lyklema, J., *J. Electroanal. Chem.,* **133**(1982), 147.
421. Pochard, I.; Doneyl, R.; Couchet, P. P. and Foissy, A., *J. Colloid Interface Sci.,* **255**(2002), 27.
422. Mishra, S. P. and Vjaya, *J. Colloid Interface Sci.,* **296**(2006), 383.
423. Zachara, J.; M. Smith, S. C. Resch, C. T. and Cowan, C. E., *Soil Sci. Soc. Am. J.,* **56**(1992), 1074.
424. Spark, K. M.; Johnson, B. B. and Wells, J. D., *Euop. J. Soil Sci.,* **46**(1995), 625, 633.
425. Barrow, N. J.; Gerth, J. and Brümmer, G. W., *J. Soil Sci.,* **40**(1989), 437.
426. Kooner, Z. S., *Environ. Geol. Water Sci.,* **20**(1992), 205.
427. Kooner, Z. S., *Environ. Geol.,* **21**(1993), 242.
428. Zachara, J. M.; Resch, C. T. and Smith, S. C., *Geochim. Cosmochim. Acta,* **58**(1994), 553.
429. Burns, R. G., *Geochim. Cosmochim. Acta,* **40**(1976), 95.
430. Dillard, J. G., *Geochim. Cosmochim. Acta,* **43**(1979), 781.
431. Gray, M. J. and Malati, M. A. *J. Chem. Technol. Biotechnol.,* **29**(1979), 127, 135.

432. Zasoki, R. J. and Burau, R. G., *Soil Sci. Soc. Am. J.*, **52**(1988), 81.

433. Balistrieri, L. S. and Murray, J. W., *Geochim. Cosmochim. Acta,* **46**(1982), 1041.

434. Catts, J. G. and Langmuir, D., *Appl. Geochem.*, **1**(1986), 255.

435. Pretorius, P. J. and Linder, P. W., *Appl. Geochem.*, **16**(2001), 1067.

436. Fu, G.; Allen, H. E. and Cowan, C. E., *Soil Sci.,* **152**(1991), 72.

437. Mishra, S. P.; Dubey, S. S. and Tiwary, D., *J. Colloid Interface Sci.*, **279**(2004), 61.

438. Murray, D. J.; Healy, T.W. and Fuerstenau, D.W., in *Adsorption from Aqueous Solution,* Adv. Chem. Ser., Am. Chem. Soc., Washington DC, vol. **79**, 1968, pp 74-79.

439. Jefferies, D. S. and Stumm, W., *Canad. Mineralogist,* **14**(1976), 16.

440. Takematsu, N., *J. Oceanographic Soc.* (Japan), **35**(1979), 36.

441. Anderson, B. J.; Jenne, E. A. and Chao, T. T., *Geochim. Cosmochim. Acta,* **37**(1973), 611.

442. Hasany, S. N. and Chaudhary, M. H., *Int'l J. Appl. Radiation & Isotopes,* **32**(1981), 899.

443. Cvjeticanin, N.; Cjveticanin, D. and Pravica, M., *J. Chromatography,* **241**(1982), 183.

444. Dahal, M. P. and Lawrence, G. A., *Adsorp. Sci. Technol.,* **13**(1996), 231.

445. Pan, G.; Qin, Y.; Xianbing, Li; Hu, T.; Wu, Z. and Lie, Y., *J. Colloid Interface Sci.,* **271**(2004), 28, 35.

446. Feng, X. H.; Zhai, L. M.; Tan, W. F.; Zhou, W.; Liu, F. and He, J. Z.,*J. Colloid Interface Sci.,* **298**(2006), 258.

447. Tripathy, S. S.; Bersillon, J.-L.and Gopal, K., *Desalination,* **194**(2006), 11.

448. Zaman, M. I.; Mistafa, S.; Khan, S. and Xing, B., *J. Colloid Interface Sci.,* **330**(2009),9.

449. Hachiya, K.; Sasaki, M.; Saruta, Y.; Mikami, N. and Yasunaga, T., *J. Phys. Chem.,***88**(1984), 23.

450. McBride, M. B., *Soil Sci. Soc. Am. J.,* **42**(1978), 27.

451. Kosmulski, M., *Colloids Surfaces,* **117**(1996), 301.

452. Tamura, H.; Katayama, N. and Furuichi, K., *J.Colloid Interface Sci.,* **195**(1997), 192.

453. Baily, J. F. and Fein, J. B., *Geochim. Cosmochim. Acta,* **60**(1996), 2929.

454. Naiya,T.K.; Bhattacharyya, A. K. and Das, S.K., *J. Colloid Interface Sci.,* **332**(2009), 14.

455. Sugiyama, S.; Kanda,Y. Ishizuka, H. and Satowa, K-I.; *J. Colloid Interface Sci.,***320**(2008), 535.

456. Mor, S.; Ravindra, K. and Bishnoi, N. R., *Bioresource Technol.*, **98**(2007), 954.

457. Sen, T. K.; and Sarzali, M. V., *Chem. Engg. J.*, **142**(2008), 256.

458. Bhattacharyya, A. K.; Mandal, S. K. and Das, S. K., *Chem. Engg. J.*, **123**(2006), 43.

459. Weerasooriya, R.; Wijesekara, H-K. D-K. and Bandara, A., *Colloids Surfaces*, A, **207**(2002), 13.

460. Yang, J. K.; Lee, S-M. and Davis, A. P.; *J. Colloid Interface Sci.*, **295**(2006), 14.

461. Barkat, M. A.; *J. Colloid Interface Sci.*, **291**(2005), 345.

462. Nano, G. V. and Strathmann, T. G.; *J. Colloid Interface Sci.*, **297**(2006), 445.

463. Li, W.; Pan, G.; Zhang, M.; Zhou, D.; Yang, Y.; Chen, H. and He, G., *J. Colloid Interface Sci.*, **319**(2008), 160.

464. Jai, K.; Pan, B.; Zhang, Q.; Zhang, W.; Jiang, C. and Zhang, M., *J. Colloid Interface Sci.*, **318**(2008), 160.

465. Subramaniam, K.; Vithayveraj, V.; Yiacounnis, S. and Tsouris, C., *J. Colloid Interface Sci.*, **268**(2003), 12.

466. Gissinger, P. B.; Alnot, M.; Licke, J. P.; Ehrhardt, J. J. and Behra, P., *J. Colloid Interface Sci.* **215**(1994), 313.

467. Missana, T.; Garcia-Gufirrez, M. and Maffiotte, C., *J. Colloid Interface Sci.*, **260**(2003), 29.

468. Angove, M. J.; Wells, J. D. and Johnson, B. B., *J. Colloid Interface Sci.*, **211**(1999), 281.

469. Bergaya, F., and Vayer, M., *Appl. Clay Sci.*, **12**(1997), 275.

470. Alberti, G. and Constantino,U., in *Solid State Supra Molecular Chemistry: Two or Three Dimensional Inorganic Networks* (Alberti, G. and Bain, T. eds.) Pergamon-Elsevier, Oxford, 1996.

471. Zhu, H. Y. and Yamanaka, S., *J. Chem. Soc., Farad. Trans.*, **93**(1997), 477.

472. Vicente, M. A. and Lambert, *J. F., Phys. Chem. & Chem. Phys.*, **3**(2001), 4843.

473. Burch, R. and Warburton, C., *J. Catal.*, **97**(1986), 503.

474. Ohtsuka, K.; Hayashi, Y. and Suda, M., *Chem. Mater.*, **5**(1993), 1823.

475. Pereira, P. R.; Pires, J. and deCarvalho, M. B., *Langmuir*, **14**(1998), 4584.

476. Christiano, S. P. and Pinnavala, T. J., *J. Solid State Chem.*, **64**(1986), 232.

477. Velverde, J. L.; Romero, A.; Romero, R.; Garcia, R. B.; Sanchez, M. L.and Ascencio, I., *Clays Clay Miner.*, **53**(2005), 613.

478. Gil, A.; Gaudia, L. M. and Vicente, M. A., Catal. *Rev.(Sci. & Eng)*, **42**(2000), 143.

479. Theocharis, C. R.; Jacob, K. J. and Gray, A. C., *J. Chem. Soc., Farad. Trans.*, **84**(1988), 1509.
480. Ravichandran, J. and Sivasankar, B., *Clays Clay Miner.*, **45**(1997), 854.
481. Suraj, G.; Iyer, C.S.P. and Lalithambica, M., *Appl. Clay Sc.*, **13**(1998), 29.
482. Bhattacharyya, K. G. and Sengupta, S., *Ind. Eng. Chem., Res.*, **46**(2007), 3734.
483. Rodrigues, M. G. F., *Ceramica*, **49**(2003), 146.
484. Bhattacharyya, K. G. and Sengupta, S., *Colloids Surfaces,* A, **277**(2006), 191.
485. Bhattacharyya, K. G. and Sengupta, S., *Adsorption Sci. Technol.*, **27**(2009), 47.
486. Angove, M. J.; Johnson, B.B. and Wells, J. D., *J. Colloid Interface Sci.*, **204**(1998), 193.
487. Angove, M. J.; Johnson, B.B. and Wells, *J. D., Colloids Surfaces,* A, **126**(1997), 137.
488. Ikhsan, J.; Johnson, B.B. and Wells, J. D., *J .Colloid Interface Sci.*, **217**(1999), 403.
489. Srivastava, P.; Singh, B. and Angove, M. *J., J. Colloid Interface Sci.*, **290**(2005), 28.
490. Miranda-Trevino, J. C. and Coles, C. A., *Appl. Clay Sc.*, **23**(2003), 133.
491. Yavuz, o.; Altunkaynak, Y. and Guzel, F., *Water Res.*, **37**(2003), 948.
492. Gu, X. and Evans, L., *Geochim. Cosmochim. Acta,* **72**(2008), 267.
493. Elliot, H. A. and Denneny, C. M., *J. Environ. Quality,* **11**(1982), 658.
494. Sengupta, S. and Bhattacharyya, K. G., *J. Hazard. Mater.*, **128**(2006), 247.
495. Bhattacharyya, K. G. and Sengupta, S., *Sepn. Purif. Technol.*, **50**(2006), 388.
496. Bhattacharyya, K. G. and Sengupta, S., *Sepn. Sci.Technol.*, **42**(2007), 3391.
497. Bhattacharyya, K. G. and Sengupta, S., *Chem. Eng. J.*, **136**(2008), 1.
498. Alkan, M.; Kalay, B.; Dogan, M. and Demirbas, O., *J. Hazard. Mater.*, **153**(2008), 867.
499. Bhattacharyya, K. G. and Sengupta, S., *Adsorption*, **12**(2006), 185.
500. Bhattacharyya, K. G. and Sengupta, S., *Adv. Colloid Interface Sci.*, **140**(2008), 92.
501. Bhattacharyya, K. G. and Sengupta, S., *J. Colloid Interface Sci.*, **310**(20007), 411.
502. Zhang, S. Q. and Hou, W. G., *Colloids Surfaces,* A, **320**(2008), 92.
503. Peigneur, F., Maes, A. and Cremers, A., *Clays Clay Miner.*, **23**(1975), 71.
504. Abollino, O.; Giacomino, A.; Malandrino, M. and Mentasti, E., *Appl. Clay Sci.*, **38**(2008), 227.
505. Donat, R.; Akdogan, A.; Erdem, E. and Cetisli, H., *J. Colloid Interface Sci.*, **286**(2005), 43.
506. Karapinar, N. and Donat, R., *Desalination,* **249**(2009), 123.
507. Khan, S. A.; Rahman, R. and Khan, M. A., *Waste Manag.*, **14**(1994), 629; **15**(1995), 271.

508. Sengupta, S.and Bhattacharyya, K.G., *J. Colloid Interface Sci.,* **295**(2006), 21.

509. Lin, S. H. and Juang, R. S., *J. Hazard. Mater.,* **B92**(2002), 315.

510. Mellah, A. and Chegrouche, S., *Water Res.,* **31**(1997), 621.

511. Naseem, R. and Tahir, S.S., *Water Res.,* **35**(2001), 3982.

512. Du, Q.; Sun, Z.; Forsling, W. and Tang, H., *J. Colloid Interface Sci.,* **187**(1997), 232.

513. deFonseca, M. G.; deOliviera, M. M.; Arakaki, L. N. H.; Espinola, J. G. P. and Airoldi, O., *J. Colloid Interface Sci.,* **285**(2005), 50.

514. Malandrino, M.; Abollino, O.; Giacomino, A.; Aceto, M. and Mentasti, E., *J. Colloid Interface Sci.,* **299**(2006), 537.

515. Ouki, S. K. and Kavannag, M., *Water Sci. Technol.,* **39**(1999), 115.

516. Sprynskyy, M.; Buszewski, B.; Terzyk, A. P. and Namiesnik, J., *J. Colloid Interface Sci.,* **304**(2006), 21.

517. Ttsitsishvili, G.; Andronikashvili, T.; Kirov, G. and Filizoval, L., *Natural Zeolites,* Ellis Horwood, U. K., 1992.

518. Erdem, E.; Karapinar, N. and Donat, R., *J. Colloid Interface Sci.,* **280**(2004), 309.

519. Galli, E.; Gottardi, G.; Mayer, H.; Preisinger, A. and Passaglia, E., *Acta Crystallog. Sect.* B, **39**(1983), 189.

520. Malliou, E.; Lozidou, M. D. and Spyrellis, N., *Sci. Total Environ.,* **149**(1994), 239.

521. Inglezakis, V. J.; Lozidou, M. D. and Grigoropoulou, H. P., J. *Colloid Interface Sci.,* 275(2004), 570.

522. Cukrovic, L.; Cerjan-Stefanovic, S. and Filipan, T., *Water Res.,* **31**(1997), 1379.

523. Mandale, K. D.; Carland, R. M., and Aplan, F. F., *Min. Eng.,* **8**(1995), 535.

524. Ackley, M. W. and Yang, R., *AIChE J.,* **37**(1991), 1645.

525. Zamzow, M. J. and Murphy, J. E., Sepn. *Sci. Technol.,* **27**(1992), 1969.

526. Carland, R. M. and Aplan, F. F., *Miner. Metall. Process.,* **11**(1995), 210.

527. Zamzow, M. J.; Eichbaum, B. R.; Sandgren, K. R., and Shanks, D. E., *Sepn. Sci.Technol.,* **25**(1990), 1555.

528. Park, J. -B.; Lee, S. -H.; Lee, J. -W., and Lee, C. -Y., *J. Hazard. Mater.,* **B95**(2002), 65.

529. Mier, M. V.; Callejas, R. L.; Gehr, R.; Cisneres, B. E. J. and Alverez, P. J. J., *Water Res.,* **35**(2001), 373.

530. Inglezakis, V. J. and Grigoropoulou, H. P., *J . Colloid Interface Sci.,* **234**(2001), 434.

531. Doula, M. K. and Ioannou, A., *Micropor. Mesopor. Mater.,* **58**(2003), 115.

532. Petrus, R. and Warchol, J., *Micropor. Mesopor. Mater.,* **61**(2003), 137.

533. Panayotova, M. I., *Water Manag.,* **21**(2001), 671.

534. Jama, M. A. and Yücel, H., *Sepn. Sci. Technol.*, **24**(1990), 1399.

535. Leinonen, H. and Lehto, J., *Waste Manag.*, **19**(2001), 45.

536. Farrah, H. and Pickering, W. F., *Aust. J. Chem.*, **29**(1976), 1167.

537. Farrah, H. and Pickering, W. F., *Aust. J. Chem.*, **29**(1976), 1177.

538. Farrah, H. and Pickering, W. F., *Aust. J. Chem.*, **29**(1976), 1649.

539. Alverez-Pueblo R. A.; Santos (Jr), D. S.; Blanco, C.; Echeverria, J. C. and Garrido, J. J., *J. Colloid Interface Sci.*, **285**(2005), 41.

540. Egozy, Y., *Clay Clay Minerals*, **28**(1980), 311.

541. Hizal, J. and Apak, R., *J. Colloid Interface Sci.*, **295**(2006), 1.

542. Lacin O., Bayrak, B., Korkut, o. and Sayan, E., *J. Colloid Interface Sci.*, **292**(2005), 330.

543. Inskeep, W. P. and Bahm, J., *Soil Sci. Soc. Am. J.*, **47**(1983), 64.

544. Salles, F.; Doullard, J -M.; Denoyel, R.; Buildstein, O.; Jullien, M.; Beurroies, I. and Damme, H. V., *J. Colloid Interface Sci.*, **333**(2009), 510.

545. Rozalin, M.; Brady, P. V. and Javier-Huerton, F., *J. Colloid Interface Sci.*, **333**(2009), 474.

546. Zhu, S.; Hou, H.; Xue, Y.; Wei, N.; Sun, Q. and Chen, X ., *J. Colloid Interface Sci.*, **315**(2007), 8.

547. Sengupta, S. and Bhattacharyya, K. G., *J. Environ. Manag.*, **87**(2008), 46.

548. Bereket, G; Arog, A. Z. and Orzel, M. Z., *J. Colloid Interface Sci.*, **187**(1977), 338.

549. Apiratikul, R. and Pavasant, P., *Chem. Eng. J.*, **144**(2008), 245.

550. Bhattacharyya, K. G and Sengupta, S., *Appl. Clay Sci.*, **41**(2008), 1.

551. Barbier, F.; Duc, G and Petit-Ramel, M., *Colloids Surfaces A*, **166**(2000), 159.

552. El-Batouti, M.; Sadek, O. M. and Assad, F. F., *J. Colloid Interface Sci.*, **259**(2003), 223.

553. Catalano, J. G and Brown(Jr), G E., *Geochim. Cosmochim. Acta*, **69**(2005), 2995.

554. Chen, L. and Lu, S., *Appl. Radiation Isotopes*, **66**(2008), 288 ; Song, X., Wang, S., Chen, L., Zhang, M. and Dong, Y., *Appl. Radiation Isotopes*, **67**(2009), 1007.

555. Bascetin, E. L. and Atun, G., *Appl. Radiation Isotopes*, **64**(2006), 957.

556. Missana, T. and Garcia-Gufirrez, M., *Physics and Chemistry of the Earth, Parts A/B/C*, **32**(2007), 539.

557. Kulik, D.A.; Aza, S. U.; Sinisyn, V. A. and Wood, S. A., *Geochim. Cosmochim. Acta*, **64**(2000), 195.

558. Bradbury, M. H. and Baeyens, B., *Geochim. Cosmochim. Acta*, **66**(2002), 2325.

559. Nagy, N. M., Konya, J. and Urbin Z., *Colloids Surfaces A*, **121**(1997), 117.

560. DalBosco, S. M.; Jimerez, R. S.; Vignado, C.; Fontanna, J.; Geraldo, B.; Figueiredo, F. C. A., Mandelli, D. and Carvalho, W. A., *Adsorption,* **12**(2006), 133.

561. Kubilay, S.; Gurkan, R., Savran, A.; and Sahan, T.; Adsorption, **13**(2007), 41.

562. Alverez -Ayuso, E.; and Garcia -Sanchez, A., *Clay Clay Minerals,* **51**(2003), 475.

563. Adebowale, K.O., Unuabonah, I, E. and Olualabi, B. I., *J. Hazard. Mater.,* **134**(2006), 130.

564. Wu, J.; Li, B.; Liao, J.; Feng, Y.; Zheng, D.; Zhou, J.; Wen, W.; Yang, Y. and Lin, N., *J. Environ. Radioactivity,* **100**(2009), 914.

565. Turan, P.; Doegan, M. and Alkan, M., *J. Hazard. Mater.,* **148**(2007), 56.

566. Arias, F. and Sen, T. K., *Colloids Surfaces A*, **348**(2009), 100.

567. Heidmann, I.;Christl, I.; Len, C. and Kretzschmer, R., *J. Colloid Interface Sci.,* **282**(2005), 270.

568. Al-Qunnaibit, M.H.; Mekhemer, W.K. and Zaghloul, A. A., *J. Colloid Interface Sci.,* **283** (2005), 316.

569. Shahwan, T.; Erten, H. N. and Unugur, S., *J. Colloid Interface Sci.,* **300**(2006), 447.

570. Sarkar, D.; Essington, M. E. and Mishra, K.C., *Soil Sci. Soc. Am. J.,*4(2000), 1968.

571. El Mouzdahir, Y.; El Mashauri; Mahboub, R.; El Ansari, A.; Gil, A.; Korili, S.A.and Vicente, M.; *Appl. Clay Sci.,* **35**(2007), 47.

572. Sharma, P.; Singh, G. and Tomar, R ., *J. Colloid Interface Sci.,* **332**(2009), 298.

573. Plaska, A. G.; Majdan, M. and Pikuo, S., *J. Colloid Interface Sci.,* **317**(2008), 409.

574. Li, X.; Chen, C. L.; Chang, P.P.; Yu, S.M.; Hu, W.S. and Wang, X.K., *Desalination,***244**(2009), 283.

575. Coles, C. A. and Yong, R.N., *Appl. Clay Sci.,* **22**(2002), 39.

576. Calvo, B.; Canoira, L.; Morante, F.; Gonzalez, J.; Elsen, J. and Alcantara, R., *J. Hazard. Mater.,* **166**(2009), 619.

577. Jha, V.K., Matsuda, M. and Miyake, M., *J. Hazard.Mater.,* **160**(2008), 148.

578. Amphlett, C. B., *Inorganic Ion Exchangers,* Elsevier, Amsterdam, 1964.

579. Mishra, D. N., *Adsorption on and Surface Chemistry of Hydroxyapatites,* Plenum Press, New York, 1974.

580. Sugiyama, S.; Ichii, T.; Matsumoto, H. and Hayashi, H., *Advances Environmental Res.,* **6**(2002), 285.

581. Elasri, S.; Laghzizil, A.; Coradin, T.; Saoiabi, A.; Alaoui, A. and Mohamed, R., *Colloids Surfaces A,* **362**(2010), 33.

582. Sugiyama, S.; Shimizu, Y.; Manabe, T.; Nakagawa, K.and Sotowa, K,I., *J. Colloid Interface Sci.,* **332**(2009), 439.

583. Yasukawa, A.; Yokoyama, T.; Kondari, K. and Ishikawa, T., *Colloids Surfaces A,* **299**(2007), 203.

584. Wu, G. and Li, L. Y., *J. Hazard.Mater.,* **152**(2008), 876.

585. Miyake, M.; Watanabe, K.; Nagayama, Y.; Nagasawa, H. and Suzuki, T., *J. Chem. Soc. Farad. Tans.,* **86**(1990), 2303.

586. Sugiyama, S.; Matsumoto, H.; Hiromu, H. and Moffat, J. B., *Colloids Surfaces A,* **169**(2000), 17.

587. Mustafa, S.; Murtaza, S.; Naeem, A. and Farina, K., *J. Colloid Interface Sci.,* **283**(2005), 287.

588. Sugiyama, S.; Fakuda, N.; Matsumoto, H.; Hayashi, H.; Shigemoto, N.; Hiraga, Y. and Moffat, J. B., *J. Colloid Interface Sci.,* **220**(1999), 324.

589. Sugiyama, S.; Matsumoto, H.; Ichii, T.; Hayashi, H.; Hiraga, Y. and Fakuda, N., *J. Colloid Interface Sci.,* **238**(20001), 183.

590. Sandrine, B.; Ange, N.; Didier, B. -A.; Eric, C. and Patric, S., *J. Hazard.Mater.,* **139**(2007), 443.

591. Radoicic, T. K., and Raicevic, S., *Chem. Eng. J.,* **166**(2010), 503.

592. Corami, A.; Mignardi, S. and Ferrini, V., *J.Colloid Interface Sci.,***317**(2008), 402.

593. Peld, M.; Tonsuaadu, K. and Bender, V., *Environ. Sci. Technol.,* **38**(2004), 5626.

594. Zhu, R.; Yu, R.; Yao, J.; Mao, D.; Xing, C. and Wang, D., *Catalysis Today,* **139**(2008), 94.

595. Corami, A.; Mignardi, S. and Ferrini, V., *J. Hazard.Mater.,* **146**(2007), 164.

596. Sheha, R. R., *J . Colloid Interface Sci.,* **310**(2007), 18

597. Yasukawa, A.; Yokoyama, T.; Kondari, K. and Ishikawa, T., *Colloids Surfaces A,* **238**(2004), 133.

598. Yasukawa, A.; Yokoyama, T.; Kondari, K. and Ishikawa, T., *Colloids Surfaces A,* **317**(2008), 123.

599. Leyva, A.G.; Marrero, J.; Smichowski, P. and Cicerone, D., *Environ. Sci. Technol.,* **35**(2001), 3669.

600. Perrone, J.; Fourest, B. and Giffout, E., *J. Colloid Interface Sci.,* **239**(2001), 303; **249**(2002), 441.

601. Clearfeld, A., *Inorganic Ion Exchange Materials,* CRC Press, Boca Raton, Florida (USA), 1982.

602. Sdlakova, L. and Pekarek, V., *J Less Common Metals,* **10**(!966), 130.

603. Alberti, G.; Bertrami, R. and Constantino, H., *J. Inorg. Nucl. Chem.,* **38**(1976), 1729.

604. Parida, K. M.; Sahu, B. B. and Das, D. P., *J. Colloid Interface Sci.,* **270**(2004), 436.

605. Finck, N.; Drot, R.; M-Bion, F.; Simoni, E. and Catalette, H., *J. Colloid Interface Sci.,* **312**(2007), 230.

606. Suarez, M.; Garcia, J. R. Rodriguez, *J., Mater. Chem. Phys.,* **8**(1983), 451.

607. Jia, K.; Pan, B.; Zhang, Q., Zhang, W., Jiang, P., Hong, C. and Zhang, M., *J. Colloid Interface Sci.,* **318**(2008), 160.

608. Sahu, B. B.; and Parida, K. M., *J. Colloid Interface Sci.,* **248**(2002), 221.

609. Sahu, B. B., Mishra, H.K. and Parida, K. M., *J. Colloid Interface Sci,* **225**(2000), 511.

610. Mustafa, S.; Naeem, A. and Rehana, N., *J. Chem. Soc. Farad. Trans.,* **89**(1993), 3843.

611. Naeem, A.; Mustafa, S.;Rehana, N.;Dilara, B. and Murtaza, S., *Environ. Technol.,***24**(2003), 779.

612. Mustafa, S.; Naeem, A. and Murtaza, S., *J. Colloid Interface Sci.,* **220**(1999), 63.

613. Fernane, F.; Mecherri, M. O.; Sharrock, P.; Houdioui, M.; Lounicci, H. and Fedoroff, M., *Materials Characterization,* **59**(2008), 554.

614. Naeem, A.; Mustafa, S. and Rehana, M., *J. Colloid Interface Sci.,* **252**(2002), 6.

615. Babel. S. and Kurniawan, T. A., *J. Hazard. Mater.,* **B97**(2003), 213.

616. Ref. 27, pp 308-313.

617. Hingston, F. J.; Atkinson, R. J.; Posner, A. M. and Quirk, J. P., *Nature,* **215**(1967), 213.

618. Hingston, F. J.; Posner, A. M. and Quirk, J. P., *J. Soil Sci.,* **23**(1972), 177.

619. Kummert, R. and Stumm, W., *J. Colloid Interface Sci.,* **75**(1980), 373.

620. Antelo, J.; Avena, M. J.; Fiol, S.; Lopez, R. and Arce, F., *J. Colloid Interface Sci.,* **285**(2005), 476.

621. Harrison, R. M.(Ed.), *Pollution: Causes, Effects and Controls,* 3rd edn., Royal Society of Chemistry, London, 1996.

622. Clark, T.; Stephenson, T. and Pearce, P. A., *Water Res.,* **31**(1997), 2559.

623. Omoke, A. I. and van Łoon, G. W., *Water Res.,* **33**(1999), 3617.

624. Boisvert, J. P.; To, T.C., Berrak, A. and Jolicour, C., *Water Res.,* **31**(1997), 1939

625. Drost, R. L., *Theory and Practice of Water and Waste Water Treatment*, John Wiley & Sons, 2000, p 475.

626. Yamada, H.; Kazyama, M.; Saito, K. and Kara, M., *Water Res.,* **20**(1986), 547.

627. Orguz, E., *J. Colloid Interface Sci.,* **281**(2005), 62.

628. Tsitoridou, R. and Georgiou, J. A., *Toxicol. Environ.Chem.,***17**(1987), 129.

629. Ageyei, N. M.; Strydom, C. A. and Potgieter, J. H., *Cement Concrete Res.,* **30**(2000), 823.

630. Can, M. Y., and Yildiz, E., *J. Hazard. Mater.,* **135**(2006), 165.

631. Wu, D.; Zhang, B.; Li, C.; Zhang, Z. and Kong, H., *J. Colloid Interface Sci.,* **304**(2006), 165.

632. Chen, J.; Kong, H.; Wu, D.; Hu, Z.; Wang, Z. and Wang, Y., *J. Colloid Interface Sci.,* **300**(2006), 49.

633. Okada, K.; Shimazu, M.; Kameshima, Y.; Nakazima, A. and MacKenzie, K. J. D., *J. Colloid Interface Sci.,* **305**(2007), 229.

634. Roques, H.; Judy, L. N. and Lbugle, A., *Water Res.,* **25**(1991), 959.

635. Karaca, S.; Gurses, A.; Ejder, M. and Acikylidiz, M., *J. Colloid Interface Sci.,* **277**(2004), 257.

636. Akay, G.; Keskinler, B.; Cakici, A. and Danis, U., *Water Res.,* **32**(1998), 717.

637. Yeoman, S.; Stephenson, T.; Lester, J. N. and Perry, R., *Environ. Pollut.,* **49**(1988), 183.

638. Georgentus, D. A. and Grigoropoulou, H. P., *J. Colloid Interface Sci.,* **315**(2007), 70.

639. Brattebe, H. and Odegaard, H., *Water Res.,* **20**(1986), 977.

640. Narkis, N. and Meiri, M., *Environ. Pollut. Series B(Chemical and Physical),* **2**(1981), 327.

641. Hano, T.; Takanashi, H.; Hirata, M.; Urano, K. and Eto, S., *Water Sci. Technol.,* 35(1977), 39.

642. Strauss, R.; Brümmer, G. W Barrow, N. J., *Europ. J. Soil sci.,* **48**(1997), 101.

643. Chitrakar, R.; Tezuka, S.; Sonada, A.; Sakane, K.; Ooi, K. and Hirotsu, T., *J. Colloid Interface Sci.,* **298**(2006), 602.

644. Moon H. and Lee, W. K., *J. Colloid Interface Sci.,* **96**(1983), 162.

645. Manceau, A., *Geochim. Cosmochim. Acta,* **59**(1995), 3647.

646. Randall, S. R.; Sherman, D. R. and Ragnarsdottir, K. V., *Geochim. Cosmochim. Acta,* **65**(2001), 1015.

647. Persson, P.; Nilsson, N.; and Sjoebrg, S., *J. Colloid Interface Sci.,* **177**(1996), 263.

648. Parfitt, R. L.; Atkinson, R. J. and Smart, R. S. C., *Soil Sci. Soc. Am. Proc.,* **39**(1975), 837.

649. Russel, J. D. and Farmer, V. C., *J. Chem. Soc. Farad. Trans.1,* **72**(1976), 1082.

650. Goldberg, S. and Sposito, G., *Soil Sci. Soc. Am. J.,* **48**(1984), 772.

651. Nilsson, N.; Lövgren, L. and Sjoberg, S., *Chem. Spectr. Bioavail.,* **4**(1992), 121.

652. Jie, X. L.; Lui, F.; Zxou, D. H.; Xu, F. L.; Wi, X. Y. and Wang, D. F., *Pedosphere,* **5**(1995), 229.

653. Nanzyo, M., *Soil Sci. Plant Nutrition,* **32**(1986), 51.

654. Manceau, A.; Nagy, K. L.; Spadini, L. and Ragnarsdottir, K. V., *J. Colloid Interface Sci.,* **228**(2000), 306.

655. Geelhoed, J. S.; Hiemstra, T. and vanRiemsdijk, W. H., *Environ. Sci. Technol.,* **32**(1998), 2119.

656. Antelo, J.; Fiol, S.; Péréz, C.; Marino, S.; Arce, F.; Gondar, D. and Lopéz, R., *J. Colloid Interface Sci.,* **347**(2010), 112.

657. Ainsworth, C.C.; Sumner, M. E. and Hurst, V. J., *Soil Sci. Soc. Am. J.,* **49**(1985), 1142

658. Colombo, C.; Barron, V. and Torrent, J., *Geochim. Cosmochim. Acta,* **58**(1994), 1261.

659. Barron, V.; Herruzo, M. and Torrent, J., *Soil Sci. Soc. Am. J.,* **52**(1988), 647.

660. Mandal, B. K.; Chowdhury, T. R.; Samanta, G.; Basu, G. K.; Chowdhury, P. P.; Chanda, C.R.; Lodh, D.; Karan, N. K.; Dhar, R.K.; Tamili, D. K.; Das, D.; Saha, K.C. and Chakraborti, D., *Current Sci.,* **70**(1996), 976.

661. Berg, M.; Tran, H. C.; Nguyen, T. C.; Pham, H.V.; Schertenlieb, R and Giger, W., *Environ. Sci. Technol.,* **35**(2001), 2621.

662. Chen, S. L.; Dzeng, S. R.; Yang, M. H.; Chiu, K. H.; Sieh, G. M.and Wai, C. M., *Environ. Sci. Technol.,* **28**(1994), 872.

663. Zhang, H.; Ma, D.; and Hu, X., *Environ. Geol.,* **41**(2002), 618.

664. Bergono, J. M.; Vincent, P.; Venturino, H. and Infante, A., *Environ. Health Prospect,* **19**(1977), 103.

665. Smith, A.H.; Goycolea, M.; Haque, R. and Biggs, M. L., *Am. J. Epidemiology,* **147**(1998), 660.

666. Arminta, M. A.; Rodriguez, R. and Cruz, O., *Bull. Environ. Contam. Toxicol.* **9**(1997), 583; Del Razo, L. M.; Arellano, M. A. and Cebrian, M. E-., *Environ. Pollut.,* **64**(1990), 143.

667. Nikoladis, N. P.; Dobbs, G. M. and Lackovic, J. A., *Water Res.,* **37**(2003), 1417; Welch, A. H., Lica, M.S. and Hughes, J.L., *Ground Water,* **26**(1988), 333.

668. Nickoon, R. T.; McArthur, J. M.; Ravenscroft, P.; Burgen, W. G. and Ahmed, K. M., *Appl. Geochem.* **15**(2000), 403.

669. Smedly, P. L. and Kinniburgh , D. G., *Appl. Geochem.* **17**(2002), 517.

670. Shevade, S. and Ford, R. G., *Water Res.* **38**(2004), 3197.

671. Wadsworth, M. E. in *Rate Process of Extractive Metallurgy* (Sohn, H. Y. and Wadsworth, M. E.. eds.) Plenum Press, New York, 1979, pp 179-180.

672. Karim, M., *Water Res.,* **34**(2000), 304.

673. Jain, C. K. and Ali, I., *Water Res.,* **34**(2000), 4304.

674. Saha, J. C.; Dikshit, A. K.; Bandyopadhyay, M. and Saha, K. C., *Crit. Rev. Environ, Sci. Technol.*, **29**(1995), 281.

675. Jekel, M. R., in *Arsenic in the Environment* -Part I (Nriagu, J. O. ed.), Wiley, New York, 1994, p 119.

676. Ghosh, M.M.and Yuan, J, R, *Environ. Prog.*, **3**(1987), 150.

677. Chakravarty, S.; Dureja, V.; Bhattacharyya, G.; Maity, S. and Bhattacharjee, S., *Water Res.*, **34**(2002), 625.

678. Bellack, E., *J. Am. Water Works Assoc.*, **63**(1971), 454.

679. Gulledge, J. H.and O'Connor, J. T., *J. Am. Water Works Assoc.*,**65**(1973), 548.

680. Do, H. D.; Green, J. F.; McLean, S. J. and Scott, K. N., *J. Am. Water Works Assoc.*, **86**(1994), 79.

681. Hering, J. G.; Chen, P. Y.; Wilkie, J. A.; Elimelech, M. J. and Liang, S., *J. Am. Water Works Assoc.*, **88**(1996), 155.

682. Hering, J. G.; Chen, P. Y.; Wilkie, J. A. and Elimelech, M.J., *Environ. Eng. ASCE,* **123**(1997), 800

683. McNeill, L. S. and Edwards, M., *J. Am. Water Works Assoc.*, **89**(1997), 75.

684. Edwards, M., *J. Am. Water Works Assoc.*, **86**(1994), 64.

685. Meng, X. G.; Korfiatis, G. P.; Christodoulatos, C. and Bang, S., *Water Res.*, **35**(2001), 2805.

686. Zouboulis, A. I. and Katsoylannis, I. A., *Sepn. Sci. Technol.*, **37**(2003), 5950.

687. Lee, Y.; Ik-Hwan, U. and Yoon J., *Environ. Sci. Technol.*, **37**(2003), 5950.

688. Chang, W.; Singh, P.; Paling, E. and Delidas, S., *Miner. Eng.*, **17**(2004), 517.

689. Maity, A.; Sharma, H.; Basu, J. K. and De, S., *J. Hazard. Mater.*, **102**(2002), 928.

690. Giménez, J.; Martinez, M.; DePablo, J.; Rovira, M. and Duro, L., *J. Hazard. Mater.*, **141**(2007), 575.

691. Guo, H.; Stoeben, D. and Berner, Z., *Appl.Geochem.*, **22**(2007), 1039.

692. Katsoylannis, I. A. and Zouboulis, A. I., *Water Res.*, **36**(2002), 5141.

693. Kundu, S. and Gupta, A. K., J. Colloid Interface Sci., 290(2005), 52.

694. Hlavay, J. and Polyak, K., *J. Colloid Interface Sci.*, **284**(2005), 71.

695. Dixit, S. and Hering, J. G., *Environ. Sci. Technol.*, **37**(2003), 4182.

696. Manning, B. A. and Goldberg, S., *Soil Sci. Soc. Am. J.*, **60**(1996), 121.

697. Lakshmipathiraj, P.; Narasimhan, B. R. V.; Prabhakar, S. and Bhaskar Raju, G., *J. Hazard. Mater.*, **136**(2006), 281.

698. Hertzog, O. K.; Loganathan, V. A.; Kennel, S. R.; Jeppu, G. P. and Barnett, M. O., *J. Colloid Interface Sci.*, **333**(2009), 6.

699. Zhang, Y.; Yang, M.; and Huang, X., *Chemosphère,* **51**(2003), 945.

700. Mahapatra, M.; Sahoo, S. K.; Anand, S. and Das, R. P., *J. Colloid Interface Sci.,* **298**(2006), 6.

701. Lin, T. -F. and Wu, J. K., *Water Res.,***35**(2001), 1049.

702. Singh, T. S. and Pant, K. K., *Sep. Purif. Technol.,* **36**(2004), 139.

703. Su, T.; Guan, X. -H.; Gu, G.and Wang, J., *J. Colloid Interface Sci.,* **326**(2009), 347.

704. Ogata, F.; Kawasaki, N., Nakamura, T. and Tanada, S., *J. Colloid Interface Sci.,* **300**(2006), 88.

705. Yu, M. -J.; Li, X.. and Ahn, W. S., *Micropor. Mesopor. Mater.,* **113**(2008), 197.

706. Zhang, G.; Lin, H., Lin, R. and Qu, J., *J. Hazard. Mater.,* **168**(2009), 820.

707. Violante, A., Pucci, M., Cozzolino, V., Zhu, J. and Pigna, M., *J. Colloid Interface Sci.,* **333**(2009), 63.

708. Dou, X. -M.; Yang, M.; He, H.; Jing, C. Y. and Wu, Z. Y., *J. Hazard. Mater.,* **179**(2010), 208.

709. Manning, B. A.; Fendorf, S. E.; Bostick, B. and Suarez, D. L., *Environ. Sci. Technol.,* **36**(2002), 976.

710. Oscarson, D. W.; Huang, P. M.; Liaw, W. K. and Hammer, W. T., *Soil Sci. Soc. Am. J.,* **47**(1983), 644.

711. Bajpai, S. and Chaudhuri, M., *J. Environ. Eng.,* **125**(1999), 782.

712. Manna, S. and Ghosh, U. C., *J. Hazard. Mater.,* **144**(2007), 522.

713. Su, G. and Puls, R. W., *Environ. Sci. Technol.,* **35**(2001), 1487.

714. Melitas, N.; Wang, J.; Kouklin, M.; O'Day, P. and Farrell, J., *Environ. Sci. Technol.,* **36**(2002), 2074.

715. Manning, B.; Hunt, M. L.; Amrhiein, C. and Yarmoff, J. A., *Environ. Sci. Technol.,***36**(2002), 5455.

716. Goldberg, S., *Soil Sci. Soc. Am. J.,* **66**(2002), 413. P

717. Conru, S.; Breeze, D.; Saada, A. and Baranger, P., *Soil Sci. Soc. Am. J.,* **67**(2003), 1127.

718. Chakraborty, S.; Wolthers, M.; Chatterjee, D. and Charlet, L., *J. Colloid Interface Sci.,* **309**(2007), 392.

719. Wang, Y.; Tiwary, D. and Wang, H., *J. Environ. Sci.,* **21**(2009), 927.

720. Altundogan, H.S.; Altundogan, S.; Türmen, F. and Bildik, M., *Waste Management,* **20**(2000), 76.

721. Fuhrman , H .G-; Tjell , J. C. and McConchie, D., *Environ. Sci. Technol.,* **38**(2004), 2428.

722. Mahapatra, D.; Mishra, D. and Park, H. K., *J. Environ. Sci.*,**20**(2008), 683.

723. Diamadopoulos, E.; Ionnidis, S. and Sakellaropoulos, G. P., *Water Res.*, **27**(1993), 1773.

724. Luengo, C.; Brigante, M. and Avena, M., *J. Colloid Interface Sci.*, **311**(2007), 354.

725. Lakshmipathiraj, P.; Narasimhan, B. R. V.; Prabhakar, S. and Bhaskar Raju, G., *J. Colloid Interface Sci.*, **304**(2006), 317.

726. Raven, K.P.; Jain, A.and Loeppert, R. H., *Environ. Sci. Technol.*,**32**(1998), 344.

727. Hsia, T. H.; Lo, S. -L.; Lin, C. -F. and Lee, D. Y., *Colloids Surfaces A*, **85**(1994),1.

728. Suarez, D. L. Goldberg, S. and Su, C., in *Mineral- Water Interfacial Reactions* (Sparks, D. L. and Grundle, T. eds.), *Am. Chem. Soc. Symp. Ser.* 715, 1998, pp 136-175.

729. Goldberg, S. and Johnston, C. T., *J. Colloid Interface Sci.*, **234**(2001), 204.

730. Ghosh, M. M., and Yuan, J. R., *Environ. Prog.*, **6**(1987), 150.

731. Manning, B. and Goldberg, S., *Environ. Sci. Technol.*, **31**(1997), 2005; Manning, B., Fendorf, S. E. and Goldberg, S., *Environ. Sci. Technol.*, **32**(1998), 2383.

732. Meng, X.; Korfiatis, G. P.; Bang, S. and Bang, K. W., *Toxicol. Lett.*, **133**(2002), 103.

733. Wilkie, J. A. and Hering, J. G., *Colloids Surfaces A*, **107**(1996), 97.

734. Jain, A. and Loeppert, R. H., *J. Environ. Qual.*, **29**(2000), 1422.

735. Gao, Y. and Mucci, A., *Geochim. Cosmochim. Acta*, **65**(2001),

736. Yougran, J.; Fan, M. and van Leeuween, J., *Environ. Sciences*, **19**(2007), 910.

737. Manning, B. and Goldberg, S., *Clays Clay Minerals*, **44**(1996), 609.

738. Roberts, L. C.; Hug, S. J.; Ruettimann, T.; Billah, M. M.; Khan, A. W. and Rahman, M.T., *Environ. Sci. Technol.*, **38**(2004), 307.

739. Meng, X., Bang, S. and Korfiatis, G.P., *Water Res.*, **34**(2000), 1255.

740. Tang, T. A.; Wang, Y. W.; Wang, D.C. and Hou, S.F., in *Selenium in the Environment*, Marcel Dekkar, New York, 1994, pp 47-68.

741. National Academy of Sciences (Food and Nutrition Board), *Human Health Nutrition Review*, **34**(1976), 347.

742. Lo, S. L. and Chen, T. Y., *Chemosphere*, **35**(1997), 919.

743. Su, G. and Suarez, D. L., *Soil Sci. Soc. Am. J.*, **64**(2000), 101.

744. Rovira, M.; Giménez, J.; Martinez, M.; Martinez-Lldo, X.; DePablo, J.; Marti, V. and Duro, L., *J. Hazard. Mater.*, **150**(2008), 279.

745. Zhang, P. and Sparks, D. L., *Environ. Sci. Technol.*, **24**(1990), 1848.

746. Peak, D. and Sparks, D. L., *Environ. Sci. Technol.*, **36**(2002), 1460.

747. Davis, J. A. and Leckie, J.O., *J. Colloid Interface Sci.*, **74**(1980), 32.

748. Duc, M.; Lefevre, G. and Fédoroff, M., *J. Colloid Interface Sci.,* **298**(2006), 556.

749. Su, T.; Guan, X. -H.; Gu, G.and Wang, J., *J. Hazard. Mater.,* **176**(2010), 466.

750. Elzinga, E. J.; Tang, Y.; McDonald, J.; Desisto, S. and Reeder, J., *J. Colloid Interface Sci.,* **340**(2009), 153.

751. Wu, C.; Lo, S. and Lin, C., *Colloids Surfaces A,* **166**(2000), 251.

752. Shi, K.; Wang, X.; Guo, J.; Wang, S. and Wu, W., *Colloids Surfaces A,* **349**(2009), 90.

753. You, Y.; Vance, F. and Zhao, H. G., *Appl. Clay Sci.,* **20**(2001), 13.

754. Das, J.; Das, D.; Dash, G. P. and Parida, K.M., *J.Colloid Interface Sci.,* **251**(2002), 26.

755. Dash, S. S. and Parida, K. M., *J. Colloid Interface Sci.,* **307**(2007), 335.

756. Bar-Yosef, B. and Meek, D., *Soil Sci.,* **144**(1987), 11.

757. Neal, R. H.; Sposito, G.; Holtzclaw, K.M. and Tarina, S. J., *Soil Sci. Soc.Am. J.,* **51**(1987), 1161.

758. Blaylock, M. J.; Tawfic, T. A. and Vance, G. F., *Soil Sci.,* **159**(1995), 43.

759. Beuer, I. and Johnson, C. A., *Environ. Sci. Technol.,* **37**(2003), 3442.

760. Hayes, K. F.; Roe, A.; Brown, G. E.; Hodgson, K.O.; Leckie, J.O.and Parks, G.A., *Science,* **238**(1987), 783.

761. Harrison, J. B. and Berkheiser, V. E., *Clays Clay Minerals,* **30**(1982), 97.

762. Jordan , N.; Lomenech, C.; Marmier, N.; Giffaut, E. and Ehrhardt, J.J., *J. Colloid Interface Sci.,* **329**(2009), 17.

763. De, A. K., *Environmental Chemistry,* 4th ed.., New Age International (P) Ltd., New Delhi, 2000, pp 229-230.

764. Rai, D.; Eary, L. E. and Zachara, J. M., *Sci. Total Environ.,* **86**(1989), 15.

765. Ainsworth, C.C.; Girvin, D. C.; Zachara, J. M.and Smith, S. C., *Soil Sci. Soc. Am. J.,* **53**(1989), 411.

766. Zachara, J.M.; Girvin, D.C.; Schmidt, R.L. and Resch, C.T., *Environ. Sci. Technol.,* **21**(1987), 589.

767. Yin, S. and Ellis, D. E., *Surface Sci.,* **603**(2009), 736.

768. Hsia, T. H.; Lo, S. L.; Lin, C. P. and Lee, D. Y., *Chemosphere,* **26**(1993), 1897.

769. Weerasooriya, R. and Tobschall, H., *Colloids Surfaces A,* **162**(2000), 167.

770. Villalobos, M., Trotz, M. A. and Leckie, J. O., *Environ. Sci. Technol.,* **35**(2001), 3849.

771. Fritzen, M. B.; Souza, A. Z.; Silva, T. A.G.; Souza, L.; Nome, R. A.; Fiedler, H. D. and Nome, F., *J. Colloid Interface Sci.,* **296**(2006), 465.

772. Mor, S.; Ravindran, K. and Bishnoi, N. R., *Bioresources Technol.,* **98**(2007), 954.

773. Wang, C. H.; Wang, J. H. and Huang, C. P., *Water Sci. Technol.,* **35**(1997), 55.

774. Bhattacharyya, A. K.; Naiya, T. K.; Mandal, S. N. and Das, S. K., *Chem. Eng. J.,* **137**(2008), 529.

775. Nagendra Rao, C. R., in *Proc. 3rd Int'l Conf. on Environment and Health* (Bunch, M.J.; Suresh, V. M.and Kumaran, T. V. eds.), held at Dept. of Chemistry, Univ.of Madras, Chennai on 12-17 December, 2003, pp386-399.

776. Sinha Ray, S. P., in *Proc. Nat'l Seminar on Survey, Conservation & Utilization of Water Resources* (Kanungo, S.B.and Ray, H. S.eds.), held at Ramakrishna Vivekananda Mission, Institute of Advanced Studies, Kolkata- 58, on 20-21 January, 2005, pp 38-46.

777. Mariappa, P., Yoganarasimhan, V. and Vasudevan, T. C. *Pollut. Res.,* **19**(2000), 165.

778. WHO, *Guidelines for Drinking Water Quality. In Health Criteria and Other Supporting Information,* 2nd edn. Vol. 2, WHO, Geneva, 1984-85.

779. Apambire, W. B.; Boyle, D.R. and Michael, F. A., *Environ. Geol.,* **33**(1997), 13.

780. Gaciri, S. J. and Davis, T. C., *J. Hydrol.,* **143**(1992), 395.

781. Czarnowski, W.; Wizesniowski, K. and Krechniak, J., *Sci. Total Environ.,* **191**(1996), 177.

782. Clifford, D. A., in *Water Quality and Treatment* (Pontius, F. W. ed.), Am. Water Works Assocn., Denver, Colorado (USA), 1990, pp 501-639.

783. Ghosh, D.; Medhi, C.R. and Purkait, M. K., *Chemosphere,* **73**(2008), 1393.

784. Zuo, Q.; Chen, X.; Li, W. and Chen, G., *J. Hazard. Mater.,* **159**(2008), 452.

785. Essadki, A.H.; Gourich, B.; Azzi, M.; Vial, C. and Delmas, H., *Chem. Eng. J.,* **164**(2010), 106.

786. Nawlakhe, W. G.; Kulkarni, D. N.; Pathak, B. N. and Bulusu, W. L., *Indian J. Environ. Health,* **17**(1975), 26.

787. Bulusu, W. L.; Sundaresan, B. B.; Pathak, B. N.; Nawlakhe, W. G.; Kulkarni, D. N. and Khergaonkar, V.P., *J. Instn. Engineers (India),* **60**(1979), 1.

788. Parthasarathi, N.; Buffle, J. and Haerdi, W., *Water Res.,* **20**(1986), 443.

789. Saha, S., *Water Res.,* **27**(1993), 1347.

790. Rubel, F. (Jr), and Woosely, R. D., *J. Am. Water Works Assocn.,* **71**(1979), 45.

791. Lounici, H.; Addour, L.; Belhocine, D.; Grib, H.; Nicolas, S.; Bariou, B. and Mameri, N., *Desalination,* **114**(1997), 241.

792. Rozic, L.; Novakovic, T.; Jovanovic, N.; Baricevic, A. T-.and Gibavcic, Z.; *J. Serb.Chem. Soc.,* 66(2001), 273.

793. Rozic, L.; Novakovic, Petrovic, S.; Cupic, Z., Gibavcic, Z. and Rozic, A., *J. Serb.Chem. Soc.*, **71**(2006), 1237.

794. Cheung, C. W.; Porter, J.F. and McKay, G., *Water Res.*, **35**(2001), 605.

795. Sujana, M. G.; Thakur, R. S. and Rao, S. B., *J. Colloid Interface Sci.*, **206**(1998), 94.

796. Nigussie, W.; Zewage, F. and Chandravanshi, B.S., *J. Hazard. Mater.*, **147**(2007), 954.

797. Kau, P. M.H.; Smith D.W. and Binning, P., *J. Contam. Hydrol.*,**28**(19970, 267.

798. Chaturvedi, A.K.; Pathak, K. C. and Singh, V. N., *Appl.Clay Sci.*,**3**(1998), 337.

799. Srimurali, M.; Pragathi, A. and Karthikeyan, J. A., *Environ. Pollut.*, **99**(1998), 285.

800. Jadava, K. P.; Tyagi, B. S. and Singh, V. N., *J. Chem. Technol. Biotechnol.*, **65**(1996), 363.

801. Tor, A., *Desalination,* **201**(2006), 267.

802. Pradas, E. G-; Sanchez, M. V-; Garcia, A. V.; Viciana, M. S-; Bueno, F. and Rodriguez, G., *J. Chem. Technol. Biotechnol.*, **42**(1988), 105.

803. Pradas, E. G-; Sanchez, M. V-; Campo, A. G-; Amate, D.U- and Viciana, M.S, *J. Chem. Technol. Biotechnol.*, **69**(1997), 173.

804. Pradas, E. G-; Sanchez, M. V-; Campo, A. G-; Amate, D. U- and Perez, F., *J. Chem. Technol. Biotechnol.*, **74**(1999), 49.

805. Pollard, S. J. T.; Christopher, J. S. and Perry, R., *J. Chem. Technol. Biotechnol.*, **50**(1991), 277.

806. Boukerroui, A. and Quali, M. S. G., *J. Chem. Technol. Biotechnol.*,**75**(2000), 773.

807. Mahramanligluo, M.; Kizilcikli, I. and Bicer, I. O., *J.Fluorine Chem.*, **115**(2002), 41.

808. Zhuang, J. I E. and Yu, G. R., *Chemosphere,* **49**(2002), 619.

809. Agarwal, M., Rai, K. and Dass, S., *J. Cleaner Produc.*, **11**(2003), 439.

810. Kamble, S. P.; Dixit, P.; Rayalu, S. S. and Labhsetwar, N. K., *Desalination,* **249**(2009), 687.

811. Zevenbergen, C.; vanReeuwijk, L. P.; Flapporti, G.; Louws, R.J. and Schuiling, R.D., *Sci. Total Environ.*, **188**(1996), 225.

812. Luther, S. M.; Paulson, L.; Dudas, M. J. and Rutherford, P. M., Can. *J. Soil Sci.,* **76**(1996), 83.

813. Reardon, E. J. and Wang, Y., *Environ. Sci. Technol.*, **24**(2000), 3247.

814. Wang, Y. and Reardon, E. J., *Appl. Geochem.*, **16**(2001), 531.

815. Chaturvedi, A. K.; Jadava, K. P.; Pathak, K. C. and Singh, V. N., *Water, Air & Soil Pollut.*, **49**(1990), 51.

816. Mavros, P.; Zouboulis, A. I. and Lazaridis, N. K., *Environ. Tech.*, **14**(1993), 83.

817. Sujana, M. G. and Anand, S., *Desalination,* 267 (2011) 222

818. Cengeloglu, Y., Kir, E. and Ersoz, M., *Sepn. Sci. Technol.,* **28**(2002), 81.

819. Sarkar, M.; Banerjee, A.; Pramanick, P. P. and Sarkar, A.; *J. Colloid Interface Sci.,* **302**(2006), 432.

820. Sujana, M.G.; Pradhan, H.K. and Anand, S.; *J. Hazard. Mater.,* **161**(2009), 120.

821. Xu, Y. M-; Ning, A. R- and Zhao, J., *J. Colloid Interface Sci.,* **235**(2001), 66.

822. Drost, R. L., *Theory and Practice of Water and Waste Water Treatment, John Wiley,* New York, 1997, p 478.

823. Maliyekkal, S. M.; Sharma, A. K. and Philip, L., *Water Res.,* **40**(2006), 3497.

824. Tripathy, S. S. and Raichur, A. M., *J. Hazard. Mater.,* **153**(2008), 1043.

825. Maliyekkal, S. M.; Shukla, S.; Philip, L. and Nambi, I. M., *Chem. Eng. J.,* **140**(2008), 183.

826. Bansiwal, A.; Pillewan, P.; Baniwale, R. B. and Rayalu, S.S., *Micropor. Mesopor. Mater.,* **129**(2020), 54.

827. Biswas, K.; Saha, S. K.; and Ghosh, U.C., *Ind. Eng.Chem.(Res),* **46**(2007), 5346.

828. Sujana, M. G.; Soma, G.; Vasumathi, N. and Anand, S., *J. Fluorine Chem.,* **130**(2009), 749.

829. Sujana, M. G. and Anand, S., *Appl. Surf. Sci.,* **256**(2010), 6956.

830. Tang, S. Y.; Guan, X.; Wang, J.; Gao, N.; McPhail, M. R. and Chusuei, C. C., *J. Hazard. Mater.,* **171**(2009), 774.

831. Kumar, E.; Bhatnagar, A.; Ji, M.; Juang, W.; Lee, S.H-; Kim, S. J.; Lee, G.; Song, H.; Choi, J. Y-; Yang, J. S. and Jeon, B. H-., *Water Res.,* **43**(2009), 490.

832. Sundaram, C. S.; Viswanathan, N. and Meenakshi, S., *J. Hazard. Mater.,* **155**(2008), 206.

833. Fan, X.; Parker, D. J. and Smith, H. M. D., *Water Res.,* **37**(2003), 4329.

834. Clifford, D.; Matson, J. and Kennedy, R., *Ind. Eng. Chem.* (Res), 15(1978), 6.

835. Bahena, J.L.R.; Cabrera, A. R.; Valdivieso, A.L- and Urbina, R.H., *Sepn. Sci.Technol.,* **37**(2002), 1973.

836. George, S.; Pandit, P.; and Gupta, A. B., *Water Res.,* **44**(2010), 3055.

837. Valdivieso, A. L-; Bahena, J. L. R.; Song, S. and Urbina, R. H., *J. Colloid Interface Sci.,* **298**(2006), 1.

838. Tiwari, P. H. and McLean, A. H., *J. Colloid Interface Sci.,* **40**(1972), 267.

839. Halter, W. F., *Geochim. Cosmochim. Acta,* **63**(1999), 3077.

840. Berube, Y. G. and DeBruyn, P. L., *J. Colloid Interface Sci.,* **27**(1968), 305.

841. Fokkink, L. G. J., Dekeizer, A. and Lyklema, J., *J. Colloid Interface Sci.*, **127**(1989), 116.

842. Hansen, H.C.B.; Wetche, T. P.; Raulund-Rasmussen, K. and Borggaard, O.K., *Clay Minerals,* **29**(1994), 341.

843. Goldberg, S.; Foster, H. S. and Heick, E. L., *Soil Sci. Soc. Am. J.*, **57**(1993),704.

844. Blesa, M. A.; Alberto, J.; Maroto, G. and Regazzoni, A.I., *J. Colloid Interface Sci.,* **99**(1984), 32.

845. Su, C. and Suarez, D. L., *Environ. Sci. Technol.,* **29**(1995), 302.

846. Bloesch, P. M.; Bell, L. C. and Hughes, J. D., *Austr. J. Soil. Res.,* **25**(1987), 377.

847. Le, X. C.; Yalcin, S. and Ma, M., *Environ. Sci. Technol.,* **34**(2000), 2342.

848. Yalcin, S. and Le, X. C., *Environ. Monitor.,* **3**(2001), 81.

849. Yamamoto, M.; Urata, K.; Murashige, K. and Yamamoto, Y., *Spectrochim Acta, Part B,* **36**(1981), 671.

850. Elizalde-Gonzalez, M.P.; Mattusch, J.; Einicke, W. D. and Wennrich, R., C., *Chem. Eng. J.,* **81**(2001), 187.

851. Londesborough, S.; Mattusch, J. and Wennrich, R., C., *Fresnius Z. Anal. Chem.,* **363**(1999), 577.

INDEX